High-Performance Thin-Layer Chromatography (HPTLC)

ManMohan Srivastava
Editor

High-Performance Thin-Layer Chromatography (HPTLC)

Springer

Editor
ManMohan Srivastava
Professor
Department of Chemistry
Dayalbagh Educational Institute
Agra-282110
India
smohanm@rediffmail.com

ISBN 978-3-642-14024-2 e-ISBN 978-3-642-14025-9
DOI 10.1007/978-3-642-14025-9
Springer Heidelberg Dordrecht London New York

Cover design: deblik, Berlin

Printed on acid-free paper

Springer is part of Springer Science+Business Media (www.springer.com)

About the Book

HPTLC: High-Performance Thin-Layer Chromatography

MM. SRIVASTAVA
EDITOR

The present edited book is the presentation of 18 in-depth national and international contributions from eminent professors, scientists and instrumental chemists from educational institutes, research organizations and industries providing their views on their experience, handling, observation and research outputs on HPTLC, a multi-dimensional instrumentation. The book describes the recent advancements made on TLC which have revolutionized and transformed it into a modern instrumental technique HPTLC. The book addresses different chapters on HPTLC fundamentals: principle, theory, understanding; instrumentation: implementation, optimization, validation, automation and qualitative and quantitative analysis; applications: phytochemical analysis, biomedical analysis, herbal drug quantification, analytical analysis, finger print analysis and potential for hyphenation: HPTLC future to combinatorial approach, HPTLC-MS, HPTLC-FTIR and HPTLC-Scanning Diode Laser. The chapters in the book have been designed in such a way that the reader follows each step of the HPTLC in logical order.

About the Editor

Dr. MM. Srivastava is Professor in Department of Chemistry of Dayalbagh Educational Institute, Agra, India and has extensive experience of twenty six years of teaching and research in Analytical and Environmental Chemistry. Prof. Srivastava is actively engaged in the research under the domain of Green Chemistry and delivered lectures in National Research Council, University of Alberta, Canada, University of Illinois, Chicago, Wisconsin and Maryland, USA. He has more than 100 research papers in journals of repute. Prof. Srivastava is recipient of Department of Science and Technology Visiting Fellowship and has recently been elected as Fellow of Royal Society, London, UK (FRSC) and Fellow of Indian Society of Nuclear Techniques in Agriculture and Biology (FNAS). He has edited books on Recent Trends in Chemistry, Green Chemistry: Environmental Friendly Alternatives and Chemistry of Green Environment.

Preface

Thin-layer chromatography is without doubt one of the most versatile and widely used separation methods in chromatography. The concept of TLC is simple and samples usually require only minimal pretreatment. It has been frequently used in pharmaceutical analysis, clinical analysis, industrial chemistry, environmental toxicology, food chemistry, pesticide analysis, dye purity, cosmetics, plant materials, and herbal analysis. The previous image of TLC regarding low sensitivity, poor resolution, and reproducibility made it stagnant and forgotten technique few years back. Now, it is the most used chromatographic technique and likely to remain so for times to come.

Today, most stages of this technique are automated and operated instrumentally in the form of modern high-performance thin-layer chromatographic system that allows the handling of a large number of samples in one chromatographic run. Speed of separation, high sensitivity, and good reproducibility result from the higher quality of chromatographic layers and the continual improvement in instrumentation. It is now capable of handling samples with minimal pretreatment, detecting components at low nanogram sensitivities and with relative standard deviations of about 1%. HPTLC is now truly a modern contemporary of HPLC and GC and continues to be an active and versatile technique in research with large number of publications appearing each year.

This edited book is the presentation of 18 in-depth national and international contributions from eminent professors, scientists, and instrumental chemists from educational institutes, research organizations, and industries providing their views on their experience, handling, observation, and research outputs on this multidimensional instrumentation. The book describes the recent advancements made in TLC which have revolutionized and transformed it into a modern instrumental technique HPTLC. The book addresses different chapters on HPTLC fundamentals, principle, theory, understanding, instrumentation, implementation, optimization, validation, automation, and qualitative and quantitative analysis; applications of HPTLC separation with special reference to phytochemical analysis, biomedical analysis, herbal drug quantification, analytical analysis, finger print analysis; and HPTLC future to combinatorial approach, potential for hyphenation, HPTLC–MS, HPTLC–FTIR, and HPTLC–scanning diode laser. The chapters in the book have been designed in such a way that the reader follows each step of the HPTLC in logical order.

Our greatest ambition for editing this book has been to familiarize and popularize the theoretical and practical aspects of working and applications of a recent, modified, versatile analytical instrument HPTLC system among students, researchers, academicians, analysts, and chemists involved in various areas of research. We wish to place on record our appreciation to Prof. VG Das, Esteemed Director, Prof. LD Khemani, Head, Department of Chemistry, Prof. Satya Prakash, Professor Emeritus, Dayalbagh Educational Institute, Dayalbagh, Agra, and all the contributors for their cooperation and encouragement extended to me. Without their enthusiasm and timely submission of their articles, this work would have not been possible. Although the bulk of material is original and/or taken from sources that the authors have been directly involved with, every effort has been made to acknowledge materials drawn from other sources.

Editor trusts that his apology will be accepted for any error, omission, and editing mistake in the manuscripts.

Agra, India ManMohan Srivastava

Contents

List of Contributors

Abdul Moheman Department of Applied Chemistry, Faculty of Engineering and Technology, Aligarh Muslim University, Aligarh 202002, India

Ajai Prakash Gupta Public Health Engineering Department, IIIM-CSIR, Jammu & Kashmir, India

Ali Mohammad Department of Applied Chemistry, Faculty of Engineering and Technology, Aligarh Muslim University, Aligarh 202002, India

Ariane Nimptsch University of Leipzig, Medical Department, Institute of Medical Physics and Biophysics, Härtelstr, 16/18, D-04107 Leipzig, Germany

Arun Sharma Food Technology Division, Bhabha Atomic Research Centre, Mumbai 400085, India

Beate Fuchs University of Leipzig, Medical Department, Institute of Medical Physics and Biophysics, Härtelstr, 16/18, D-04107 Leipzig, Germany

Bharat G. Patel A. R. College of Pharmacy and G. H. Patel Institute of Pharmacy, Sardar Patel University, University of Leipzig Gujarat, Vallabh Vidyanagar 388 120, India

Claudia Cimpoiu Faculty of Chemistry and Chemical Engineering, Babes Bolyari University, Cluj Napoca, Romania

Devanand B. Shinde Department of Chemical Technology, Dr. Babasaheb Ambedkar Marathwada University, Aurangabad 431001, India

G. Subramanian Department of Pharmaceutical Quality Assurance, Manipal College of Pharmaceutical Sciences, Manipal, Karnataka 576104, India

Gunawan Indrayanto Plant Biotechnology Research Group and Assessment Service Unit Faculty of Pharmacy, Airlangga University, Surabaya 60286, Indonesia

Janki N. Thakker Department of Biotechnology, PD Patel Institute of Applied Science, Charutar University of Science & Technology, Education Campus Changa, 388421 Gujarat, India

Joachim Franzke ISAS—Institute for Analytical Sciences, Bunsen-Kirchhoff-Straße 11, 44139, Otto-Hahn-Straße 6b, 44227 Dortmund, Germany

Jurgen Schiller University of Leipzig, Medical Department, Institute of Medical Physics and Biophysics, Härtelstr, 16/18, D-04107 Leipzig, Germany

Kathrin Nimptsch University of Leipzig, Medical Department, Institute of Medical Physics and Biophysics, Härtelstr, 16/18, D-04107 Leipzig, Germany

Kristin Teuber University of Leipzig, Medical Department, Institute of Medical Physics and Biophysics, Härtelstr, 16/18, D-04107 Leipzig, Germany

Lukasz Ciesla Department of Inorganic Chemistry, Faculty of Pharmacy, Medical University of Lublin, Lublin, Poland

M.A. Khan Centre for Transgenic Plant Development, Department of Biotechnology, Faculty of Science, Jamia Hamdard, Hamdard Nagar, New Delhi 110062, India

M.Z. Abdin Centre for Transgenic Plant Development, Department of Biotechnology, Faculty of Science, Jamia Hamdard, Hamdard Nagar, New Delhi 110062, India

Machindra J. Chavan Department of Pharmacognosy, Amrutvahini College of Pharmacy, Sangamner, S.K. Dist-Ahmednagar (M.S) 422 605, India

ManMohan Srivastava Department of Chemistry, Faculty of Science, Dayalbagh Educational Institute, Dayalbagh, Agra 282110, India

Mauji Ram Centre for Transgenic Plant Development, Department of Biotechnology, Faculty of Science, Jamia Hamdard, Hamdard Nagar, New Delhi 110062, India

Michael Edler ISAS—Institute for Analytical Sciences, Bunsen-Kirchhoff-Straße 11, 44139 Otto-Hahn-Straße 6b, 44227 Dortmund, Germany

Monika Waksmundzka Hajnos Department of Inorganic Chemistry, Faculty of Pharmacy, Medical University of Lublin, Lublin, Poland

Mrunali R. Patel Indukaka Ipcowala College of Pharmacy, Sardar Patel University, New Vallabh Vidyanagar, 388 121 Gujarat, India

N. Vadera Department of Pharmaceutical Quality Assurance, Manipal College of Pharmaceutical Sciences, Manipal, Karnataka 576104, India

Norman Ahlmann ISAS—Institute for Analytical Sciences, Bunsen-Kirchhoff-Straße 11, 44139 Otto-Hahn-Straße 6b, 44227 Dortmund, Germany

P. Musmade Department of Pharmaceutical Quality Assurance, Manipal College of Pharmaceutical Sciences, Manipal, Karnataka 576104, India

Paweł K. Zarzycki Section of Toxicology and Bioanalytics, Koszalin University of Technology, Śniadeckich 2, 75-453 Koszalin, Poland

Pinakin Dhandhukia Ashok and Rita Patel Institute of Integrated Study & Research in Biotechnology and Allied Sciences, New Vallabh Vidyanagar, 388 121 Gujarat, India

Prabhakar Jha Department of Botany, Faculty of Science, Jamia Hamdard, Hamdard Nagar, New Delhi 110062, India

Prasad S. Variyar Food Technology Division, Bhabha Atomic Research Centre, Mumbai 400085, India

Pravin S. Wakte Department of Chemical Technology, Dr. Babasaheb Ambedkar Marathwada University, Aurangabad 431 001 (M.S), India

Rashmin B. Patel A. R. College of Pharmacy and G. H. Patel Institute of Pharmacy, Sardar Patel University, Vallabh Vidyanagar, 388 120 Gujarat, India

Rosmarie Süß University of Leipzig, Medical Department, Institute of Medical Physics and Biophysics, Härtelstr, 16/18, D-04107 Leipzig, Germany

Shipra Ahuja Department of Pharmaceutics, Jamia Hamdard University, Hamdard Nagar, New Delhi 110062, India

Song Peng ISAS—Institute for Analytical Sciences, Bunsen-Kirchhoff-Straße 11, 44139 Otto-Hahn-Straße 6b, 44227 Dortmund, Germany

Tomasz Tuzimski Department of Physical Chemistry, Faculty of Pharmacy, Medical University of Lublin, Lublin, Poland

Part I
Introduction

Chapter 1
An Overview of HPTLC: A Modern Analytical Technique with Excellent Potential for Automation, Optimization, Hyphenation, and Multidimensional Applications

MM. Srivastava

Abstract High performance thin layer chromatography (HPTLC) is a sophisticated instrumental technique based on the full capabilities of thin layer chromatography. The advantages of automation, scanning, full optimization, selective detection principle, minimum sample preparation, hyphenation, etc. enable it to be a powerful analytical tool for chromatographic information of complex mixtures of inorganic, organic, and biomolecules. The chapter highlights related issues such as journey of thin-layer chromatography, basic principle, protocol, separation, resolution, validation process, recent developments, and modifications on TLC leading to the HPTLC, optimization, process control, automation, and hyphenation. It explains that HPTLC has strong potentials as a surrogate chromatographic model for estimating partitioning properties in support of combinatorial chemistry, environmental fate, and health effect studies.

Analytical chemists work to improve the reliability of existing techniques to meet the demands for better chemical measurements which arise constantly in our society. They adapt proven methodologies to new kinds of materials or to answer new questions about their composition. They carry out research to discover completely new principles of measurement and are at the forefront of the utilization of recent discoveries for practical purposes. Modern analytical chemistry is dominated by instrumental analysis. Analytical chemists focus on new applications, discoveries and new methods of analysis to increase the specificity and sensitivity of a method. Many methods, once developed, are kept purposely static so that data can be compared over long periods of time. This is particularly true in industrial quality assurance, forensic, and environmental applications. Analytical chemists are also equally concerned with the modifications and development of new

MM. Srivastava
Department of Chemistry, Faculty of Science, Dayalbagh Educational Institute, Dayalbagh, Agra 282110, India
e-mail: smohanm@rediffmail.com

MM. Srivastava (ed.), *High-Performance Thin-Layer Chromatography (HPTLC)*,
DOI 10.1007/978-3-642-14025-9_1, © Springer-Verlag Berlin Heidelberg 2011

instrument. The types of instrumentation presently being developed and implemented involve analytical tools including vibrational, rotational, optical, absorption, colorimetric and scattering spectroscopy, mass spectrometry, chromatography, electro chemicals, acoustics, laser, chemical imaging, light-induced fluorescence, light scattering, etc.

At this point, we will talk about chromatographic techniques. Chromatography, defined as the group techniques used for the separation of a complex mixture of compounds by their distribution between two phases, was invented in 1901 by Russian botanist Mikhail Semyonovich Tswet, during his research on plant pigments. No other separation method is as powerful and applicable as in chromatography. It is the most versatile and widespread technique employed in modern analytical chemistry. The fact has genuine reasons. First, very sensitive methods of detection are available to all types of chromatography and small quantities of material can be separated, identified and assayed. Second, chromatographic separations are relatively fast and an analysis can be completed in a short interval of time. Another advantage of chromatography is its relative simplicity and ease of operation compared with other instrumental techniques. Finally, if the established procedure is well controlled and the apparatus is well maintained, good accuracy and precision can be achieved.

Thin-layer chromatography, among various chromatographic techniques, score high over other chromatographic techniques where altogether a new problem, one might not have encountered or solved. It is a valuable tool for reliable identification providing chromatographic fingerprints.

The feature that distinguishes TLC from other physical and chemical methods of separation is that two mutually immiscible phases are brought in to contact while one phase is stationary and the other mobile. A sample is loaded on the stationary phase and is carried by the mobile phase. Species in the sample undergo repeated interaction between the mobile and stationary phase. When both phases are properly selected, the sample components are gradually separated into bands or zones. Figure 1.1 explains the facts involving the separation of the sample.

The common method of development in thin-layer chromatography employs capillary forces to transport the mobile phase through the layer. These weak forces arise from the decrease in free energy of the solvent as it enters the porous structure of the layer. For fine particle layers, capillary forces are unable to generate sufficient flow to minimize the main sources of zone broadening. Firstly, the mobile-phase velocity varies as a function of time and migration distance. Secondly, the mobile-phase velocity is established by the system variables and is otherwise beyond experimental control. This results in a slow and variable mobile-phase velocity through the layer with separation times that is longer than required. Separated zones are broader than they would be for a constant and optimum mobile-phase velocity and the zone capacity limited by the useful range of mobile-phase velocities. Multiple developments with an incremental increase in the development length and a decreasing solvent strength gradient is the basis of separations by automated multiple developments (AMDs). Results from phenomenological models indicate that further improvements over those already realized are

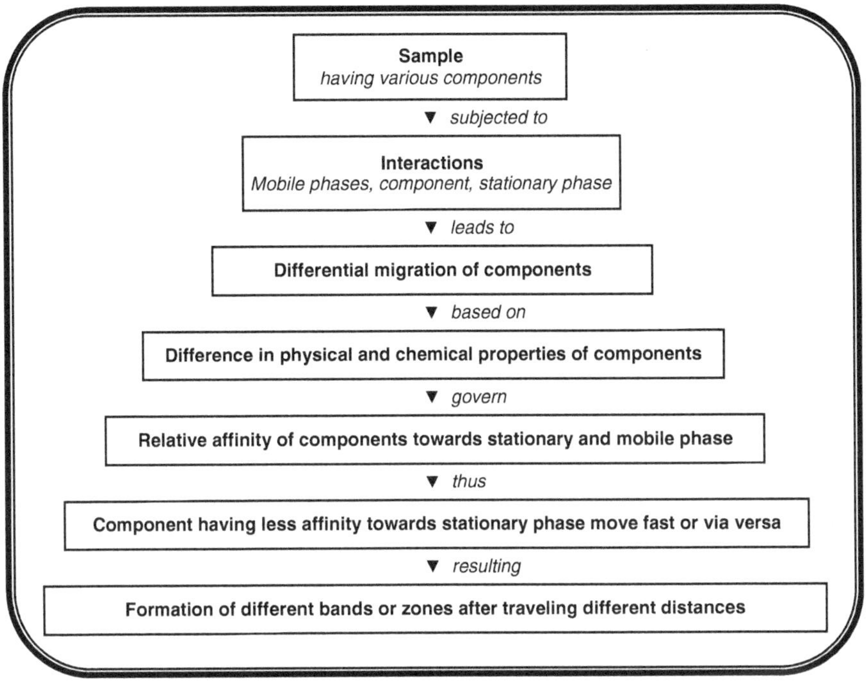

Fig. 1.1 Separation of bands on thin-layer chromatographic plate

unlikely for capillary flow systems and there is no solution to the significant increase in separation time. The magnitude and range of capillary flow velocities fundamentally limit separations in thin-layer chromatography. Faster separations with an increase in zone capacity require a higher mobile-phase velocity than in capillary flow as well as a velocity that is independent of the solvent front migration distance.

The attractive features of TLC are low-cost analysis of samples requiring minimal sample clean up and allows a reduction in the number of sample preparation steps. TLC is also preferred for the analysis of substances with poor detection characteristics requiring post-chromatographic treatment for detection. Thin-layer chromatography retains a historic link with the characterization of dyes and inks and the control of impurities in industrial chemicals. It is used for the identification of drugs and toxic substances in biological fluids, unacceptable residue levels, maintaining a safe water supply by monitoring natural and drinking water sources for crop projecting agents used in modern agriculture, and confirmation of label claims for content of pharmaceutical products. It remains one of the main methods for class fractionation, speciation and flavor potential of plant materials. It continues to be widely used for the standardization of plant materials used as traditional medicines. It is frequently selected as the method of choice to study the metabolism and fate of radiolabeled compounds in the body and environment.

Journey of Thin-Layer Chromatography

In order to separate inorganic ions, Meinhard and Hall (1949) used a starch binder to give some firmness to the layer and described as surface chromatography. Advances were made by Kirchner et al. (195l) who used the now conventional ascending method using a sorbent composed of silicic acid. Reitsema (1954) used much broader plates and was able to separate several mixtures in one run. However, from 1956 a series of papers from Stahl appeared in the literature introducing thin-layer chromatography as an analytical procedure. Since then, silica gel nach Stahl became well known as a stationary phase. Plaster of Paris (calcium sulfate) was used as a binder and TLC began to be widely used. First book on thin layer chromatography was published by Kurt Randerath (1962), followed by those of Stahl and co-workers and second edition of Stahl's book (1969). These authors showed the wide versatility of TLC and its applicability to a large spectrum of separation problems and also illustrated how quickly the technique had gained acceptance throughout the world. Stahl (1965) could quote over 4,500 publications on TLC works. Stahl's publication highlighted the importance of factors such as the controlling of the layer thickness, the layer uniformity, the binder level, and the standardization of the sorbents as regards pore size, volume, specific surface area and particle size. Commercialization of the technique began in 1965 with the first precoated TLC plates and sheets. TLC quickly became very popular with about 400–500 publications per year appearing in the late 1960s. It was recognized as a quick, relatively inexpensive procedure for the separation of a wide range of sample mixtures. It soon became evident that the most useful sorbents was silica gel, particularly with an average pore size of 60 Å. Modifications to the silica gel began with silanization to produce reversed-phase layers. This opened up a far larger range of separation possibilities based on a partition mechanism, compared with adsorption. Until to this time, quantitative TLC was fraught with experimental error. However, the introduction of commercial spectro densitometric scanners enabled the quantification of analytes directly on the TLC layer. Initially, peak areas were measured manually, but later, integrators achieved this automatically.

Halpaap (1973) was the first to recognize the advantage of using a smaller average particle size of silica gel (5–6 mm) in the preparation of TLC plates. He compared the effect of particle size on development time, R_f values and plate height. Commercially the plates were first called nano-TLC plates but soon changed to the designation HPTLC plates with the recognition that HPTLC has added a new dimension to TLC. It was demonstrated that less amount of mobile phase, precision (tenfold) and reduction in analysis time (similar factor) could be achieved. The first major HPTLC publication was made by Zlatkis and Kaiser (1977). Halpaap and Ripphahn described their comparative results with the new 5.5-cm HPTLC plates versus conventional TLC for a series of lipophilic dyes. Reversed-phase HPTLC was reported by Halpaap et al. (1980). Jost and Hauck (1982) reported an amino (NH2−) modified HPTLC plate which was soon followed by cyano-bonded (1985) and diol-bonded (1987) phases. The era of 1980s also saw improvements in spectro-densitometric scanners with full computer control including options for peak purity and the measurement of full UV/visible spectra for all separated components. AMD made its appearance because of the pioneering work of Burger (1984). This improvement enabled a marked increase in the number and resolution of the separated components.

Recent Developments

The multiple developments and its combination with other analytical techniques have dramatically increased the use of thin-layer chromatography for the characterization of complex mixture. TLC has strong potential as a surrogate chromatographic model for qualitative and quantitative analysis. To convert these opportunities in to the practice, several modifications have been carried out on the conventional TLC system.

Over-Pressured Layer Chromatography

Forced flow separations in the overpressured development chamber involves the sealing of the layer on its open side by a flexible membrane under hydraulic pressure and a pump is used to deliver the mobile phase to the layer. A constant mobile-phase velocity independent of the solvent front migration distance is obtained as long as the hydraulic pressure applied at the membrane maintains an adequate seal with the layer. When a solvent is forced through a dry layer of porous particles sealed from the external atmosphere, the air displaced from the layer by the solvent usually results in the formation of a second front (β front). The space between the α and β fronts is referred to as the disturbing zone and consists of a mixture of solvent and gas bubbles. In practice, the disturbing zone can be eliminated or minimized by predevelopment of the layer with a weak solvent in which the sample does not migrate. The solvent dislodges trapped air from the layer before starting the separation and consists of a mixture of solvent and gas bubbles.

Planar Electrochromatography

Electro-osmosis provides a suitable alternative transport mechanism to pressure driven flow in open tubular and packed capillary chromatography. Electro-osmotic flow in packed capillary columns is the basis of capillary electrochromatography. The plug-like flow profile reduces *trans*-axial contribution to band broadening as well as providing a constant and optimum mobile-phase velocity. In addition, the mobile-phase velocity is independent of column length and average particle size up to the limits established by double-layer overlap. The general interest created by the rapid development of capillary electro chromatography as a useful separation method has trickled over to thin layer chromatography. Electroosmotically driven flow could provide an effective solution to the limitations of capillary flow. The current status of electroosmotically driven flow in thin-layer chromatography is probably more confusing. Recent studies have brought some enlightenment to this technique. Enhanced flow is caused by forced evaporation of the mobile phase from a solvent-deficient region at the top of the layer. Because of drainage in vertically mounted layers, electrical resistance is highest at the top of the layer and the increase in heat production drives the evaporation of solvent, pulling additional solvent through the layer. In an open system like thin-layer chromatography, evaporation of mobile phase from the layer surface competes with electro osmotic flow along the layer. The voltage, pH, and buffer concentration must be optimized to minimize either excessive flooding or drying of the layer to avoid degradation of the separation quality. These processes are probably better controlled by enclosing the layer and improving the thermostating of the system. Since high pressures are not involve, mechanisms for enclosing the layer could be relatively simple compared to pressure-driven forced flow and new approaches suggest that effective temperature control is possible. Thinner layer may also help to contain temperature gradients in combination with adequate thermostating.

Image Analysis

Slit-scanning densitometry is the dominant method of recording thin-layer separations for interpretation and quantification. This technology is now relatively mature although limited to absorption and fluorescence detection in the UV–visible range. It has adequately served the needs of thin-layer chromatography for the last two decades. Evolution of slit-scanning densitometry is now largely progressive and major changes in operation and performance seem unlikely. A possible exception is the development of scanners employing a fiber optic bundle for illumination of sample zones and collection of reflected light in conjunction with a photodiode array detector for simultaneous multi-wavelength detection and spectral recording. This approach simplifies data acquisition for some applications and affords the possibility of facile application of modern chemometric approaches for data analysis. This approach may improve the quality of available data from thin-layer separations, but does not overcome the principal limitations of slit-scanning densitometry.

For video densitometry, optical scanning takes place electronically, using a computer with video digitizer, light source, monochromators, and appropriate optics to illuminate the plate and focus the image onto a charge-coupled device (CCD) video camera. The main attractions of video densitometry for detection in thin-layer chromatography are fast and simultaneous data acquisition, a simple instrument design without moving parts, increase in sensitivity, longer acquisition times and compatibility with data analysis. Video densitometry cannot compete with slit scanning densitometry in terms of sensitivity, resolution and available wavelength-measuring range.

Two-Dimensional Separations

Multidimensional separations employing two or more coupled orthogonal separation systems represent the preferred approach in chromatography to obtain a high peak capacity for the separation of complex mixtures. Two-dimensional separations are easily performed using planar separation systems. Even capillary flow separations can be expected to afford a zone capacity of a few hundreds rising to a few thousands for forced flow developments. In most cases, the two-solvent systems differ only in their intensity for a given set of intermolecular interactions and are not truly complementary. Such systems are responsible for the low success of two-dimensional separation systems to provide a significant increase in the separation potential apparent in many applications. Recent reports are encouraging and recognize the importance of the orthogonal nature of the retention mechanisms if a high separation capacity is to be achieved. Bilayer plates with a smaller reversed-phase strip along one edge of the plate adjacent to a larger silica gel layer have provided the most popular approach for the implementation of two-dimensional separations with a high separation capacity. Chemically bonded layers can also be used in the reversed-phase and normal phase mode and allow the use of buffers as a further means of adjusting selectivity. The awaited breakthrough in general detection for two-dimensional planar separations is likely to come from video densitometry. Data acquisition is straight forward since the whole plate is imaged simultaneously, but a problem remains with quantification that has still to be addressed.

High-Performance Thin-Layer Chromatography (HPTLC)

HPTLC allows fast, inexpensive method of analysis in the laboratory as well as in field. Modern quantitative HPTLC, when properly performed by well-trained analysts, can be advantageous compared to high-performance liquid-column chromatography in many analytical situations. The modern HPTLC technique, combined with automated sample application and densitometric scanning, is sensitive and completely reliable, suitable for use in qualitative and quantitative analysis. HPTLC is a valuable tool for reliable identification because it can provide chromatographic fingerprints that can be visualized and stored as electronic images. To fully take advantage of this unique feature inherent to HPTLC, reproducible results and images must be ensured. Special advantages of HPTLC include high sample throughput and low cost per analysis; multiple samples and standards can be separated simultaneously, and sample preparation requirements are often minimal because the stationary phase is disposable. Other advantages include static, off-line detection of zones using a great variety of complementary post-chromatographic universal and selective detection methods that are often applied sequentially, and storage of the separation, containing all sample components, on the layer for identification and quantification at a later time by in situ or elution methods (Fig. 1.2). At the present time, all steps of the TLC process can be computer controlled.

TLC	FEATURES	HPTLC
Manual	Technique	Instrumental
Lab Made/ Pre-coated	Layer	Pre-coated
Circular (2-4 nm dia)	Shape of sample	Rectangular (6mm L X 1mm W)
Uncontrolled/ Solvent dependent	Size of sample	Controlled Solvent independent
2-10%	Vol. precision	0.5%-2%
Not Compliant	Good Lab Practice	Compliant
1 TO 10 μL	Vol. range	0.1 to 500 μL
No	PC connectivity	Yes
No	Method storage	Yes
No	Validation	Yes
Capillary/ pipette	Sample holder	Syringe
No	Quantitative analysis	Yes
254 or 366 nm, visible	Wavelength range	190 or 800 nm, Monochromatic
No	Spectrum analysis	Yes
By analyst	Analysis judgment	By machine

Fig. 1.2 Advancements made on TLC leading to the development of HPTLC

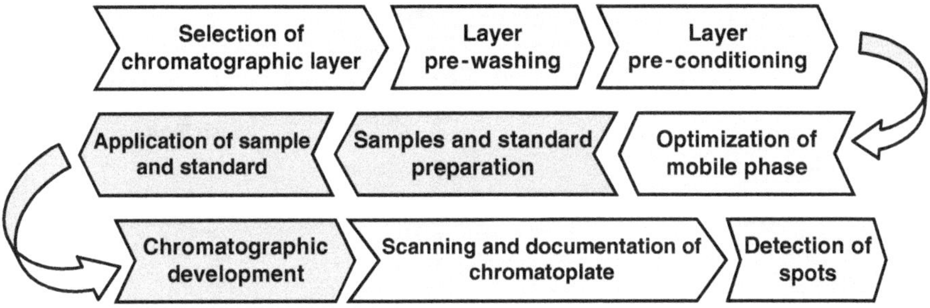

Fig. 1.3 Schematic procedure for HPTLC method development

The use of highly sensitive (CCD) cameras has enabled the chromatographer to electronically store images of chromatograms for future use and for direct entry into reports at a later date.

HPTLC-based separations involves several steps shown in Fig. 1.3. The details of each step have been discussed in the preceding chapters.

HPTLC: Separation and Resolution

To which extent various components of a formulation are separated by a given HPTLC system is the important factor in quantitative analysis. It depends on the following factors:

- Type of stationary phase
- Type of precoated plates
- Layer thickness
- Binder in the layer
- Mobile phase
- Solvent purity
- Size of the developing chamber
- Saturation of chamber
- Sample's volume to be spotted
- Size of the initial spot
- Solvent level in the chamber
- Gradient
- Relative humidity
- Temperature
- Flow rate of solvent
- Separation distance
- Mode of development

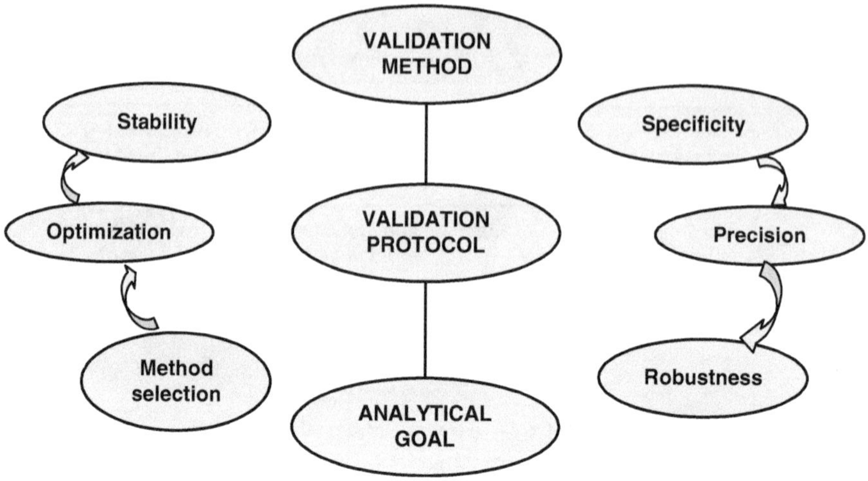

Fig. 1.4 Validation process involved in HPTLC

HPTLC: Validation Process

Validation should not be seen separately from the development of a method. The entire process can be visualized with the scheme in Fig. 1.4. It starts from a clearly defined analytical goal, method selection, optimization, and development, which is called prevalidation considerations before arriving at the elaboration of a validation protocol and is the starting point of the actual validation. After performing all the experiments described in the validation protocol, obtained data are evaluated and compared with the acceptance criteria. If all criteria are met, the method can be regarded as valid. In a less-formal approach, some validation data may be incorporated from experiments, which were conducted previously as part of the method development.

The above approach is widely accepted for validation of qualitative HPTLC methods for identification during routine use. It is possible that the validation method in different situations may require some changes in the standard validation protocol. Such changes may include restrictions with respect to relative humidity, waiting times, precision, etc. The validation protocol is a key instrument for structuring, regulating and documenting the validation processes, depending on the quality management system. The following elements must be included:

Selectivity

Ability of the developed analytical method is to detect analyte quantitatively in the presence of other components which are expected to be present in the sample matrix. Results are expressed as Resolution. If the expected impurities or related substances arc available, they should be chromatographed along with the analyte to check the system suitability, retention factor, tailing factor, and resolution.

Sensitivity

Ability of the method within a given range to obtain test results in direct proportion to the concentration of analyte in the sample – calibration curve for the analyte.

Precision

Precision provides an indication of random error. Its results should be expressed as relative standard deviation (RSD) or coefficient of variation (COV). Precession is observed in terms of *replication*: precision under same conditions, same analyst, same apparatus, short interval of time and identical reagents using the same sample; *measurement of peak area*: RSD should not be greater than 1%, based on seven times measurement of same spot; *peak position*: RSD should not be greater than 2% based on seven times repositioning the instrument after each measurement; *sample application*: equal volume applied as seven spots and RSD should not be greater than 3% and under different conditions, different analyte, different laboratory, and different days and reagents from different sources using the same sample. RSD should not be greater than 10% within laboratory reproducibility.

Accuracy

Accuracy of an analysis is determined by systematic error involved. It is defined as closeness of agreement between the actual value and mean analytical value obtained by applying the test method a number of times. The accuracy is acceptable if the difference between the true value and mean measured value does not exceed the RSD values obtained for repeatability of the method. This parameter is very important for formulated pharmaceutical dosage forms as it provides information about the recovery of the analyte from sample preparation and effect of matrix. It the recovery rate is found to be 100%, it implies that the proposed analytical method is free from constant and proportional systematic error. A blank matrix and known impurities must be available to test the accuracy of the method.

Ruggedness

This is one of the most important parameters for validation of HPTLC method. Experiments are usually recommended to evaluate ruggedness of a HPTLC method like *sample preparation*: composition, quantity of solvent, pH, shaking time, temperature and number of extractions; *sample application*: volume applied, spot shape and size, band and spot stability; *separation*: at least on three different plates; *chromatographic conditions*: chamber saturation, eluent composition, eluent volume, temperature, humidity and development distance; *spot visualization*: post-chromatographic derivatization, spraying, dipping, reaction temperature and time; *quantitative evaluation*: drying of plates, detection and wavelength.

Once the analytical method is developed, it should be performed independently by three analysts well conversant with practical aspects of the technique, analyzing the same sample under same experimental conditions to check reproducibility of the method.

Limit of Detection

Lowest amount of analyte that can be detected is not greater than 10% of the individual impurity limit. If this is not possible, then amount of analyte to be applied has to be increased. Limit of detection (LOD) is determined on the basis of signal to noise ratio. Mean of 15 noise peak areas and their absolute SD values are determined. LOD is the amount of applied sample producing a peak area which is equal to the sum of mean blank area and three times standard deviation.

Stability

Analyte should not decompose during development of the chromatogram and should be stable in solution and on the sorbent for at least 30 and 15 min, respectively. The intensity of the spot on the chromatogram should be constant for at least 60 min while optimization of the extraction/purification procedure and one must keep in mind the chemical properties and purity of the extraction solvent. Chemical reaction of the solvents and their impurities may produce extra spot/peak, thus leading to false assay values. Other important factor is pH of the aqueous phase used for extraction/purification which may lead to hydrolysis, oxidation and isomerization. The complete removal of organic solvent should be avoided.

HPTLC: Optimization and Process Control

A standard methodology is applied for optimization. Sample preparation, in most cases, a 5-min sonication with methanol, followed by centrifugation and using the supernatant as test solution, yields satisfactory results. Derivatization is optimized with the goal of convenience, safety, and reproducibility. Botanical Reference Materials (BRM) of known adulterants are used to ensure sufficient specificity of the method. Small modifications of the mobile phase composition are applied to fine-tune separation. Each step of the optimization process is documented for complete traceability. The optimization of the chromatographic mobile phase proved to be possible when the number of experimental determinations of separation parameters for each compound is obtained for more than one distinct compositions of mobile phase, at least equal with the number of variable use in the mathematical model. A mobile phase optimization program based on an original mathematical approach is to be developed for its performances by applying on three sets of compounds. The original optimization procedure starts from the idea that into a mixture of three solvents the quantitative measure of the choused chromatographic parameter is dependent on composition of mobile phase through an equation of dependency with six or seven parameters, taking into consideration the molar fraction of the solvents. The optimization procedure is included in a program and applied on three sets of previously studied compounds through high-performance thin-layer chromatography with three solvents. The mobile-phase optimization process proved to be able to provide accurate, precise, and reproducible method of characterization and analysis of chromatographic parameters.

HPTLC: Automation

For the past 50 years, both automatic and automated instruments have been used to monitor and control process stream, such as density, viscosity, and conductivity. It is necessary to distinguish between the characteristics of automatic and automated devices. According to the current definitions of the International Union of Pure and Applied Chemistry (IUPAC), both devices are designed to replace, refine, extend or supplement human effort and facilities in the performance of a given process. The unique feature of automated devices is the feedback mechanism, which allows at least one operation associated with the device to be controlled without human intervention. An automatic photometer might continuously monitor the absorbance of a given component in a process stream, generating some type of alarm if the absorbance exceeds a preset value. By contrast, an automated system could transmit absorbance values to a control unit that adjusts process parameters (temperature and amount of additional reagent) to maintain the concentration of the measured component within preset limits. In spite, of this fundamental difference, the terms automatic and automated are often interchanged.

The use of automated sample processing, analytics and screening technology for profiling absorption, distribution, metabolism, excretion, and physicochemical properties is becoming more widespread. The use and application of these technologies is both diverse and innovative. High throughput screening technologies have been utilized enabling the profiling of an increased number of compounds. Although the drivers for using these technologies are common, different approaches can be taken. Control Systems, Safe, efficient, and economical operations of chemical processes are ever more dependent in the use of online analyzers. The use of analytical measurements of component properties in near real time for process control during manufacturing is becoming more common. The combination of online analyzers and advanced control technologies holds an enormous economic potential. As a result, the number of existing applications of HPTLC is growing steadily.

Advances in science and technology have raised an increasing demand for control analyses and posed various challenges to analytical chemists such as the need to develop new methods exhibiting as much selectivity, sensitivity, sample and reagent economy, throughput, cost-effectiveness, simplicity, and environmental friendliness as possible. The large of number of samples, with which analysts can be confronted, imposes the use of expeditious automatic methods. Despite the major conceptual and operational differences between partly and fully automated methods, the two are frequently confused. Thus, a fully automated method allows the whole analytical process to be completed with no intervention from the analyst; also, it can by itself make the decision as to whether the operating conditions should be altered in response to the analytical results. All methods are deemed automated simply because one or several steps of the analytical process are performed in an automated manner. However, an automated method should be capable of completing all steps including sampling, sample preparation and dissolution, interference removal, aliquot withdrawal, analyte measurement, data processing, result evaluation, and decision making, and also of restarting the whole process in order to adapt it to the particular needs of a new sample if needed.

A fully automatic method is very difficult to develop especially for solid samples, the first steps in the analysis of which can rarely be performed in an inexpensive manner. Usually, the operations posing the greatest difficulties among those involved in such steps are those requiring some mechanical handling, automation of which is only possible in most cases by using robot arm adapted to the particular chemical operations to be performed. Because this equipment is too expensive for most analytical applications, fully automated methods for the analysis of solid samples are very scant and largely restricted to the control of manufacturing processes. The automation of analyses involving fluid samples is facilitated by their usually adequate homogeneity and easy mechanical handling by the use of peristaltic or piston pumps, or some other liquid-management devices. This is not the case with solid samples, analysis of which frequently involves their prior conversion into liquids by dissolution. The dissolution step is the bottleneck of analytical processes involving solid samples as it is frequently slow and must be performed manually. The earliest automatic methods used dedicated devices suited to the particular application. This restricted their scope to very specific uses such as the control of manufacturing processes or in those cases where the number of samples to be analyzed was large enough to justify the initial effort and investment required. The computer-controlled techniques have introduced a great number of advantages to HPTLC systems mainly a dramatic decrease of the needed sample and reagents volumes, and have allowed the introduction of the concept of unit laboratory operations.

HPTLC: Hyphenation

Over the past several years, the concept of hyphenation has gained rapid growth in the pharmaceutical industry because of its ability to produce a large number of compounds with a wide range of structural diversity in a short time. The combinational approach (hyphenation) has received a significant recognition compared to a traditional one-compound-at-a-time approach.

The various steps having potential for the advancements on the thin-layer chromatography are methods to provide a constant and optimum mobile-phase velocity, video densitometry for recording multidimensional chromatograms, in situ scanning, and monitoring for selective detection. These improvements dramatically increased the use of thin-layer chromatography in the form of HPTLC. Today, thin-layer chromatography has been successfully hyphenated with high-performance liquid chromatography (HPLC), mass spectroscopy (MS), Fourier transform infra-red (FTIR), and Raman spectroscopy to give far more detailed analytical data on separated compounds. Even the UV/visible diode array technique has been utilized in TLC to determine peak purity or the presence of unresolved analytes.

Liquid Chromatography–Thin-Layer Chromatography (LC–TLC)

The most general interface for coupling column liquid chromatography to thin-layer chromatography (LC–TLC) is based on different modification to the spray-jet applicator. Flow rates typical for mobile phase can be applied to the layer. A splitter in the transfer line to the spray-jet applicator is required to accommodate higher flow rates from wider-bore columns. The column eluent is nebulized by mixing with nitrogen gas and sprayed as an aerosol onto the layer. The spray head is moved horizontally on one line within a defined bandwidth. Contemporary interest in LC–TLC remains weak. The main problems are more on the detection and data handling side than separations. It is simpler to obtain mass spectral information from the solution phase using liquid chromatography–mass spectrometry (LC–MS) than to either quantify or identify separated bands by thin layer chromatography–mass spectrometry (TLC–MS).

High-Performance Thin-Layer Chromatography–Mass Spectrometry (HPTLC–MS)

The combination of chromatographic separations with mass spectrometric detection is considered an indispensable tool for problem solving in analytical chemistry and increasingly for routine analytical methods. Mass spectrometric detection brings an added level of information, complementary to the chromatographic process that improves the certainty of identification and the specificity of detection. Mass spectral information can generally be obtained from sample sizes typical of common analytical methods. HPTLC–MS is mainly a research tool available to a small number of research groups. The evolution of HPTLC–MS has been slow compared with LC–MS. The challenge was to develop an automated system for in situ acquisition of mass spectral data directly from layers with retention of the spatial integrity of the chromatographic separation. This is certainly not a simple problem but is a problem of some importance, since it restricts the range of applications that HPTLC is considered suitable.

For more than 20 years, efforts have been made to hyphenate HPTLC with mass spectrometry, similar to that of HPLC and MS. Dr. Luftmann Head of the Mass Spectrometry Department at the Institute of Organic Chemistry of the University of Munster, Germany, developed an interface (ChromeXtractor) which allows such HPTLC–MS hyphenation. Dr. Morlock, assistant professor at the Institute of Food Chemistry of the University of Hohelnheim in Stuttgart, Germany, modified ChromeXtractor and demonstrated the performance of this versatile interface in comparison to other technical solutions for hyphenation. The substance of interest is eluted directly form the HPTLC plate and is transferred online into the mass spectrometer.

Component mixtures, even with heavy matrix load, can be separated cost efficiency on HPTLC plates. If the target zone is not visible, it can be marked either under UV 254 nm or UV 366 nm, by extrapolation of the adjacent zone made visible by derivatization. The HPTLC–MS interface is operated in semiautomatic mode which means that after manual positioning of the zone the piston is lowered at the push of a button. Moving a lever starts the solvent flow through the layer and extracts the zone. Previously, data acquisition has to be started by flow injection analysis/direct flow infusion/placebo injection or the direct data acquisition window, followed by the cleaning procedure and the HPTLC–MS interface can be used for next analysis. Hyphenating HPTLC with MS appears to hold considerable promise for those analysts who previously have had reservations towards the use of planar chromatography. The hyphenation opens a new dimension for the technique and makes it more prestigious from the scientific view. Recently, HPTLC–MS has been successfully used for the identification and quantification of amino acid in peptides, fast identification of unknown impurities, problem-solving technique in pharmaceutical analysis, identification of botanicals, screening for bioactive natural products in sponges, determination of ginkgolides A, B, and C and bilobalide in *Ginkgo bilodes* and identification of *Hoodia gordonii* a popular ingredient of botanical slimming products.

High-Performance Thin-Layer Chromatography–Infrared Spectroscopy (HPTLC–IR)

In recent years, much effort has been devoted to the coupling of high-performance thin-layer chromatography (HPTLC) with spectrometric methods. It is because of the robustness and simplicity of use of HPTLC and the need for detection techniques that provide identification and determination of sample. Infrared (IR) is one of the spectroscopic methods that have been coupled with HPTLC. IR spectroscopy has a high potential for the elucidation of molecular structures and characteristic absorption bands. Almost all chemical compounds yield good IR spectra that are more useful for the identification of unknown substances and discrimination between closely related substances. The HPTLC and FTIR coupling can be divided into two groups – indirect and direct methods. Indirect coupling involves either the transfer of the substance from a TLC spot to a nonabsorbing IR material (KBr or KCl) or in-situ measurement of excised HPTLC spots when the spectra are recorded directly from the plate. The direct methods are based on the direct hyphenated HPTLC–FTIR technique introduced in 1989 by Glauninger and co-workers. Until then, the combination of HPTLC and ultraviolet–visible (UV–VIS) spectroscopy was the only on-line coupling method available in planar chromatography. The information content of UV–VIS spectra is rather poor and rarely enables unambiguous identification of a substance; furthermore, a chromophore is needed for UV detection. The HPTLC–FTIR spectra make possible the detection and quantification of even non-UV absorbing substances on HPTLC plates. These reasons make this hyphenated technique more universally applicable. The direct on-line coupled HPTLC–FTIR offers advantages relative to other hyphenated techniques (HPTLC–Raman spectroscopy, HPTLC–PA, and HPTLC–MS), such as: the ease of operation and the optimized operational aspects of on-line coupling. The HPTLC–FTIR coupled method has been widely used in modern laboratories for qualitative and quantitative analysis. The potential of this method is demonstrated by its application in various fields of analysis such as drug, forensic, food, environmental, and biological analysis, etc. The hyphenated HPTLC–FTIR technique will continue to be developed in the future with the aim of taking full advantage of this method's capabilities.

HPTLC: Laser

For the purpose of investigation, atmospheric pressure-matrix-assisted laser desorption/ionization is chosen for a certain applications depending on the analytes. Combination of laser desorption and APCI was recently developed in which desorption (laser) and ionization (APCI) were well decoupled. This combination was easily incorporated into HPTLC/MS system. Such a system benefits from the high spatial resolution of the laser, simple transfer of analyte molecules, compatibility with modern mass spectrometric systems and less fragmentation under atmospheric pressure. One drawback of such a system is that the cost for a traditional pulsed laser system is relatively high which somehow counteracts the advantage of TLC in the low costs. The size of the laser system is also not ideal for a miniaturization of the whole analytical system.

The initial efforts were carried out on a graphite plate (photon-absorbing material). A continuous wave diode laser replacing traditional pulsed lasers was employed for this purpose. The thermally desorbed analyte was ionized in the gas phase by a corona discharge device at atmospheric pressure and detected mass spectrometerically. Both essential processes, desorption and ionization of analyte molecules are separated. The technique was subsequently applied to thin-layer chromatography to realize the combination of TLC and mass spectrometry. Thus, a graphite suspension was employed to couple the laser energy and improve the desorption efficiency. The power density for desorption was decreased by two orders of magnitude. In addition, a TLC plate-scanning device was developed, by which the chromatography on a TLC plate can be recovered and rapid screening for numerous analytes on a TLC plate was obtained. The device can also be applied for the identification of unknown compounds and overlapping sample spots. Finally, a quantification method for this system was developed. An internal standard was added into the mobile phase to yield a "background" signal, which was used as a reference signal for the quantification. If TLC plates with embedded graphite particles would become commercially available in the future, laser diode desorption-APCI-MS analysis would be facilitated and the pretreating time shortened.

HPTLC: Multidimensional Applications

HPTLC is now frequently used in the identification of hydrocarbon, alcohols, phenols, carbohydrates, ethers, epoxides, organic acids and lipids, organic peroxides, steroids, steroid glycosides, saponins, terpenoide glycoside, alkloides, nitro and nitroso compounds, amino acids and peptides, proteins, enzymes, nucleic acids, organic sulfur and phosphorus compounds, organometallic compounds, vitamins, growth regulators, antibiotics, pesticides and agrochemicals, synthetic and natural dyes, plastic and their intermediates and also in pharmaceutical, environmental, toxicological, forensic, and food chemical applications. TLC is a routine tool in the monitoring of synthesis processes, HPTLC offers several advantages over the present methods; such as fast, simple, and inexpensive analysis of many samples simultaneously; a disposable stationary phase, possibility to use a number of nondestructive detection methods; and cost-effective reagents. In recent years, HPTLC research has entered the chiral-separation field using a number of chiral selectors and chiral-stationary phases.

Notes

HPTLC is a most versatile technique and is known for uniformity, purity profile, assay values and precision and accuracy of results. It can handle several samples of even divergent nature and composition. It is accepted as a time-saving and most economical machine practically with minimum trouble shootings. It speeds up analysis work which is usually not possible with other parallel chromatographic techniques available. The scope of hyphenation of HPTLC with other analytical techniques appears to hold considerable promise for the analysts who previously have had reservation towards the use of planar chromatography. Its hyphenation with mass/infra red/laser spectroscopy, etc. opens a new dimension which makes it the most prestigious among the analytical chemists in the present perspective. Undoubtedly, HPTLC is a modern analytical separation method with extensive versatility, although already much utilized, is still with great potential for future development into areas where research apparently is only just beginning.

Suggested Readings

Adegbola AR, Dare AA and Lasisi EO. Spray reagents for the visualization and detection of sesame oil unsaponiiables on thin-layer chromatograms. Chromatographia, 2008, 68, (1–2), 151–155.

Ahmad S, Rizwan M, Abeaparveen R, Mujeeb M and Aquil M. A validated stability indicating TLC method for determination of forskolin in crude drug and pharmaceutical dosage form. Chromatographia, 2007, 67, (5–6), 441–447.

Ahuja S (1992). Chromatography and Pharmaceutical Analysis. In Chromatography of Pharmaceuticals; published by American Chemical Society: Washington, DC.

Bagócsi B, Végh Z and Ferenczi F. Optimization of the visualization of steroids separated by OPLC. J Planar Chromatogr, 2008, 21, 107–112.

Bard AJ, Faulkner LR (2000). Electrochemical Methods: Fundamentals and Applications, 2nd edition, published by John Wiley & Sons, New York.

Bauer GK, Pfeifer AM, Hauck HE and Kovar K. A. Development of an optimized sorbent for direct HPTLC-FTIR on-line coupling. J Planar Chromatogr-Mod TLC, 1998, 11, 84–89.

Bezuidenhout W and Brett MJ. Ultrathin layer chromatography on nanostructured thin films. J Chromatogr A, 2008, 118, (1–2), 179–185.

Burger K., Z. Anal. Chem., 1984, 318, 228.

Burger K. Instrumental Thin-Layer Chromatography/Planar Chromatography, Proceedings of the International Symposium, Brighton, UK, 1989, 33–44.

Cha S, Yeung ES. Colloidal graphite-assisted laser desorption/ionization mass spectrometry and MSn of small molecules. 1. Imaging of cerebrosides directly from rat brain tissue. Anal Chem, 2007, 79, 2373–2385.

Christensen JH, Mortensen J, Hansen AB and Andersen O. Chromatographic preprocessing of GC-MS data for analysis of complex chemical mixtures. J Chromatogr A, 2005, 1062, (1), 113–123.

Cserhati T, Forgacs E, Morais MH and Ramos AC. TLC-FTIR of color pigments of chestnut sawdust. J Liq Chromatogr Rel Technol, 2001, 24, 1435–1445.

Davld N, Rose WI Becker and Karen A. Sassic. Time optimization in thin-layer chromatography. Anal Chem. 1982, 54, 1955–1959.

Ellis LA and Roberts DJ. Chromatographic and hyphenated methods for elemental speciation analysis in environmental media. J Chromatogr A, 1997, 774 (1–2), 3–19.

Fuchs B, Schiller J, Süß R, Zscharnack M, Bader A, Müller P, Schürenberg M and Becker M. Analysis of stem cell lipids by offline HPTLC-MALDI-TOF MS Detlev Suckau. Anal Bioanal Chem, 2008, 392, 849–860.

Gombosuren N, Nova Z, Kotschy A, Mincsovics E and Dibo G. A multidimensional overpressured layer chromatographic method for the characterization of tetrazine libraries. J Biochem Biophys Methods, 2007, 69, 239–249.

Gopu C, Aher S, Mehta H, Paradkar A and Mahadik K. Simultaneous determination of cinnamaldehyde, eugenol and piperine by HPTLC densitometric method Phytochem Anal, 2008, 19, 116–121.

Guetens G, and De Boeck G, Highley MS, Wood M, Maes RA, Eggermont AA, Hanauske A, de Bruijn EA, Tjaden UR. Hyphenated techniques in anticancer drug monitoring. II. Liquid chromatography-mass spectrometry and capillary electrophoresis-mass spectrometry. J Chromatogr A, 2002, 976, (1–2), 239–247.

Guetens G, De Boeck G, Wood M, Maes RA, Eggermont AA, Highley MS, van Oosterom AT, de Bruijn EA and Tjaden UR. Hyphenated techniques in anticancer drug monitoring I. Capillary gas chromatography-mass spectrometry. J Chromatogr A, 2002, 976, (1–2), 229–238.

Halpaap H. J. Chromatogr, 1973, 78, 77–78.

Halpaap H, Krebs KF and Hauck HE. J. HRC and CC, 1980, 3, 215–240.

Hauck HE and Jost W, 'Instrumental High Performance, Thin-Layer Chromatography', Proceedings of 2nd International Symposium, Kaiser RE (ed.), Interlaken, Switzerland, 1982, 25–37.

Hhodisan Y, Ccasoni D, Beldean-Ggalea MS and Cimpoiu C. Identification and quantification of tocopherols in vegetable oil by thin-layer chromatography. J Planar Chromatogr, 2008, 21, 213–215.

Saxena S, Jain DC, Gupta MM, Bhakuni RS, Mishra HO and Sharma RP. High-performance thin-layer chromatographic analysis of hepatoprote diterpenoids from *Andrographis paniculata.* Phytochem Anal, 2000, 11, 34–36.

Holt RM, Newman MJ, Pullen FS, Richards DS and Swanson AG. High-performance liquid chromatography/NMR spectrometry/mass spectrometry: further advances in hyphenated technology. J Mass spectrometry, 1997, 32, (1), 64–70.

Janecki DJ, Novotny AL, Woodward SD, Wiseman JM and Nurok D. A preliminary study of the coupling of desorption electrospray ionization-mass spectrometry with pressurized planar electrochromatography. J Planar Chromatogr, 2008, 21, 11–14.

Jäntschi L, Bolboacă DS, Ungureşan ML. Mobile Phase Optimization in Three Solvents High Performance Thin-Layer Chromatography: Methodology and Evaluation. 6th European Conference on Computational Chemistry, September 3–7, 2006, Slovakia.

Kaiser RE. Instrumental Thin-Layer Chromatography/Planar Chromatography', Proceedings of the International Symposium, Brighton, UK, 1989, 251–262.

Kirchner JG, Miller JM and Keller GE. Anal. Chem., 1951, 23, 420.

Kirchner JG. (1978) Thin-Layer Chromatography, 2nd edition, Techniques in Chemistry, vol. XIV, published by Wiley-Interscience, Chichester, UK.

Kowalska T, Sherma J (2007). Thin-layer Chromatography in Chiral Separations and Analysis. Chromatographic Science Series, vol. 98, published by CRC Press, Boca Raton, New York.

Kwiecien A, Krzek J and Biniek L. TLC-densitometric determination of azithromycin in pharmaceutical preparations. J Planar Chromatogr, 2008, 21, 177–181.

Leichert M, Eckhardt HS, Klein KF and Spangenberg F. A simple and reliable method for quantification of glucosamine in nutritional supplements. J Planar Chromatogr, 2008, 21, 55–59.

Mavle R, Katkar H, Daundkar B, Malve M and Krishnamurthy R. Thin layer chromatographic technique for detection and identification of endosulfan insecticide with m-dinitrobenzene reagent. J Planar Chromatogr, 2008, 21, 197–198.

Meinhard JE and Hall NF. Anal. Chem., 1949, 21, 185.

Mmirakor V, Vaidya V, Menon S and Champanerker P. HPTLC method for determination of colchicines in a pharmaceutical formulation. J Planar Chromatogr, 2008, 21, 187–189.

Naik S, Desai SK, Nanda RK and Narayanan MS. Fermentation, isolation, purification, and biological activity of SJA-95, a heptaene polyene macrolide antibiotic produced by the Streptomyces sp. Strain S24. Arzneim-Forsch/Drug Res, 2007, 56, 171–179.

Orinák A, Talian I, Efremov EV, Ariese F and Oriáakovác R. Diterpenoic acids analysis using a coupled TLC-surface-enhanced Raman spectroscopy system. Chromatographia, 2008, 67–74, 15–19.

Patel BH, Suhagia BN, Patel MM and Patel JR. High-performance liquid chromatography and thin-layer chromatography for the simultaneous quantitation of rabeprazole and mosapride in pharmaceutical products. J Chromatogr Sci, 2008, 46, 10–14.

Patel RK and Prajapati AM. Development and validation of a visible absorption densitometry method for quantitation of conessine in *Holarrhena antidysenterica* (Kurchi). J AOAC Int, 2008, 91.

Peterson BL and Cummings BS. A review of chromatographic methods for the assessment of phospholipids in biological samples. Biomed Chromatogr, 2006, 20, 227–243.

Polak B, Halka A and Dzido T. Pressurized planar electrochromatographic separation of the enantiomers of tryptophan and valine. J Planar Chromatogr, 2008, 21.

Poole CF. Thin layer chromatography: challenges and opportunities. J Chromatogr, 2003, 1000, 963–984.

Prichard E and Barwick V (2007). Quality Assurance in Analytical Chemistry. Analytical Techniques in the Sciences; published by Wiley-Interscience: Hoboken, NJ.

Rahul G, Leena B, Asfak V and Mrinalini D. A validated high performance thin layer chromatographic method for simultaneous estimation of oloxacin and satranidazole in pharmaceutical dosage form. J Pharm Res, 2007, 6, (4).

Randerath K. In Thin-Layer Chromatography, Academic Press, London, UK, 1962.

Reich E, Schibli A and Debatt A. Validation of HPTLC methods for the identification of botanicals in a cGMP environment. J AOAC Int, 2008, 91, 10–20.

Reitsema RH. Anal. Chem., 1954, 26, 960.

Salo PK, Pertovaara AM, Salo VMA, Salomies HEM and Kostiainen RK. High-performance thin-layer chromatography method for assessment of the quality of combinatorial libraries, and comparison with liquid chromatography ultraviolet mass spectrometry. J Comb Chem, 2003, 5, 223–232.

Salo PK, Vilmunen S, Salomies H, Ketola RA and Kostiainen R. Two-dimensional ultra-thin-layer chromatography and atmospheric pressure matrix-assisted laser desorption/ionization mass spectrometry in bioanalysis. Anal Chem, 2007, 79, 2101–2108.

Sean X. Peng, Charles Henson, Michael J. Strojnowski, Adam Golebiowski, and Sean R. Klopfenstein. Automated high-throughput liquid extraction for initial purification of combinatorial libraries. Anal Chem, 2000, 72, 261–266.

Somsen GW and Visser T. (2003) Liquid Chromatography/Infrared Spectroscopy. In Encyclopedia of Analytical Chemistry; Meyers RA, ed.; John Wiley & Sons Ltd: Chichester.

Soponar F, Mot AC and Sârbu C. Quantitative determination of some food dyes using digital processing of images obtained by thin-layer chromatography. J Chromatogr A, 2008, 1188, (2), 295–300.

Stahl E. (1965) Thin-Layer Chromatography – A Laboratory Handbook, published by Springer-Verlag, Berlin, Germany.

Stahlmann S and Kovar KA. Analysis of impurities by high-performance thin-layer chromatography with Fourier transform infrared spectroscopy and UV absorbance detection in situ measurement: chlordiazepoxide in bulk powder and in tablets. J Chromatogr A, 1998, 813, 145–152.

Tang T, Hong Wu. An image analysis system for thin-layer chromatography quantification and its validation. J Chromatogr Sci, 2008, 46, (6), 560–564.

Tyihák E, Mincsovics E, Kátay G, Király-Véghely Z, Móricz AM and Ott PG. An unlimited possibility of biochemical interactions in the adsorbent layer after chromatographic separation. J Planar Chromatogr, 2008, 21, 15–20.

Urakova I, Pozharitskaya O, Shikov A, Kosman V and MakaRov V. Comparison of high performance TLC and HPLC for separation and quantification of chlorogenic acid in green coffee bean extracts. J Sep Sci, 2008, 1, 207–241.

Waksmundzka-hajnos M and Józwiak GW. Special modes of development in preparative-layer chromatography of extracts of alkaloids from *Fumaria oficinalis*. J Planar Chromatogr, 2008, 21, 61–66.

Wall PE. (2005). Thin-layer Chromatography a Modern Practical Approach. RSC Chromatography Monographs Salesand Customer Care. Smith, RM, ed.; Royal Society of Chemistry, Thomas Graham House, Science Park, Milton Road, Cambridge.

Wang PG (2008). High-Throughput Analysis in the Pharmaceutical Industry; Critical Reviews in Combinatorial Chemistry; CRC Press, Boca Raton, FL.

Wilkins CL. Hyphenated techniques for analysis of complex organic mixtures, Science, 1983, 222, (4621), 291–296.

Workman J, Koch M, Lavine B and Chrisman R. Process analytical chemistry. Anal Chem, 2009, 81, 4623–4643.

Yuanling L, Mark EB, Todd Y and Bonner DM. quantitative analysis of aflatoxins by high-performance thin-layer chromatography utilizing a scientifically operated charge-coupled device detector. Anal Chem, 1996, 68, 3885–3891.

Zlatkis A and Kaiser RE. (1977) HPTLC High Performance Thin-Layer Chromatography, published by Elsevier, Amsterdam, Netherlands.

Part II
Fundamentals, Principle and Advantages of HPTLC

Chapter 2
Fundamentals and Theory of HPTLC-Based Separation

Prasad S. Variyar, Suchandra Chatterjee, and Arun Sharma

Abstract High-performance thin-layer chromatography (HPTLC) is a form of thin-layer chromatography (TLC) that provides superior separation power using optimized coating material, novel procedures for mobile-phase feeding, layer conditioning, and improved sample application. It promotes for higher separation efficiencies, shorter analysis time, lower amounts of mobile phase, and efficient data acquisition and processing. The major parameters that influence separation of the constituents within a mixture are the partition coefficients, retention factor (R_f), and capacity factor of the individual constituents on the plate, selectivity of the mobile and stationary phase to the solutes, and the plate height that decide the separation efficiency as well as resolution of the individual constituents within a mixture. The partition coefficient is defined as the molar concentration of the analyte in the stationary phase to that in the mobile phase. R_f, a fundamental qualitative value, is expressed as the ratio of migration distances of an individual components of a mix relative to the mobile phase. Capacity factor (k) is a fundamental characteristic of a substance that determines its qualitative chromatographic behavior. It can be expressed as the ratio of the retention time of the substance in the stationary to that in the mobile phase and is influenced by the chemical nature of the two phases. The separation number (SN) that influences separation power of HPLC is defined as the highest possible number of components that are completely separated in a mixture under a gradient-free isocratic TLC. The efficacy of separation of two components of a mixture in a chromatogram is termed as resolution and is influenced by the selectivity of the components between the stationary and the mobile phase, mobile phase flow rate influenced by particle size and solvent strength that influence capacity factors.

P.S. Variyar (✉), S. Chatterjee, and A. Sharma
Food Technology Division, Bhabha Atomic Research Centre, Trombay, Mumbai 400085, India

MM. Srivastava (ed.), *High-Performance Thin-Layer Chromatography (HPTLC)*,
DOI 10.1007/978-3-642-14025-9_2, © Springer-Verlag Berlin Heidelberg 2011

Planar Chromatography

Chromatography is a physical method of separation in which the components to be separated are distributed between two phases, one of which is a stationary phase over which the other mobile phase moves in a specific direction. Planar chromatography is a mode of chromatography in which the stationary phase is spread on a flat, planar surface. TLC is an important planar chromatographic technique that is widely used as a cost-effective method for rapid analysis of simple mixtures. It requires minimum sample cleanup as it uses a disposable stationary phase and has a high sample throughput because of its ability to analyze several samples in parallel. It allows greater flexibility and simplicity in sample evaluation because of the possibility of sequential detection by complementary techniques, postchromatographic derivatization for identification and quantification, archiving a separation for evaluation at a later time, and accessibility of the sample because of the planar format. It can be applied to samples where analyte lack a convenient chromophore making detection by other methods difficult and can be applied for analyzing samples with minimal prepurification. In some respects, the techniques of TLC and HPLC are complementary. Optimum resolution in TLC is obtained when R_f is about 0.25, and it is interesting to note that this corresponds to a retention factor of 3.0, which is in the optimal range found for column. Liquid chromatographic column methods provide better efficiencies, better mass-transfer properties and higher operating velocities. HPLC can also be fully automated making this method more popular for most analytical applications. By contrast, TLC is the most popular method because of its low cost, simplicity, and flexibility.

Theoretical Considerations

Separation Efficiency

In TLC, the flow of the mobile phase is not controlled as in the column methods. The position of the mobile phase at time t as it moves through a sorbent layer when governed by capillary forces is thus described by $(Z_f)^2 = \kappa t$, where Z_f is the distance moved by the mobile phase from the sample origin and κ is the velocity constant. Velocity constant is dependent on external factors including saturation level of vapor phase in contact with the stationary phase. It is *al*so related to the properties of the mobile and stationary phases by the equation

$$\kappa = 2K_o d_p (\gamma/\eta) \cos \theta, \tag{2.1}$$

where K_o is the permeability constant, d_p is the average particle diameter, γ is the surface tension of the mobile phase, η is the viscosity of the mobile phase, and θ is the contact angle. Assuming a uniform particle size distribution, the velocity constant increases linearly with the average particle size and thus the solvent front velocity is greater for coarse-particle layers. The velocity constant also depends linearly on the ratio of the surface tension of the solvent to its viscosity. Solvents that maximize this ratio are preferred for TLC. For silica gel layers, the contact angle for all common solvents is close to zero as all solvents wet it adequately. For reversed-phase layers containing bonded, long-chain alkyl groups the contact angle of the solvent increases rapidly with increasing water content of the mobile phase. At about 30–40% (v/v) water, cos θ becomes smaller than 0.2–0.3 (Wall 2005). The mobile phase is virtually unable to ascend the thin-layer plate, and chromatography becomes impossible. To improve solvent compatibility, modern reversed-phase TLC plates are prepared from large-size particles (10–15 μm) or from sorbents with a defined degree of modification that is lower than that of exhaustively silanized sorbents. Solvent compatibility is much less of a problem for polar, bonded sorbents such as 3-aminopropylsilanized and 3-cyanopropylsilanized sorbents, which are wetted by all solvents including pure water. Consequently, solvent with high viscosity and surface tension will migrate at much slower rate than that with low viscosity and surface tension. It is therefore advantageous to mix solvents with low κ values with those with higher values to improve migration rate, but care will need to be exercised to ensure that the solvents are miscible and that the appropriate polarity is maintained.

The mobile-phase velocity for the fine-particle layer declines rapidly with the migration distance until it becomes so slow that diffusion causes the spots to broaden. The coarse-particle layer is more permeable than the fine-particle layer, and both the solvent velocity and the efficiency are higher at longer plate lengths. For fine-particle layers with a development length of 5–7 cm, it should be possible to obtain up to about 5,000 theoretical plates, but it is nearly impossible to exceed this number using capillary flow-controlled development. It is futile to use solvent migration distances greater than this for HPTLC plates since the mobile-phase velocity declines to the point where zone broadening exceeds the rate of zone center separation. For coarse-particle layers (d_p = 15 μm), a development length of about 15 cm is required to obtain around 5,000 theoretical plates and, though higher numbers are possible, they will lead to long separation times. When the development length is optimized, the separation performance of conventional and high-performance TLC are not very different, the virtue of HPTLC being that it requires a shorter migration distance to achieve a given efficiency resulting in faster separations and more compact zones which are easier to detect by scanning densitometry (Fig. 2.1).

A popular approach for improving resolution under capillary flow-controlled conditions is to use multiple developments. Either one-dimensional or two-dimensional separations are possible in planar chromatography (Geiss 1972). Mobile-phase velocity can also be controlled by external means, such as in forced-flow development. The parameters of (2.1) above then become unimportant provided that sufficient pressure is available to provide the desired velocity. The limited range of mobile-phase velocities under capillary controlled flow prevent the optimum performance of a given stationary phase. This combined with shorter separation times is the rational theoretical reason to prefer forced-flow development in TLC.

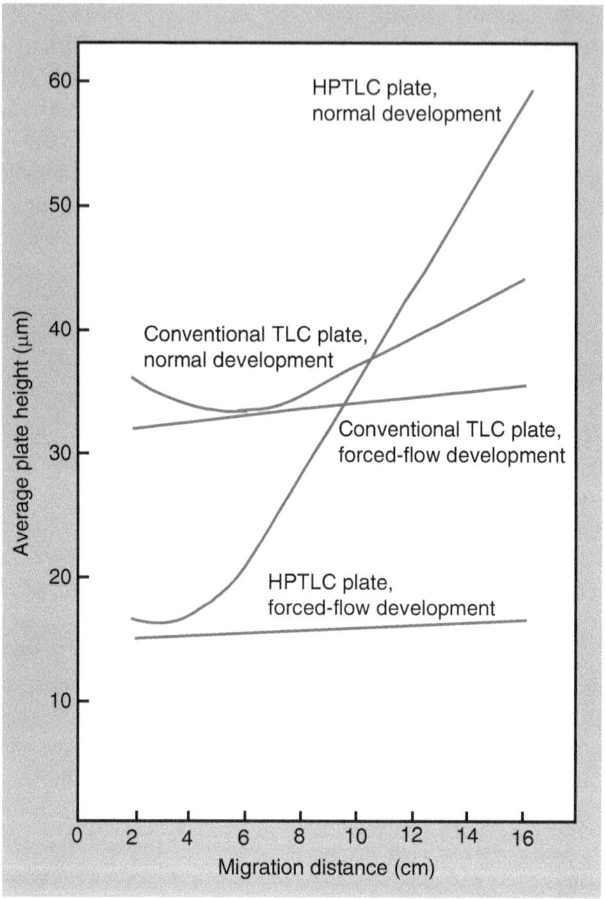

Fig. 2.1 Variation of efficiency (average plate height) of fine- and coarse-particle layers as a function of migration distance and development technique. Adapted from Poole and Poole (1989)

The use of forced flow clearly lowers the plate height, especially for the HPTLC plates. In addition, analysis times can be decreased, and gradient elution can be used. We have thus seen how smaller particle sizes that are used in HPTLC affect separation efficiencies as a result of changes in solvent flow rate. We shall now discuss some of the major parameters that influence chromatographic system efficiency with special reference to TLC and in particular HPTLC.

Partition Coefficient

In a chromatographic system, the mobile phase passes over through the stationary phase. The components of the mixture ideally equilibrate or differentially partition between the two phases. This results in differential rates of migration of the components of the mixture while passing through the system. Various components of the mixture are thus retarded in proportion to their interaction with the adsorbent. At any given time, a particular analyte molecule is either in the mobile phase, moving along at its velocity, or in the stationary phase and not moving at all in the downstream direction. The sorption–desorption process occurs many times as the molecule moves through the bed, and the time required to do so depends mainly on the proportion of time it is sorbed to the time it is held immobile. The ratio of the equilibrium concentration of an analyte in the stationary phase divided by its equilibrium concentration in the mobile phase is described by the distribution constant K_a and is represented by the equation

$$K_a = C_S/C_M, \tag{2.2}$$

where C_S is the concentration of the analyte in the stationary phase and C_M is its concentration in the mobile phase. It is this ratio that controls the rate of migration of an analyte. As analyte proceeds through the system at a given temperature, it partitions itself between the two phases and is retained in the system in proportion to its affinity for the stationary phase. A solute with large K_a has a great affinity for stationary phase and will travel slowly through the layer. Thus, for a mixture containing two constituents A and B, if A has a lower partition coefficient than B, the former will spend less time in the stationary phase and thus will migrate faster than the latter. For an ideal situation, the ratio of C_S to C_M will follow a linear correlation. However, if the spot/band is overloaded, the relationship of C_S to C_M becomes nonlinear. Overloading of sample usually results in a slight increase in R_f value. This increases with loading concentration. Nonlinear curves are undesirable because they not only cause the R_f value to vary but also can impair critical separations because of "tailing" effects and introduce unnecessary errors in quantification.

Retention/Retardation Factor

The position of any solute spot in TLC is characterized by its retention/retardation factor R_f. It is a fundamental qualitative value and is expressed as

$$R_f = \text{Distance travelled by the analyte } (Z_s) \,/\, \text{distance travelled by the solvent front} (Z_f).$$

R_f values range from 1.0 for analyte migrating to the solvent front to 0.0 for an analyte strongly retained at the point of application (Fig. 2.2).

It is sometimes more convenient to report as $R_f \times 100$. The reproducibility of R_f values depends on many factors, such as quality of the sorbent, humidity, layer

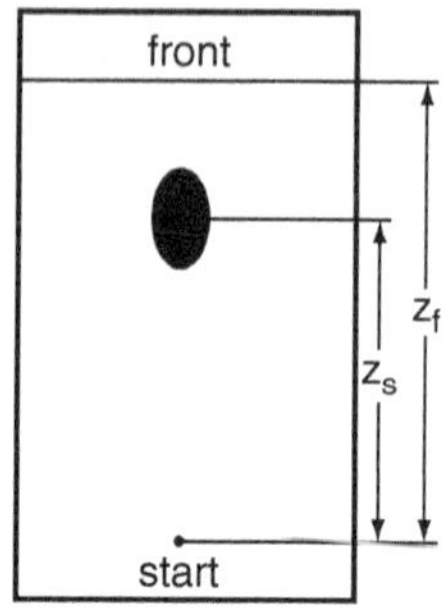

Fig. 2.2 The thin-layer chromatographic parameters used in calculation of R_f. Adapted from Sherma and Fried (2003)

thickness, development distance, and ambient temperature. Overloading of sample usually results in a slight increase in R_f value. System errors affect this fundamental qualitative TLC value when the exact position of the solvent front cannot be located. In this case, the R_f values become systematically too large. The loss of mobile phase or the "piling up" of mobile-phase components already present results in the values becoming too small. These R_f-values are not real and cannot be used to calculate k, the capacity factor. Real values can be obtained when no gradient exists along the separation path, no loss of mobile phase occurs, the correct position of the solvent front can be measured without errors, and by excluding any influences resulting from prevaporization. Consequently, a HPTLC chamber for obtaining real R_f values has almost no gas phase, does not exhibit a temperature gradient, the layer is in a horizontal position, and is fully symmetrical. This is the only effective way of eliminating any disturbance of the phase ratios. In other approaches, complex equipment is used to compensate and regulate the many variables which influence the R_f value. R_f, also known as the retardation factor in planar chromatography (such as TLC), has the same meaning, as the retardation factor R in column techniques. A comparison of HPLC and TLC for systems with equivalent stationary and mobile phases can be made by comparing their respective k values, each calculated with the appropriate retardation factor, R or R_f. In conventional thin-layer chromatography, or especially for the techniques of overrun (continuous) or multiple developments, where the solvent front is not measurable, Rx values can be recorded. Rx is defined by the equation

$$Rx = \text{Distance traveled by solute/distance traveled by standard substance.}$$

Unlike R_f, Rx values can be greater than 1. Neither R_f nor Rx values are true constants, but Rx values are more reproducible than absolute R_f values and should be preferred for purposes of identification when comparing sample mobilities to tabulated values.

The optimum method for obtaining tentative identification of a substance is to spot the sample and a series of reference compounds on the same chromatogram. In this way, mobilities of all compounds are compared under the same experimental conditions, and a match in R_f values between a sample and standard is evidence for the identity of the sample. Experimental conditions should be chosen so that the compound to be identified moves to a point near the center of the layer ($R_f = 0.5$) and resolution between spotted standards is as good as possible. If R_f values on silica gel are higher than desired, the polarity of the mobile phase is reduced. For R_f values that are too low, the polar component of the mobile phase is increased. If the spots of interest from the sample do not correspond with the standards, they are either not the same or their mobilities have been affected by accompanying extraneous material from the sample. A correlation of the R_f values between circular and linear HPTLC was proposed by Geiss and is expressed as R_f linear = (R_f circular)^2. This relation was found to be completely valid when the starting point is exactly in the center of the circular chromatogram. If the substance is applied at a distance from the center point, the chromatogram will resemble a run in linear TLC.

Capacity Factor

The capacity factor of a substance is defined as "the ratio of its retention time in the stationary phase to that in the mobile phase."

$$k = t_{\mathrm{s}}/t_{\mathrm{m}}. \tag{2.3}$$

This is the simplest and most fundamental formulation of the qualitative chromatographic behavior of a substance that measures the degree of retention. A measurement of capacity factor will help to determine whether retention shifts are due to the stationary phase (capacity factor is changing with retention time changes) or due to the mobile phase (capacity factor remains constant with retention time changes).

k is related to R_f by the equation

$$R_f = 1/k + 1 \text{ or } k = (1 - R_f)/R_f, \tag{2.4}$$

k can also be described by the ratio of total number of moles of analytes in each phase. It is primarily controlled by the strength of the mobile phase, nature of the stationary phase, and the temperature at which separation is performed. If a substance does not migrate, i.e., not even a trace of the substance can be detected in the mobile phase, then its elution would last $\infty x t_{\mathrm{m}}$; the R_f-value is zero. A substance which migrates half of the separation length ($R_f = 0.5$) would require a time of $t_{\mathrm{m}} + t_{\mathrm{m}}$ for its elution. Its retention time in the mobile phase would be the same as the retention time in the stationary phase, and therefore $k = t_{\mathrm{s}}/t_{\mathrm{m}} = 1$. A substance which migrates with the front is not retained by the stationary phase. Its R_f-value is 1 and $k = t_{\mathrm{s}}/t_{\mathrm{m}} = 0/1 = 0$. When retention factor is less than one, elution of the analyte is so fast that accurate determination of its retention time is very difficult. High retention factors (greater than 20) mean that elution takes a very long time. Ideally, the retention factor for an analyte is between 1 and 5. Thus, higher the capacity factor, the longer the retention time.

Spot Capacity

The separation number or spot capacity is defined as the maximum number of substances, which are completely separated between $R_f = 0$ and $R_f = 1$, provided that the separation conditions are isocratic. In quantitative chromatography, the separation number is defined as the highest possible number of completely separated substances, between $k = 0$ and $k = 10$. Two substances are completely separated, when the distance between the two adjacent peak maxima and the sum of both peak widths at half-peak height are same. It thus measures the efficiency of separation. Separation number is a quantity measurable with sufficient accuracy, simple, relatively independent of the chromatographic technique, easy to handle, and suitable for comparison and optimization.

Separation number is given by the expression

$$\mathrm{SN} = (Z_\mathrm{f}/b_0 + b_1) - 1. \tag{2.5}$$

Z_f = migration distance of the front
b_o = extrapolated width of the spot at $R_f = 0$
b_1 = extrapolated width of the spot with $R_f = 1$

In a TLC densitogram, widths are measured at half-height. A typical capillary-controlled HPTLC has a separation number of 10–20 while a capacity of 40–80 is possible for forced-flow TLC. A spot capacity of 100–400 and 500–3,000 is indicated for two-dimensional TLC with capillary flow and forced flow, respectively (Poole and Poole 1995).The accuracy of the R_f-value determination depends on SN and is interrelated by the formula

$$\Delta R_f = 1/\mathrm{SN} + 1, \tag{2.6}$$

where $\Delta R_f = R_f$ differences between two adjacent, but completely separated spots.

Plate Height

The most common measure of the efficiency of a chromatographic system is the plate number, an analogy with distillation, originally called the number of theoretical plates contained in the chromatographic column system. Lower separation efficiency of TLC in comparison with liquid chromatography has resulted in the concept of theoretical plates being less frequently applied to this separation technique. A successful attempt in recent years to enhance efficiency of the technique has, however, led to the application of theoretical plate concept to thin-layer chromatography. Broadening of a chromatographic spot can be simply expressed in terms of the theoretical plate number N of the given chromatographic system:

$$N = 16 Z_f Z_s / W_s, \tag{2.7}$$

where, Z_f and Z_s are the migration lengths of the mobile phase and solute, respectively, and W_s is the chromatographic spot width in the direction of the mobile-phase migration.

N can also be calculated by the equation:

$$N = 16 (Z_s / W_s)^2. \tag{2.8}$$

The calculation of plate number based on this equation is demonstrated in Fig. 2.3.

Although the numerical values of N obtained for different solutes on the same chromatographic plate coincide fairly well, they usually differ significantly for another plate type. For this reason, the quantity N can be regarded as an approximate measure of the separating efficiency of chromatographic plates. It is proportional to the migration length of the mobile phase Z_f, such that, when Z_s/W_s ratio remains constant, an increase in Z_f results in an increase of N and better separation. This proportionality of N and Z_f is given by the relationship

$$N = Z_f / H, \tag{2.9}$$

where H is the so-called HETP value (i.e., height equivalent of a theoretical plate). The quantity H, or simply the plate height, measures the efficiency of a given

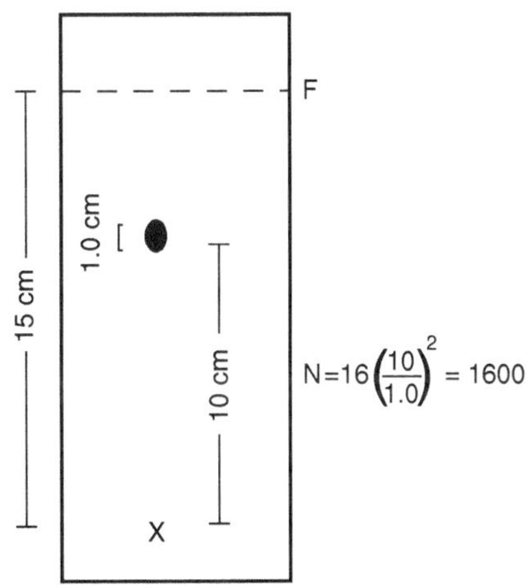

Fig. 2.3 Calculation of plate number, X = origin, F = solvent front. Adapted from Fried and Sherma (1999)

chromatographic system per unit length of the migration distance, Z_f, of the mobile phase. Small H values mean more efficient chromatographic systems and larger N values. The main goal of efforts to enhance performance of thin layers is the attainment of small H values and maximum N values. As in other chromatographic techniques, the efficiency of a given TLC system is better (i.e., H is smaller) for smaller particles of stationary phases or supports, lower mobile-phase flow rates, less-viscous mobile phases, and smaller solute molecules.

The plate height or number of plates is dependent on diffusion processes which result in a spreading of the sample spot. This is true for all chromatographic techniques. The determination of the number of plates is meaningful for the substance with $R_f = 1$, i.e., the substance which migrated the longest possible distance. One of the most important chromatographic relationships, the Van Demeter equation, attempts to estimate the relative contributions of eddy and molecular diffusion, and of the effects of mass transfer, on H. It is an empirical equation, originally established for column chromatographic techniques but valid also for thin-layer chromatography. The Van Demeter relationship can be written as

$$H = \mathrm{Aw}^{0.33} + \mathrm{B/u} + \mathrm{Cu} + \mathrm{Du}, \qquad (2.10)$$

where, u is the flow rate of the mobile phase and A, B, C, and D are the equation constants, measuring contributions of the different spot-broadening processes to the quantity H. The effects of eddy diffusion and mass transfer on the flowing mobile phase are described jointly by A. The molecular diffusion is reflected in B, while C and D correspond to the effects of mass transfer in the stagnant mobile and stationary phases, respectively. The constants A, B, C, and D depend mostly on the parameters of the microporous solid, but they are also influenced by the nature of the solute and the mobile phase and by the working temperature of the chromatographic system.

Resolution

The main goal of chromatography is separation of constituents in a mixture. Chromatographic spots of two adjacent solutes may, however, overlap to a smaller or greater degree in many cases. The separation between two spots is measured by the quantity R_s, and is called resolution (Miller 2005). The R_s of two adjacent chromatographic spots 1 and 2 is defined as being equal to the distance between the two spot centers divided by the mean spot widths (Fig. 2.4)

$$R_s = z_2 - z_1/0.5(w_1 + w_2). \qquad (2.11)$$

The quantity R_s serves to define separation. When $R_s = 1$, the two spots are reasonably well separated. R_s values larger than 1 indicates better separation while

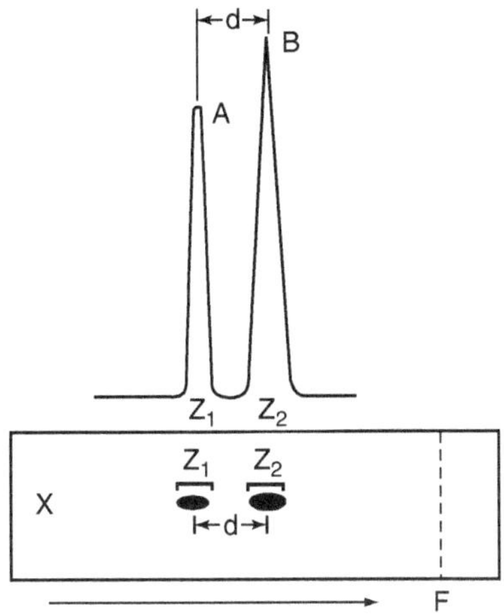

Fig. 2.4 Chromatographic resolution determined from spot or densitometric scan. Adapted from Fried and Sherma (1999)

smaller than 1 suggest a poor separation. Spot overlap becomes more disturbing when the concentration of solute in one spot is much greater than that in the other.

R_s is also given by the equation

$$R_s = R_{f(2)} - R_{f(1)}/0.5(w_1 + w_2), \qquad (2.12)$$

where $R_{f(1)}$ and $R_{f(2)}$ are the R_f values of chromatographic spots 1 and 2, respectively.

Assuming Gaussian concentration profiles of two closely spaced (i.e., overlapping) chromatographic spots and mean R_f value for both of them ($R_{f(1)} \approx R_{f(2)} = R_f$), the above equation can be written as

$$R_s = 0.25 \overset{(I)}{(K_2/K_1 - 1)} \overset{(II)}{(R_f N)^{1/2}} \overset{(III)}{(1 - R_f)}, \qquad (2.13)$$

where K_1 and K_2 are distribution coefficients of solutes 1 and 2 between the stationary and mobile phases ("distribution" is used in a general sense and means partition, adsorption, or any other phenomenon, depending on the retention mechanism of a particular chromatographic technique). Equation (2.12) is the thin-layer chromatographic version of a fundamental chromatographic relationship that allows discussion of spot resolution in terms of the influence of K_2/K_1, N, and R_f. K_2/K_1 monitors interdependence between the stationary and mobile phases, R_f monitors elution strength of the mobile phase, and N depends on the length of the mobile-phase migration and on the plate height i.e., Z_f and H, respectively.

Selectivity

Selectivity of separation is seldom referred to in the case of thin-layer chromatography, although no serious reason can be given for avoiding this term. To the contrary, selectivity of separation is a useful chromatographic notion, no matter which particular technique, column or planar, is being considered. In the case of thin-layer chromatography, the separation factor *a* can be defined as $\alpha = K_2/K_1$ (14) which remains in full conformity with the definition used for the column technique. In fact, the quantity α makes use of part of term I in (2.13) describing the resolution R_s of two overlapping chromatographic spots. It can be stated that with greater difference between distribution coefficients of solutes 1 and 2 ($K_1$and K_2), greater selectivity of separation (α) and better resolution (R_s) are observed. With $K_1 = K_2$, the two chromatographic spots entirely overlap ($\alpha = 1$) and the respective spot resolution R_s is nil. Several options for increasing α are available, and these can be ranked in order of decreasing impact; change in mobile-phase composition, change in mobile-phase pH, change in stationary phase, change in temperature, and special chemical effects.

Notes

For the optimization of chromatographic separation, the conditions for separation on a specified coating material can be optimized by the right choice of mobile phase, i.e., its chemical composition. This can aid in achieving satisfactory separation within a minimum migration distance i.e., in shortest possible time. The minimum migration distance, necessary to obtain a separation (according to R_f-values) can be calculated when SN is known. The separation number, the result of a combined action between mobile phase and stationary phase, is determined by the plate height value. The plate height is dependent on the coating material and by starting spot width which in turn is a function of the sample loading (dosage) technique. A satisfactory resolution requires a minimum mobile-phase migration distance Z_f, i.e., a minimum analysis time and a good coating material (low plate height). Best results are obtained when the system has a high selectivity for the two substances to be separated or has a sufficiently high separation number and an optimum dosage technique. Combined with a maximum selectivity, it guarantees an excellent separation. Slight inadequacies in the sample loading can considerably reduce the quality of separation, even with excellent coating material. The effect of spread starting spot of a substance with a high R_f value are less pronounced than those of a substance with low R_f value. The experimental parameters discussed above are thus all important factors that should be optimized to achieve effective separation.

References

Fried B and Sherma J (1999). Thin Layer Chromatography, 4th Edition, published by Marcel Dekker Inc., New York.

Geiss F (1972). Die Parameter der Dunnschicht Chromatographie, eine modern Einfuhrung in Grundlagen und Praxis, published by Friedr. Vieweg + Sohn\Braunschweig, now Wiesbaden, Germany.

Miller JM (2005). Chromatography-Concepts and Contrasts, 2nd Edition, published by John Wiley & Sons, Inc., Hoboken, New Jersey.

Poole CF and Poole SK. Modern thin layer chromatography. Anal. Chem., 1989, 61, (22), 1257–1269.

Poole CF and Poole SK. Multidimensionality in planar chromatography. J. Chromatogr., 1995, 703, 573–612.

Sherma J and Fried B (2003). Handbook of Thin Layer Chromatography, published by Marcel Dekker Inc., New York.

Wall PE (2005). Thin-layer Chromatography – A Modern Practical Approach, published by The Royal Society of Chemistry, Cambridge, UK.

Chapter 3
Experimental Aspects and Implementation of HPTLC

Rashmin B. Patel, Mrunali R. Patel, and Bharat G. Batel

Abstract High-Performance Thin-Layer Chromatography (HPTLC) is a sophisticated instrumentation technique. It has been reported in many publications to provide excellent separation and qualitative and quantitative analysis of a wide range of compounds, such as herbal and botanical dietary supplements, nutraceuticals, traditional western medicines, traditional Chinese medicines, and Ayurvedic (Indian) medicines (Sharma 2008). Comparative studies have often found that HPTLC is superior to High-Performance Liquid Chromatography (HPLC) in terms of total cost and time required for analysis. HPTLC is an off-line process in which the various stages are carried out independently. Important features of HPTLC are the ability to analyze cruder samples containing multicomponents; application of large number of sample and a series of standards using the spray-on technique; choice of solvents for the HPTLC development is wide as the mobile phases are fully evaporated before the detection step; processing of standards and samples identically on the same plate, leading to better accuracy and precision of quantification; different and universal selective detection methods, and in situ spectra recording in sequence to obtain positive identification of fractions; storage of total sample on layer, without time constrains. In addition, it minimizes exposure risks and significantly reduces disposal problems of toxic organic effluents; thereby, reduce possibilities of environment pollution. In view of this, HPTLC-based methods could be considered as a good alternative as they are being explored as an important tool in routine analysis. This chapter provides detailed information regarding HPTLC-based analytical method development (Renger 1993; Renger 1998; Patel and Patel 2008; Patel et al. 2010).

High-Performance Thin-Layer Chromatography (HPTLC) is a sophisticated and automated form of thin-layer chromatography. Modern instrumental HPTLC

R.B. Patel (✉) and B.G. Batel
A.R. College of Pharmacy and G.H. Patel Institute of Pharmacy, Sardar Patel University, Vallabh Vidyanagar 388120, Gujarat, India
e-mail: rashmru@gmail.com

M.R. Patel
Indukaka Ipcowala College of Pharmacy, Sardar Patel University, New Vallabh Vidyanagar 388121, Gujarat, India

MM. Srivastava (ed.), *High-Performance Thin-Layer Chromatography (HPTLC)*,
DOI 10.1007/978-3-642-14025-9_3, © Springer-Verlag Berlin Heidelberg 2011

is a powerful analytical method equally suitable for qualitative and quantitative analytical tasks. This has been demonstrated by many publications and research projects. HPTLC is a flexible, versatile and economical process in which the various stages are carried out independently. The advantages of this off-line arrangement as compared with an on-line process, such as column high-performance liquid chromatography (HPLC), have been outlined and include the following (Sherma and Fried 1996; Sethi 1996; Sherma 2007):

- Technically, it is simple to learn and operate.
- Several analysts work simultaneously on the system.
- Lower analysis time and less cost per analysis.
- Low maintenance cost.
- Visual detection possible – as it is an open system.
- Availability of a great range of stationary phases with unique selectivity for mixture components. Chromatographic layer (sorbent) requires no regeneration as TLC/HPTLC plates are disposable.
- Ability to choose solvents for the mobile phase is not restricted by low UV transparency or the need for ultra-high purity. Corrosive and UV-absorbing mobile phases can be employed.
- No prior treatment for solvents like filtration and degassing.
- There is no possibility of interference from previous analysis as fresh stationary and mobile phases are used for each analysis. No carry over, hence no contamination.
- Repetition of densitometric evaluation of the same sample can be achieved under different conditions without repeating the chromatography to optimize quantification, since all sample fractions are stored on the TLC/HPTLC plate.
- Samples rarely require cleanup.
- High sample throughput since several samples can be chromatographed simultaneously.
- Lower expenditure of solvent purchase and disposal since the required amount of mobile phase per sample is small. In addition, it minimizes exposure risks of toxic organic effluents and reduces possibilities of environment pollution.
- Accuracy and precision of quantification is high because samples and standards are chromatographed and measured under the identical experimental conditions on a single TLC/HPTLC plate.
- Sensitivity limits of analysis are typically at nanogram (ng) to picogram (pg) levels.
- Use of different universal and selective detection methods.

HPTLC is a modern adaptation of TLC with better and advanced separation efficiency and detection limits. The table below compares HPTLC and TLC (Table 3.1).

HPTLC Methodology

Set the analytical objective first that may be quantification or qualitative identification or separation of two components/multicomponent mixtures or optimization of analysis time before starting HPTLC. Method for analyzing drugs in multicomponent dosage forms by HPTLC demands primary knowledge about the nature of the

Table 3.1 Comparison between HPTLC and TLC on the basis of parameters (Patel and Patel 2008)

Parameters	HPTLC	TLC
Technique	Automated/instrumental	Manual
Mean particle size	5–6 μm	10–12 μm
Layer thickness	100 μm	250 μm
Plate height	12 μm	30 μm
Efficiency	High due to smaller particle size generated	Less
Separations	3–5 cm	10–15 cm
Analysis time	Shorter migration distance and the analysis time is greatly reduced	Slower
Solid support	Wide choice of stationary phases like silica gel for normal phase and C8, C18 for reversed-phase modes	Silica gel, alumina, and Kiesulguhr
Development chamber	New type that require less amount of mobile phase	More amount
Sample spotting	Auto sampler	Manual spotting
Sample volume	0.1–0.5 μl	1–5 μl
Starting spot's diameter	1–1.5 mm	3–6 mm
Separated spot's diameter	2–5 mm	6–15 mm
Sample tracks per plate	≤36 (72)	≤10
Scanning	Use of UV/visible/fluorescence scanner scans the entire chromatogram qualitatively and quantitatively and the scanner is an advanced type of densitometer	Not possible
Separation time	3–20 min	20–200 min
Detection limits (absorption)	100–500 pg	1–5 ng
Detection limits (fluorescence)	5–10 pg	50–100 pg

sample, namely, structure, polarity, volatility, stability, and the solubility parameter. Method development involves considerable trial and error procedures. The most difficult problem usually is where to start, with what kind of mobile phase.

Selection of stationary phase is quite easy, that is, to start with silica gel which is reasonable and nearly suits all kind of drugs. Mobile phase optimization is carried out by using three level techniques. First level involves use of neat solvents and then by finding some such solvents which can have average separation power for the desired drugs. Second level involves decreasing or increasing solvent strength using hexane or water for respective purposes. Third level involves trying of mixtures instead of neat solvents from the selected solvents of first and second level which can further be optimized by the use of modifier like acids or bases.

Analytes are detected using fluorescence mode or absorbance mode. But, if the analytes are not detected perfectly than it needs change of stationary phase or mobile phase or need the help of pre or post chromatographic derivatization.

Optimization can be started only after a reasonable chromatogram which can be done by slight change in mobile-phase composition. This leads to a reasonable chromatogram which has all the desired peaks in symmetry and well separated. Procedure for HPTLC method development is outlined as follow (Figure 1). (Sherma and Fried 1996; Sethi 1996).

SCHEMATIC PROCEDURE FOR HPTLC METHOD DEVELOPMENT

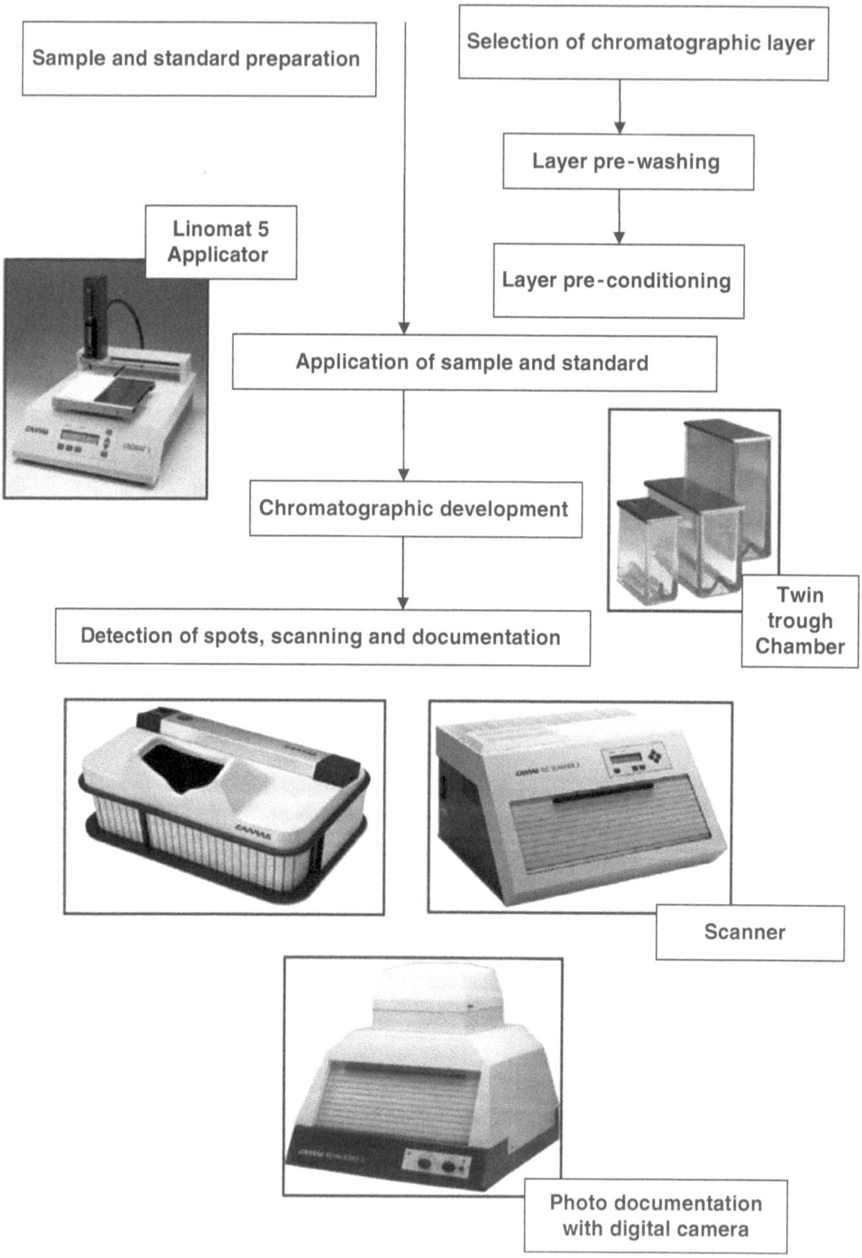

Stationary Phase

HPTLC can be regarded as the most advanced form of modern TLC. It uses HPTLC plates featuring small particles with a narrow size distribution. As a result, homogenous layers with a smooth surface can be obtained. HPTLC uses smaller plates (10 × 10 or 10 × 20 cm) with significantly decreased development distance (typically 6 cm) and analysis time (7–20 min). HPTLC plates provide improved resolution, higher detection sensitivity, and improved in situ quantification and are used for industrial pharmaceutical densitometric quantitative analysis. Normal phase adsorption TLC on silica gel with a less polar mobile phase, such as chloroform– methanol, has been used for more than 90% of reported analysis of pharmaceuticals and drugs. Lipophilic C-18, C-8, C-2; phenyl chemically-modified silica gel phases; and hydrocarbon- impregnated silica gel plates developed with a more polar aqueous mobile phase, such as methanol–water or dioxane–water, are used for reversed-phase TLC. Other precoated layers that are used include aluminum oxide, magnesium silicate, magnesium oxide, polyamide, cellulose, kieselguhr, ion exchangers, and polar modified silica gel layers that contain bonded amino, cyano, diol, and thiol groups. Optical isomer separations that are carried out on a chiral layer produced from C-18 modified silica gel impregnated with a Cu (II) salt and an optically active enantiomerically pure hydroxyproline derivative, on a silica layer impregnated with a chiral selector such as brucine, on molecularly imprinted polymers of α-agonists, or on cellulose with mobile phases having added chiral selectors such as cyclodextrins have been reported mostly for amino acids and their derivatives. Mixtures of sorbents have been used to prepare layers with special selectivity properties. HPTLC plates need to be stored under appropriate conditions. Before use, plates should be inspected under white and UV light to detect damage and impurities in the adsorbent. It is advisable to prewash the plates to improve the reproducibility and robustness of the results (Sethi 1996; Sherma 2007).

Layer Prewashing

Plates are generally handled only at the upper edge to avoid contamination. Usually plates are used without pretreatment unless chromatography produces impurity fronts due to contamination of the plate. For reproducibility studies and quantitative analysis, layers are often prewashed using 20 ml methanol (generally, methanol is used as a prewashing solvent; however, a mixture of methanol and ethyl acetate or even mobile phase of the method may also be used) per trough in a 20 × 10 cm twin-trough chamber (TTC). Up to two 20 × 10 cm or four 10 × 10 cm plates can be developed back-to-back in each trough of the TTC. Remove the plate and dry it for 20 min in a clean drying oven at 120°C. Equilibrate plate with laboratory atmosphere (temperature, relative humidity) in a suitable container providing protection from dust and fumes.

Table 3.2 Sample application parameters on HPTLC plate

Parameter	HPTLC
Distance from lower edge of plate for use in TTC	8 mm
Distance from lower edge of plate for use in horizontal development chamber (HDC)	5 mm
x-position of first track	15 mm
Minimum space between bands/spots	2 mm
Maximum diameter of application spot	5 mm
Band length	8 mm
Maximum number of tracks on a 10 × 10 cm plate	7
Maximum number of tracks on a 20 × 10 cm plate	16

Mobile Phase

The selection of mobile phase is based on adsorbent material used as stationary phase and physical and chemical properties of analyte. General mobile-phase systems that are used based on their diverse selectivity properties are diethyl ether, methylene chloride, and chloroform combined individually or together with hexane as the strength-adjusting solvent for normal-phase TLC and methanol, acetonitrile, and tetrahydrofuran mixed with water for strength adjustment in reversed-phase TLC. Separations by ion pairing on C-18 layers are done with a mobile phase such as methanol–0.1 M acetate buffer (pH 3.5) containing 25 mM sodium pentanesulfonate (15.5:4.5). Accurate volumetric measurements of the components of the mobile phase must be performed separately and precisely in adequate volumetric glassware and shaken to ensure proper mixing of the content. Volumes smaller than 1 ml are measured with a suitable micropipette. Volumes up to 20 ml are measured with a graduated volumetric pipette of suitable size. Volumes larger than 20 ml are measured with a graduated cylinder of appropriate size. To minimize volume errors, developing solvents are prepared in a volume that is sufficient for one working day.

Sample Preparation and Application

Proper sample preparation is an important prerequisite for the successful HPTLC separation. If the analyte concentration is sufficiently high, pharmaceutical dosage forms can often be simply dissolved in a solvent that will completely solubilize the analyte and leave excipients or extraneous compounds undissolved to yield a test solution that can be directly spotted for HPTLC analysis. The sample preparation is not as demanding as for other chromatographic techniques. However, several steps may be necessary such as grinding, sonication, filtration, extraction, centrifugation, and concentration procedure if low concentration of analyte is present in sample. Since layers are not reused, it is often possible to apply cruder samples. Solvent for dissolving sample should be nonpolar and volatile (i.e., methanol, ethanol, or chloroform).

Automated instruments are available for sample application; especially for quantitative HPTLC, apply samples using the spray-on technique with Automatic TLC sampler (ATS) 4 or Linomat 5. In case of spot application by contact, use the Nanomat or the ATS 4 and dissolve the sample in the solvent of lowest suitable solvent strength. Sample application in the form of narrow bands provides the highest resolution and sensitivity attainable with a given chromatographic separation method; however, the spreading of the sample band during application must be as small as possible. The sample is typically contained in a syringe, which is emptied by a motor. Delivery speed and volume are electronically controlled. A stream of an inert gas such as nitrogen around the tip of the syringe atomizes the sample and creates a band on the TLC/HPTLC plate if either the syringe or the plate is moving linearly (Table 3.2).

Development of Chromatogram

Although chromatogram development is the most decisive step in the TLC procedure, important parameters are often not given the attention they deserve. Separations obtained in HPTLC are affected by the vapor phase, which depends on the type, size, and saturation condition of the chamber during development. The interactions of these three phases as well as other factors, such as temperature and relative humidity, must be controlled to obtain reproducible TLC separations. HPTLC plates are developed in flat-bottom chambers, twin-trough chambers, or horizontal-development chambers. In general, saturated twin-trough chambers fitted with filter paper offer the best reproducibility. For development of plates in a saturated TTC, initially prepare the appropriate volume of mobile phase. Place a correctly sized piece of filter paper in the rear trough of TTC and carefully pour prepared mobile phase into chamber so that the filter paper is thoroughly wetted and adheres to rear wall of TTC. Tilt TTC to the side (about 45°) so that the solvent volume in both troughs equalizes. Set chamber on bench, replace the lid and let chamber equilibrate for 20 min. Mark the desired developing distance (70 mm from lower edge of plate) with a pencil on the right edge of the plate. Slide off the lid to the side and place the plate into the front trough in such a way that the layer and filter paper should face each other and the back of the plate is resting against front wall of TTC. Replace the lid and develop plate to the mark. Remove plate from chamber and dry it (vertically in the direction of chromatography) for 5 min in a stream of cold air. After each development, remaining mobile phase and filter paper are discarded. Prior to being prepared for the next run, the chamber is dried and, if necessary, also cleaned.

Detection

After removal of the mobile phase from the developed plate by heating, zones are detected on the layer by their natural color, natural fluorescence, quenching of fluorescence, or as colored, UV-absorbing, or fluorescent zones after reaction with a reagent (postchromatographic derivatization). Zones with fluorescence or quench fluorescence are viewed in cabinets that incorporate short-wave (254 nm) and long-wave (366 nm) UV lamps. Detection under UV light is the first choice as it is nondestructive. An important advantage of the off-line operation of TLC is the flexibility afforded by the use of multiple methods for zone detection and identification. For example, the layer can be viewed under long- and short-wave UV light, followed by one or more chromogenic, fluorogenic, or biological detection methods. Many hundreds of reagents and detection methods have been described in various literature sources.

Derivatization

Derivatization is necessary in most cases to visualize the analytes of interest. Derivatization can be performed either by immersing the plates or by spraying the plates with a suitable reagent. For better reproducibility, immersion is the preferred derivatization technique. To induce or optimize the derivatization reaction, it may be necessary to heat the plates. Also, the conditions and time must be specified for this step.

Immersing

Tank of immersion devices is charged with 200 ml of reagent. Place the plate in the holder of the immersion device, set parameters according to method and press start. Let excess reagent drip off the plate, wipe off the back of the plate with paper towel. Remove the plate from plate holder and dry it with cold air (vertically in the direction of chromatography).

Spraying

Charge the bottle of the sprayer with up to 50 ml of reagent. Place the plate in spray cabinet against a filter paper. Spray plate with horizontal and vertical motion until it is homogeneously covered with reagent. The plate is dried with cold air.

Heating

Turn on plate heater and adjust temperature. Place plate on plate heater when the temperature is stable. After the time specified by the method, remove the hot plate from heater.

Quantification

Most modern HPTLC quantitative analysis are performed in situ by measuring the zones of samples and standards using a chromatogram spectrophotometer usually called a densitometer or scanner with a fixed sample light beam in the form of a rectangular slit. Generally, quantitative evaluation is performed with the TLC Scanner 3 using winCATS software. It can scan the chromatogram in reflectance or in transmittance mode by absorbance or by fluorescent mode; scanning speed is selectable up to 100 mm/s. Spectra recording is fast. Calibration of single and multiple levels with linear or nonlinear regressions are possible. When target values are to be verified, such as stability testing and dissolution profile, single level calibration is suitable. Concentration of analyte in the sample is calculated by considering the sample initially taken and dilution factors.

Documentation

Each developed plate is documented using digital documentation system under UV light at 254 nm, UV light at 366 nm, and white light. If a type of light does not produce usable information, that fact must be documented. If a plate is derivatized, images are taken prior and after derivatization.

Validation of Method

The process of validating an analytical method cannot be separated from its development, as the analyst will not know whether the developed method and its performance parameters are acceptable until validation has been performed. The suitability of any analytical procedure for its intended use in analysis must be based on objective validation data. Validation is an important step in determining the reliability and reproducibility of the method because it is able to confirm that the intended method is suitable to be conducted on a particular system. The necessity for validation in analytical laboratories is derived from regulations such as International Conference on Harmonization (ICH) and current good manufacturing practices (cGMP), good laboratory practices (GLP), and the good clinical practices (GCP). Other regulatory requirements are found in quality and accreditation standards such as the International Standards Organization (ISO) 9000 series, ISO 17025, the European Norm (EN 45001), United States Pharmacopoeia (USP), Food and Drug Administration (FDA), and Environmental Protection Agency (EPA). The reliability of analytical data is critically dependent on three factors, namely, the reliability of the instruments, the validity of the methods, and the proper training of the analysts. Validation parameters typically monitored are: selectivity; stability before, during, and after development; linearity of the calibration graph; range of levels within which the analyte can be quantified; limits of detection and accurate and precise quantification; and accuracy (indication of systematic errors), precision (indication of random errors), sensitivity (ability to measure small variations in concentration), and ruggedness (results of the method when used by different analysts in a variety of locations). Each step of the analysis must be validated through error analysis and a suitability test, and includes sample preparation, application of samples, HPTLC separation, detection procedures, and quantification (Table 3.3) (Koll et al. 2003; Reich et al. 2008; Patel et al. 2009; Patel et al. 2010).

Table 3.3 Type of analytical procedures and required validation characteristics (Ferenczi-Fodor et al. 2001; Patel et al. 2010)

Type of analytical procedure Characteristics	Identification	Assay/in vitro release study/ dissolution study/content/potency	Testing for impurities	
			Quantitative	Limit test
Linearity	–	+	+	–
Range	–	+	+	–
Specificity[a]	+	+	+	+
Accuracy	–	+	+	–
Precision				
Repeatability	–	+	+	–
Intermediate precision	–	+[b]	+[b]	–
Limit of detection	–	–	–[c]	+
Limit of quantification	–	–	+	–

+ Characteristic normally evaluated, – characteristic normally not evaluated
[a]Lack of specificity of one analytical procedure could be compensated by other supporting analytical procedure
[b]If reproducibility has been performed, intermediate precision is not needed
[c]Evaluated in certain conditions

Specificity

The specificity of the developed method is established by analyzing the sample solutions in relation to interferences from formulation ingredients. The spot for the sample is confirmed by comparing retardation factor (R_f) values of the spot with that of the standard.

Sensitivity

Sensitivity of the method is determined with respect to limit of detection (LOD) and limit of quantification (LOQ). Noise can be determined by scanning blank spot (solvent) six times. Series of concentrations of drug solutions are applied on plate and analyzed to determine LOD and LOQ. LOD is calculated as three times the noise level, and LOQ is calculated as ten times the noise level. LOD and LOQ are experimentally verified by diluting the known concentrations of sample until the average responses are approximately 3–10 times the standard deviation (SD) of the responses for six replicate determinations.

Linearity and Calibration Curve

Linearity of the method is evaluated by constructing calibration curves at different concentration levels. Calibration curve is plotted over a different concentration range of analyte. The calibration curve is developed by plotting peak area vs. concentrations with the help of the win-CATS software.

Accuracy

Accuracy of the method is evaluated by carrying the recovery study at three levels. Recovery experiments are performed by adding three different amounts of standard drug, i.e., 80, 100, and 120% of the drug, to the preanalyzed formulations, and the resultant is reanalyzed six times.

Precision

Precision is evaluated in terms of intraday and interday precisions. Intraday precision is determined by analyzing sample solutions of analyte from formulations at three levels covering low, medium, and higher concentrations of calibration curve for five times on the same day. Interday precision is determined by analyzing sample solutions of analyte at three levels covering low, medium, and higher concentrations over a period of 7 days. The peak areas obtained are used to calculate mean and % RSD (relative SD) values.

Repeatability

Repeatability of measurement of peak area is determined by analyzing different amount of analyte covering low, medium, and higher ranges of the calibration curve seven times without changing the position of plate. Repeatability of sample application is assessed by spotting samples covering similar range of calibration curve seven times and analyzing them once.

Robustness

By introducing small changes in mobile-phase composition, its volume, chamber saturation time, and slight change in the solvent migration distance, the effects on the results are examined. Robustness of the method is determined in triplicate and the mean and % RSD of peak area are calculated. Parameters that are affected by the changes in chromatographic conditions are retardation factor (R_f) and peak purity.

Retardation Factor

Retardation factor (R_f) is defined as the amount of separation due to the solvent migration through the sorbent layer as shown in the formula. It depends on time of development and velocity coefficient or solvent front velocity.

$$R_f = \frac{\text{Migration distance of substane}}{\text{Migration distance of solvent front from origin}}.$$

Peak Purity

The purity of the peak is determined by comparing the spectra at three different levels: peak start (s), peak maximum (m), and peak end (e). During the purity test, the spectrum taken at the first-peak slope is correlated with the spectrum of peak maximum [r (s, m)] and the correlation of the spectra taken at the peak maximum with the one from the down slope or peak end [r (m, e)] which is used as a reference spectra for statistical calculation. An error probability of 1% only is rejected if the

Table 3.4 Basic acceptance criteria for evaluation of validation experiments (Ferenczi-Fodor et al. 2001; Patel et al. 2010)

Characteristics – parameter		Impurity testing Acceptance criteria	Assay
Linearity			
Residual plot		No trend	No trend
Correlation coefficient		$r \geq 0.99$	$r \geq 0.998$
Y-axis intercept		≤25%	≤2%
RSD residuals			≤1.5%
	Impurity level ≤0.5%	≤10%	
	Impurity level ≥0.5%	≤5%	
Range		From limit of quantification to 120% of the specified limit of impurity	80–120%
Precision			
Repeatability			
	Impurity level 0.1–0.2%	RSD ≤ 20%	RSD ≤ 2% ($n \geq 6$)
	Impurity level 0.2–0.5%	RSD ≤ 10%	
	Impurity level ≥0.5%	RSD ≤ 5%	
Intermediate precision		1.5 × RSD of repeatability	RSD ≤ 3% ($n \geq 6$)
Accuracy			RSD ≤ 3%
	Impurity level ≤0.5%	RSD ≤ 10% ($n = 3$)	
	Impurity level ≥ 0.5%	RSD ≤ 5%	
Limit of detection		RSD ≤ 10–20%	–
Limit of quantification		RSD ≤ 20–50%	–

test value is greater than or equal to 2.576 as a reference spectra for statistical calculation (Patel et al. 2008). Basic acceptance criteria for evaluation of various validation paramerers are summerized in Table 3.4.

Notes

HPTLC method is having active application in qualitative and quantitative analysis of a wide range of compounds, such as herbal and botanical dietary supplements, nutraceuticals, traditional western medicines, traditional Chinese medicines, and Ayurvedic (Indian) medicines. It is used in assaying radiochemical purity of radiopharmaceuticals; the determination of the pigments that a plant contains; detection of pesticides or insecticides in food; analyzing the dye composition of fibers in forensics, identifying compounds present in a given substance; to check starting raw materials (plant extracts, extracts of animal origin, fermentation mixtures), intermediates (crude products, reaction mixtures, mother liquors and secondary products), pharmaceutical raw materials (identification, purity testing, assay, separation of closely related compounds, stability testing), and formulated products (identification, purity testing, assay, stability testing under storage and stress, content uniformity test, dissolution test); and for the detection and identification of drugs and their metabolites in biological media such as urine, plasma, or gastric fluid (pharmacological, toxicological, pharmacokinetic, metabolic, bioequivalence, forensic, and compliance and pharmacodynamic studies). Validation requirements for HPTLC procedures are highly diversified, depending upon the actual type of analysis, as HPTLC can be used in wide range of applications (Renger 1993; Sethi 1996; Renger 1998; Sherma 2008).

References

Ferenczi-Fodor K, Vegh Z, Nagy-Turak A, Renger B and Zeller M. Validation and quality assurance of planar chromatographic procedures in pharmaceutical analysis. J AOAC Int, 2001, 84, 1265–1276.

Koll K, Reich E, Blatter A and Veit M. Validation of standardized high-performance thin-layer chromatographic methods for quality control and stability testing of herbals. JAOAC Int, 2003, 86, 909–915.

Patel RB, Patel MR, Shankar MB and Bhatt KK. Simultaneous estimation of acetylsalicylic acid and clopidogrel bisulfate in pure powder and tablet formulations by high-performance column liquid chromatography and high-performance thin-layer chromatography. J AOAC Int, 2008, 91, 750–755.

Patel RB and Patel MR. An introduction to analytical method development for pharmaceutical formulations. Pharma Rev, 2008, 6, (4). Available Online: http://www.pharmainfo.net/reviews/introduction-analytical-method-development-pharmaceutical-formulations.

Patel RB, Patel AB, Patel MR, Shankar MB and Bhatt KK. Estimation of alprazolam and sertraline in pure powder and tablet formulations by high-performance liquid chromatography and high-performance thin-layer chromatography methods. Anal Let, 2009, 42, 1588–1602.

Patel MR, Patel RB, Parikh JR and Patel BG. HPTLC method for estimation of tazarotene in topical gel formulations and *in vitro* study. Anal Methods, 2010, 2, 275–281.

Reich E, Schibli A and Debatt A. Validation of high-performance thin-layer chromatographic methods for the identification of botanicals in a cGMP environment. J AOAC Int, 2008, 91, 13–20.

Renger B. Quantitative planar chromatography as a tool in pharmaceutical analysis. J AOAC Int, 1993, 76, 7–13.

Renger B. Contemporary thin layer chromatography in pharmaceutical analysis. J AOAC Int, 1998, 81, 333–339.

Sethi PD (1996). High performance thin layer chromatography, quantitative analysis of pharmaceutical formulations, published by CBS Publishers, New Delhi, India.

Sherma J (2007). Chromatographic Methods of Analysis: Thin Layer Chromatography. In Encyclopedia of Pharmaceutical Technology; Swarbrick, J, ed. 3rd edition, Informa Healthcare, pp 538–550.

Sherma J. Modern thin layer chromatography. J AOAC Int, 2008, 91, 1142–1144.

Sherma J and Fried B (1996) Handbook of Thin Layer Chromatography, 2nd edition, published by Marcel Dekker, Inc., New York.

Chapter 4
High-Performance Thin-Layer Chromatography: Excellent Automation

Dilip Charegaonkar

Abstract Product innovations from CAMAG, Switzerland, have made thin-layer chromatography (TLC) a practical laboratory tool for both qualitative and quantitative analysis. Instrumentation was developed to permit more precise auto spotting of the sample. Sample preparation is simplified because precoated clean plates are disposable. Multiple samples and standards can be run simultaneously under identical conditions in automated multiple development chambers. Mobile phase need not be compatible with detector. Chromatogram is inspected under UV 366 nm in a TLC visualizer UV cabinet to observe fluorescence for easy detection. Fluorescence visualization is usually very specific, almost like a fingerprint, when applicable. No other chromatographic technique can directly express the result as a color image and make it available for visual evaluation. These improvements in the technique result in higher separation power, resolution, accuracy, reproducibility, and faster analysis. HPTLC is the most simple separation technique available today for the analyst and can be considered a time machine that can speed up work and allow quite a number of processes at a time, usually not possible with other analytical techniques.

Automation of TLC to HPTLC

Of all the methods used in chromatography, thin-layer chromatography (TLC) is the simplest, easiest to perform and where the samples are "visible." After Ismailor and Shraiber invented it as "Drop Chromatography," first Kirchner (1978) and then Stahl (1965) made the technique very popular for routine analysis, in all branches of Chemistry. Modern TLC is widely known and practical as high-performance thin-layer chromatography (HPTLC), which can only be performed on precoated layers, using instrumentation and mainly for the purpose of quantification. Hence, here the terminology TLC and HPTLC is used interchangeably. To teach the principle of

MM. Srivastava (ed.), *High-Performance Thin-Layer Chromatography (HPTLC)*,
DOI 10.1007/978-3-642-14025-9_4, © Springer-Verlag Berlin Heidelberg 2011

chromatography, almost all over the world, TLC is used. The primary reasons for this choice are visibility of the sample during chromatography, simplicity to perform, and ultra-low-cost apparatus for demonstration.

HPTLC are the fastest chromatography method, since chromatography of samples is done in parallel. Being offline i.e., each step of the procedure is performed independently, makes TLC/HPTLC is not only faster but flexible enough for one HPTLC System to analyze different samples in parallel. Consumption of stationary and mobile phase is directly proportional to the number of samples being analyzed. Cost per analysis is very low. One 20 × 10 cm plate can accept about 20 samples and requires 15 ml of mobile phase. Disposable stationary phase in TLC/HPTLC has two distinct advantages in sample preparation. One, it is possible to do sample cleanup on the plate itself. The other is that sample cleanup may not be that critical as residue left behind is inconsequential. Fatty matrix can be cleaned up on the plate itself, after sample application. Investment in equipment is another feature of TLC/HPTLC where it scores. One can start with a simple basic setup and invest in phases to an ultra-sophisticated fully automatic HPTLC gradient System with multiple detectors. Visibility of the sample throughout the chromatographic analysis i.e., after sample application and chromatograph development, in situ derivatization is unique to HPTLC. Post-chromatography derivatization (PCD) is very simple and routinely possible in any lab. After recording data, a plate can be derivatized to get additional information. PCD is done for several reasons viz. to detect compounds with a specific functional group or to lower detection limits by up to several orders of magnitude of target fractions or for universal detection of all organic compounds present or to visualize the sample by our eyes. Although Silica gel is by far the most widely used adsorbent (stationary phase), many other adsorbents have been used as the separation medium e.g., reverse phases, bonded phases, alumina, Kieselguhr, etc. Most Solvents can be used in the mobile phase, as the layer is disposable. Gas phase too plays an important role in TLC/HPTLC in the developing chamber's vapor saturation, its pH, and humidity in the developing chamber. TLC and HPTLC can be used for qualitative, semiquantitative, and quantitative analysis. It can also be used for the identification of industrial fractions after chromatographic separations as well for the identification of herbal extracts, complex mixtures by "HPTLC fingerprint." Most labs use TLC/HPTLC for impurity analysis, assay, or comparison with similar samples, screening of unknown samples or of a large number of samples. Quality control, analytical R&D, process monitoring, and environmental labs find TLC/HPTLC as a useful tool for everyday analysis.

Chromatogram development in TLC/HPTLC can be done with the plate in vertical or horizontal position. It can also be done by linear, circular or anticircular movement of the mobile phase. Development can also be done in one dimension, with one mobile phase (isocratic) or repeatedly in the same direction with differing mobile phases (gradient). 2D chromatography is very useful for high-resolution separations using a different mobile phase in each direction. Chromatographic stability of samples can be studied by 2D technique. However, isocratic linear development in vertical mode is practiced more than 95% of the time.

Protocol

Preparation of Plate

Precoated layers: TLC plates can be made in any lab with suitable apparatus. However such layers do not adhere well to the glass support. Precoated plates that use small quantities of very high molecular weight polymer as binder overcomes most limitations of a home-made layer. Precoated layers are reasonably abrasion resistant, very uniform in layer thickness, reproducible, preactivated, and ready to use. They are available with glass or aluminum or polyester support. Aluminum foil plates are less expensive to buy, cheaper, can be cut, and therefore easy to carry around or transport or mail. Glass plates are the best for highest quality of results. Most often, layers containing a fluorescent indicator F 254 are used. This enables the visualization of samples in a UV cabinet very simply, instantly, and in a nondestructive manner.

Commonly used size of plates in TLC is 20 × 20 cm and in HPTLC 20 × 10 cm or 10 × 10 cm is widespread. For reproducible results, conditions for chromatography need to be standardized. Research has established that the following practical conditions should be used as default unless proved otherwise (Fig. 4.1).

Plate Size

20 × 10 cm or 10 × 10 cm.

Plate support – glass or aluminum with florescent indicator.

Sample Application Position

Distance from left and right side edges – 15 mm.

Distance from bottom edge – 8 mm.

Length of bands – 6 mm.

Center to center distance between bands – 10 mm.

Fig. 4.1 TLC plate

Chromatogram Development

Chamber Saturation – 20 min in a chamber lined with filter paper on three sides.

Chamber type – twin-trough chamber.

Grease for sealing – not to be used.

Opening the lid for plate insertion – Slide the lid. Do not lift it.

Layer saturation – 5 min (keep an aliquot of the mobile phase in one trough. After 15 min of chamber saturation, keep the plate in the second trough for 5 min.)

Layer facing the chamber, not the wall.

Development distance – 70 mm.

Mobile-phase front detection – by CCD.

Derivatization

Scan speed – 20 mm/s

Center of scan beam and of sample band should overlap.

Always record spectra of all samples, except when not required in method.

By immersion technique.

Dip speed.

Dip time: 1 s

Image Documentation

With an industrial camera (and not a consumer camera) (Fig. 4.2).

Auto Spray Loading

The main requirement of sample application are precise sample volumes, precise positioning for accurate volumes, and sharpest possible "bands" by spray-on method using an inert gas. Circular spots have numerous disadvantages while "line" or "band" application is advantageous (Fig. 4.3).

TLC/HPTLC is an "open" chromatography method. An automated developing chamber that uses a normal twin-trough chamber for chromatographic development but automates all other steps such as plate insertion, addition of conditioning and mobile phases, monitoring saturation and humidity inside the chamber, drying the plate uniformly, after development, etc. is very desirable (Figs. 4.4 and 4.5).

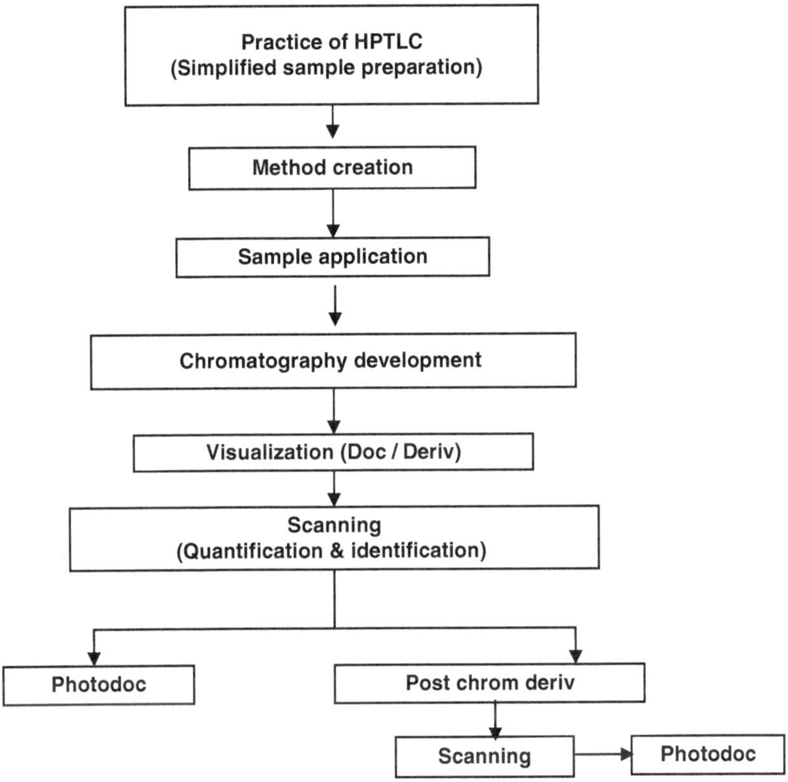

Fig. 4.2 Schematic diagram of HPTLC method for separation

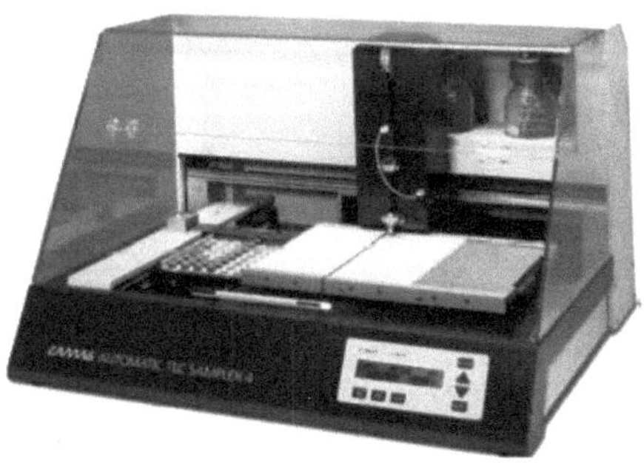

Fig. 4.3 Automatic TLC sampler (with permission from Camag)

Automated Development

After the development of chromatogram, it is inspected under UV 366 nm in a UV cabinet to observe fluorescence, if any. The plate is observed at 254 nm to observe the sample fractions are dark spots on a fluorescent background. Fluorescence visualization

Fig. 4.4 Camag automatic development chamber (with permission from Camag)

Fig. 4.5 Camag UV cabinet (with permission from Camag)

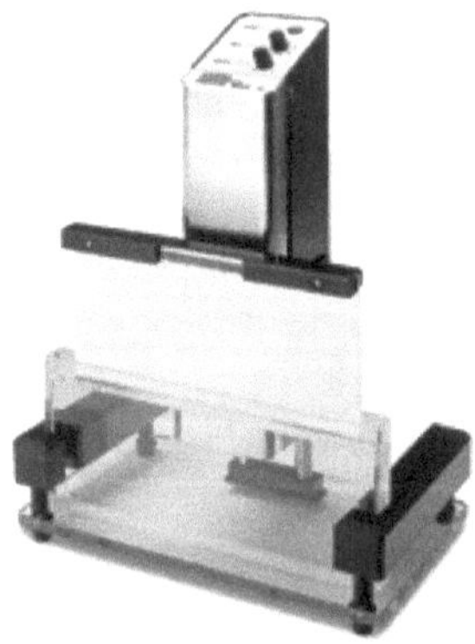

Fig. 4.6 Immersion device (with permission from Camag)

is usually very specific, almost like a fingerprint, when applicable. In situ derivatization is easily achieved in TLC/HPTLC, as the fractions are stored on the plate (Figs. 4.6 and 4.7).

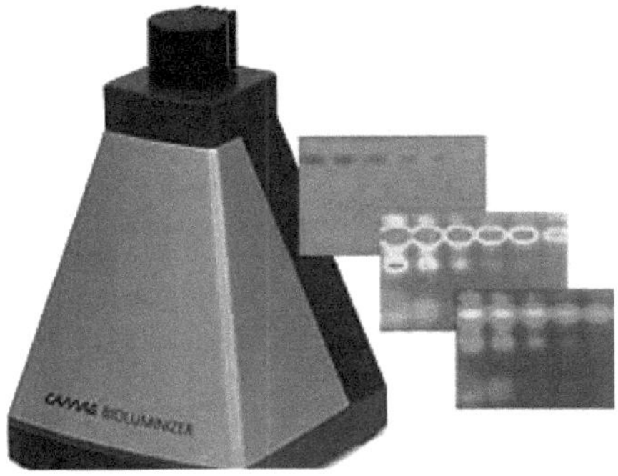

Fig. 4.7 Bioluminizer

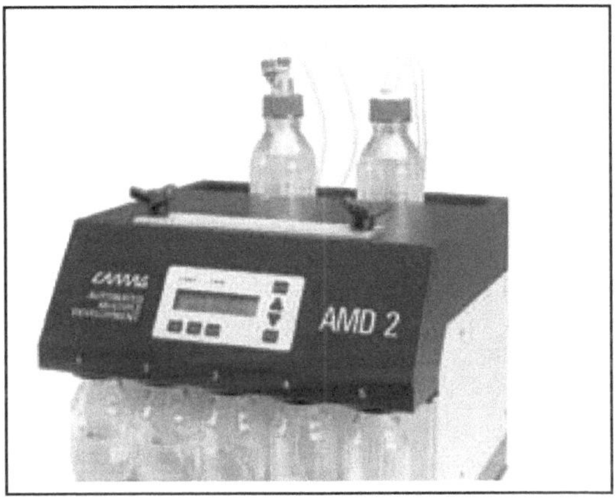

Fig. 4.8 Automatic multiple developments

Several detectors are available. Quantification is usually carried out using UV in absorbance or by fluorescence. Recently, bioluminescence detection has become possible. For reliable quantitative analysis, a software-controlled "complete HPTLC system" is a must. A complete system comprises the HPTLC system manager software, sample applicator, chromatogram development chambers, a chromatogram-visualizing device, a scanner, and small related accessories.

Isocratic separations are very powerful and very commonly used. However certain complex samples such as polymers, herbal origin, foods, etc. require much greater solving power. This is achieved in HPTLC by multiple developments in the same direction, in a stepwise manner. The first development is for a very short distance e.g., 8 mm, and then this migrating distance is increased by, say, 3 mm, for every step. Between two steps, the plate is dried and a slightly less-polar solvent is usually introduced. A typical 20- or 25-step program takes 2 h or more (Fig. 4.8).

The preferred method of choice for calibration in TLC/HPTLC is external standards since the sample application volumes (and, therefore, the weights) are extremely precise when applied on the plate.

Table 4.1 Comparison between conventional and advance TLC kits

Feature	Conventional TLC kit	Advance TLC kit
Sample application	Spot, manual	Band wise, instrumental
Chromatogram development	Twin-trough chamber	Twin-trough chamber
Post-chromatogram visualization	UV cabinet	Modified UV cabinet for photo recording
Image documentation	Not possible	Possible
Micro-prep chrom	Difficult, poor quality	Easy, high quality
Upgradation to HPTLC	Not easy	Easy, just scan the plate
Advantages		High resolution, photo documentation. Chromatogram is suitable for scanning, micro-prep.

Conventional TLC kits are unfit for use today. Most depend on lab-made layers. A comparison of the two basic kits is given below (Table 4.1).

Digital Camera-Based Image Documentation

UV Cabinets are now being replaced with improved design UV Cabinets which allow digital camera to be fixed for recording images of the plate. Small labs prefer this device although it does not conform to GLP. Today, HPTLC is a primary requirement for any laboratory involved in herbal analysis for establishing the identity of plant extracts by comparison with Botanical Reference Material (BRM) extracts to detect substitutes or adulterants, studies of formulations, etc. Forensic analysts have long ago stated that their starting points are a microscope for physical inspection and TLC for chemical inspection!

Software-Induced Scanning

An "Entry Level" HPTLC system is already very advanced and can do most of the routine jobs. It can scan for quantification in absorbance and fluorescence modes and record UV–Vis absorbance spectra, in situ. Depending on end-user requirements, gradient chamber and/or a photo documentation device and a bioluminescence detector may be added or a fully automatic system could be procured. Hyphenation with MS or IR or NMR can be achieved with a suitable commercially available interface (Fig. 4.9).

A recently available device interfaces HPTLC with MS. This interface enables the extraction of the chosen fraction from the layer and feeds it directly into the MS. This opens great new possibilities for an analytical lab. The analysis output from

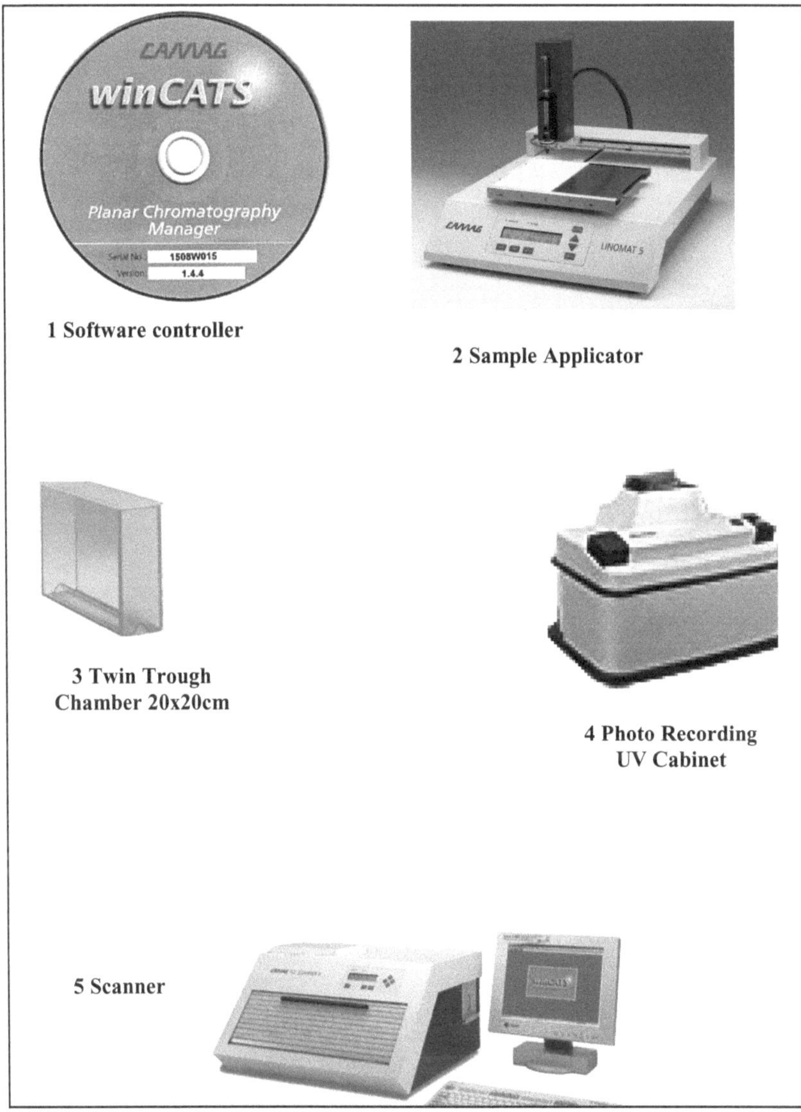

Fig. 4.9 An entry level complete HPTLC system with automation

LC–MS can be greatly increased, when compiled to TLC/HPTLC. Any specified fraction from a plate can be analyzed. Other fractions present can be ignored. Optimization of MS parameters for a particular molecule can be optimized using TLC. LC–MS and TLC–MS are complimentary techniques (Fig. 4.10 and Table 4.2).

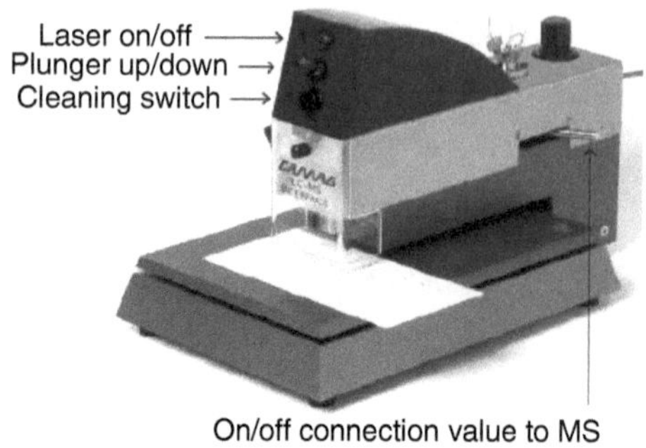

Fig. 4.10 TLC–MS

Table 4.2 Features of HPTLC and HPLC

Feature	HPLC	HPTLC
Stationary phase	Liquid/solid	Solid
Mobile phase	Liquid	Liquid
Conditioning phase	None	Gas
Samples should be		Nonvolatile
Results	By detector	Detector + eyes
Analysis	On-line	Off-line
Resolution	Very high	Moderate to high
Chromatography system	Closed	Open
Separating medium	Tubular column	Planar layer (plate)
Analysis in parallel	No. Only one at a time	Yes. Up to 100 samples
High pressure required?	Yes	No
Time per sample	2–60 min	1–3 min
Data obtained from chromatography	Little to very high (detector dependent)	High to very high (detector dependent)
Post-chromatography derivatization	Limited possibilities Cumbersome	Simple, possible for every sample, gives additional information
Fraction collection/micro-preparative chromatography	Requires prep. scale chromatograph and fraction collector	Simple, no special requirements
Sensitivity	High to ultra-high	Moderate to ultrahigh
Fluorescence data	Possible, optional	Possible, built-in
Abs. spectra for identification	Yes (PDA)	Yes
Detectors	UV, Fluor, Electrochem light scatter, MS, etc.	UV–Vis, bioluminescence, MS
Chromatogram image documentation	No	Yes, at 254 and 366 nm and visible
Sample cleanup	Thorough column reusable	Not so imp Layer disposable
Chromatographic fingerprint	Yes, but limited	Yes, comprehensive
Cost per analysis	Very high	Low
Eqpt. maintenance	Very high	Low
Analyst's skills required	High to very high	Low (TLC) to high (HPTLC)

Notes

This technique has most successfully been used in the fields of education, forensic science, herbal analysis, organic synthesis, and foods, to name a few. The introduction of TLC–MS interface will lead to a paradigm shift in the role of TLC/HPTLC in all labs. Recent improvements in the automatic development chamber will also lead to reproducible results independent of the laboratory's environment.

References

Applied Thin Layer Chromatography – Best Practice and Avoidance of Mistakes, Second, Revised & ISBN – 978-3-527-3155 3-5

Kirchner JG. (1978) Thin-Layer Chromatography, 2nd edition, Techniques in Chemistry, vol. XIV, published by Wiley-Interscience, Chichester, UK.

Kowalska T and Sharma J. Thin Layer Chromatography in Chiral Separation & Analysis. vol. 98, ISBN – 10-0-8493-4369-0.

Kowalska T and Sharma J. Preparative Layer Chromatography, vol. 95 – published by, Taylor & Francis, ISBN – 0-8493-4039-7.

Laszlo R. Quantitative Thin Layer Chromatography & its Industrial Application. vol. 36, ISBN – 0-8247-7597-X.

Reich E and Schibli A. High Performance Thin Layer Chromatography for the Analysis of Medicinal Plants; published by Thieme Medical Publishers, Inc, The Americas, ISBN – 1-58890-409-1, ISBN – 978-1-58890-409-6.

Sethi PD and Charegaonkar D. Identification of Drugs in Pharmaceutical Formulations by TLC, 2nd edition, ISBN – 81-239-0635-8.

Sethi PD. HPTLC – Quantitative analysis of Pharmaceutical Formulations, ISBN – 81-23*-0439-8.

Sharma J and Fried B. Handbook of Thin Layer Chromatography. 2nd edition, published by CRC Press.

Stahl E. (1965) Thin-Layer Chromatography – A Laboratory Handbook, published by Springer-Verlag, Berlin, Germany. (Replace with this ref.)

Wagner H and Bladt S. Plant Drug Analysis – A Thin Layer Chromatography AHas, 2nd edition, ISBN – 3-540-58676-8.

Waksmundzka M, Sharma J and Kowalska T. Thin Layer Chromatography in Phytochemistry. vol. 99, ISBN – 1-4200-4677-2.

Wall PE. Thin Layer Chromatography – A Modern Practical Approach. RS.C, ISBN – 0-85404-535-X.

Part III
Applications of HPTLC Separation

Chapter 5
Multidimensional and Multimodal Separations by HPTLC in Phytochemistry

Lukasz Ciesla and Monika Waksmundzka-Hajnos

Abstract HPTLC is one of the most widely applied methods in phytochemical analysis. It is due to its numerous advantages, e.g., it is the only chromatographic method offering the option of presenting the results as an image. Other advantages include simplicity, low costs, parallel analysis of samples, high sample capacity, rapidly obtained results, and possibility of multiple detection. HPTLC provides identification as well as quantitative results. It also enables the identification of adulterants. In case of complex samples, the resolving power of traditional one-dimensional chromatography is usually inadequate, hence special modes of development are required. Multidimensional and multimodal HPTLC techniques include those realized in one direction (UMD, IMD, GMD, BMD, AMD) as well as typical two-dimensional methods realized on mono- or bi-layers. In this manuscript, an overview on variable multidimensional and multimodal methods, applied in the analysis of phytochemical samples, is presented.

The amount of herbal drugs used worldwide has risen dramatically in the recent years. Herbal medicinal products, traditional Chinese medicines (TCM), nutraceuticals, and natural health products are only few from among the variety of plant drugs present in the market (Reich and Schibli 2008). Botanicals are believed to be very safe and at the same time not very effective drugs, but nothing could be farther from the truth. Many herbal medicines may cause severe side effects when overdosed or not properly prepared. It is also of great importance to control the amount of active ingredients within the herbal drugs, as their proper amount is responsible for the curing or health-promoting effects (Cieśla and Waksmundzka-Hajnos 2009a). However, the quality control of herbal medicines is completely different

L. Ciesla and M. Waksmundzka-Hajnos (✉)
Department of Inorganic Chemistry, Faculty of Pharmacy, Medical University of Lublin, Lubin, Poland
e-mail: monika.hajnos@am.lublin.pl

MM. Srivastava (ed.), *High-Performance Thin-Layer Chromatography (HPTLC)*,
DOI 10.1007/978-3-642-14025-9_5, © Springer-Verlag Berlin Heidelberg 2011

from the analytical evaluation of synthetic drugs. The entire plant extract should be regarded as an active component; thus, methods established for quality control of single active substance are insufficient in case of highly complex samples such as plant preparations or any other plant-derived products.

Multidimensional chromatography is often a good choice in case of very complex samples, offering many advantageous features in the analysis of medicinal plants (Cieśla and Waksmundzka-Hajnos 2009b). In multidimensional separation, a sample is first subjected to separation via one method, and then the separated compounds are further resolved by at least one additional independent method (Poole et al. 1989). According to Giddings, two conditions should be fulfilled to classify a chromatographic method as a multidimensional one (Giddings 1990). First of all, the separation mechanisms of the applied steps must be orthogonal, and secondly, the resolution gained in the first dimension may not be lost in any of the subsequent steps. The condition of orthogonal systems means the use of systems characterized by dissimilar retention mechanisms (Daszykowski et al. 2008). There are several techniques applied in the identification of such systems, for example, Daszykowski et al. used principal component analysis and hierarchical clustering methods for identification of similar and orthogonal systems for 2D-TLC separations of flavonoids (Daszykowski et al. 2008).

Multidimensionality can be realized in gas chromatography; however, in case of liquid chromatography, it is not a simple task. Planar chromatography gives a possibility of performing multidimensional separations with the use of several techniques. 2D-TLC has several unique features and thus is quite often a choice in the analysis of complex natural mixtures (Cieśla and Waksmundzka-Hajnos 2009b). First of all, multidimensional TLC is the only real multidimensional method in which, after the first separation in the first direction, all compounds can be passed to a second direction (Nyiredy 2001). Sophisticated equipment is not required, as in case of LC $\times$ LC separations, what is more plate is used once only, there is no need to worry about the adsorbed constituents that may cause column contamination. There is no need to perform complicated clean-up procedures, and multiple detection can be used to analyze the wide spectrum of compounds, which is impossible to realize in the sequential mode of HPLC. The sample preparation step does not have to be modified even if one wants to focus on different substance classes present in the extract. Proper chromatographic systems have to be chosen in order to focus on the desired constituent group (Cieśla and Waksmundzka-Hajnos 2009a). Multidimensional planar chromatography has also several noticeable disadvantages. The amount of the analyzed samples is considerably reduced in MD-PC (multidimensional planar chromatography), when compared to one-dimensional technique, as only one sample per plate can be analyzed. Multidimensional techniques are usually more time-consuming than one-dimensional methods, and if there is no significant improvement, when compared to one-dimensional mode, they simply do not pay off. Sometimes, change in chromatographic conditions may bring satisfying results without the need to apply multidimensional separation, as in the case of fatty oils' resolution, according to the European Pharmacopoeia

(Reich and Schibli 2008). In case of multidimensional methods, there is always a possibility of artifact formation, due to chemisorption or decomposition during chromatography. There are also several limitations as far as the solvents used are considered. Very polar and nonvolatile solvents should be avoided as they are difficult to be removed from the adsorbent, e.g., dimethyl sulfoxide, acetic acid, trimethylamine, as well as ion-pair reagents and nonvolatile buffer components (Poole and Poole 1995). In case of quantitative analysis, HPLC still remains a better alternative to TLC. Special modes of development can be classified as following

a) Repeated multiple development techniques in one direction
   - Unidimensional Multiple Development (UMD)
   - Incremental Multiple Development (IMD)
   - Gradient Multiple Development (GMD)
   - Bivariant Multiple Development (BMD) and its automated version – Automated Multiple Development (AMD)
b) Multidimensional techniques
   - Comprehensive 2D planar chromatography realized on one adsorbent or on bilayer plates
   - Targeted (selective) 2D planar chromatography – only chosen spots, separated after the first development, are subjected to further analysis
   - Modulated 2D planar chromatography – first eluents of decreasing strength are applied in the perpendicular direction and the sample is developed several times with solvent mixture of different selectivity at constant eluent strength
   - Graft TLC – the analyzed compounds are transferred from the first adsorbent to another and redeveloped in orthogonal system
   - Combination of MD-PC methods
   - Coupling of two chromatographic techniques realized in on- or off-line modes, e.g., HPLC–TLC, TLC–GC, TLC–MS, etc.

Techniques commonly applied in the analysis of phytochemical samples are discussed in the subsequent sections.

Multiple Development Techniques

In the chromatography of complex mixtures, the main problem is to improve resolution (Waksmundzka-Hajnos and Cieśla 2009). This might be achieved by increasing separation efficiency. Besides, for the analysis of complex mixtures, it is advantageous to obtain (if possible) a large spot capacity. The aforementioned goals can be achieved with the application of multiple development techniques in one direction. The observed improvement in separation for multiple development techniques, when compared to traditional approach, is due to the spot reconcentration mechanism. Each time the solvent front traverses the sample, it compresses the zone in the direction of development (Fig. 5.1) (Poole and Poole 1995). Due to the

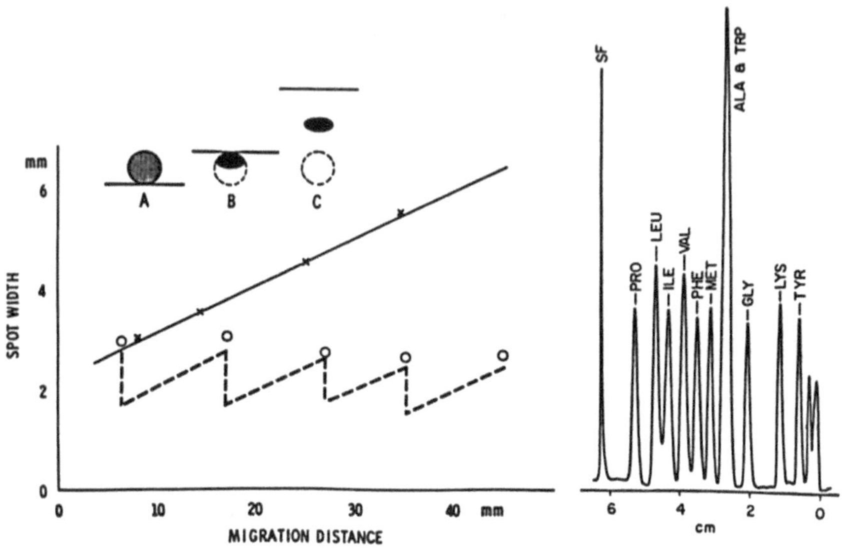

Fig. 5.1 The illustration of spot reconcentration mechanism. With permission from Poole et al. (1989)

fact that lower edge of the spot is overtaken earlier than the upper one, by the mobile phase front, spot becomes narrower. The constructional constrains causes slit-scanning densitometers are difficult to be used to scan the plate after two-dimensional separation. The application of multiple development techniques in one direction relieves the problems associated with detection (Poole et al. 1989). There are several factors that can be varied and adjusted in every single development to obtain desired separation, e.g., plate length, time of development, composition of the mobile phase, the number of developments, etc. What is more, detection can be repeated after every single step, using different detection conditions and detection modes. Constituents with comparable polarity are difficult to separate with the use of multiple development techniques, as they tend to migrate together from the origin and need long development distance for satisfying resolution. Multiple chromatographic developments in one direction should rather be reserved for those samples where only modest increase in resolution is needed (Poole and Poole 1995).

Multiple development techniques are used not only to enhance the chromatographic result but also to focus the application zone into the shape of a thin band to improve the resolution of the separation. These techniques may also serve as methods to separate the analyte from the matrix, and a very large sample can be concentrated into a thin starting zone. For removing the matrix, two modes can be applied – more polar analytes remain at the start, while nonpolar excipients migrate with the mobile phase (as in the case of the analysis of furanocumarin fraction from different fruits of the *Apiaceae* family (Waksmundzka-Hajnos and Wawrzynowicz

1992)), or the very polar matrix (e.g., carbohydrates) remains at the start, while the sample components are separated.

Unidimensional Multiple Development

UMD is a technique that consists of the repeated development of the same plate, with a mobile phase of constant composition, for the same distance (Fig. 5.2) (Nyiredy 2001). Immediately after the mobile phase has reached the specified developing distance, and before the next development, the plate must be removed from the chamber and quickly dried. The most common means of drying is cold air from a hair dryer, especially in case of rather volatile mobile phases. In case of more polar mobile phases, warm air may be used. However, in case of very volatile or labile constituents (e.g., essential oils), the warm air should not be used as it may lead to compound losses. The drying step is also critical and thus should always be validated, in case of quantitative analysis.

This mode of multiple developments is recommended in case of constituents having low R_F values in the investigated chromatographic system. It is due to the fact that differences between R_F values are increased for compounds in the lower R_F range and reduced for those in the upper range (Poole et al. 1993). For obtaining good results, when UMD is to be applied, mobile phase of low eluent strength should be used. The retention factor of a solute after n identical development steps (R_{Fn}), during UMD, can be calculated from the equation:

$$R_{Fn} = 1 - (1 - R_F)^n$$

where R_F is the retardation factor of the solute after a single development.

UMD has been successfully applied for the analysis of quaternary alkaloid fraction from *Chelidonium majus* (Waksmundzka-Hajnos et al. 2002) or the alkaloid fraction from *Fumaria officinalis* (Jóźwiak et al. 2000). In both cases, the band

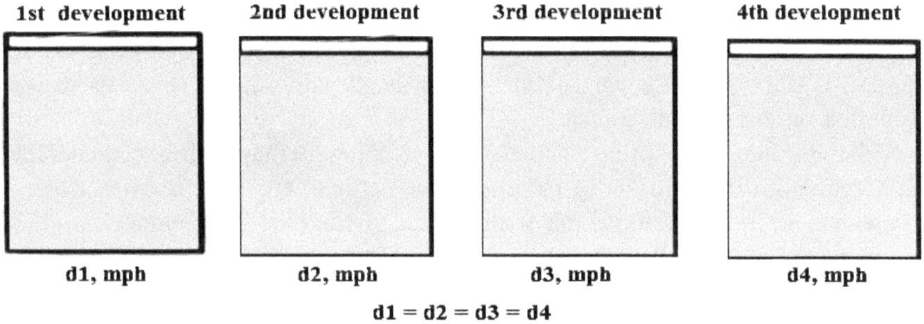

Fig. 5.2 Schematic representation of UMD technique. Symbols: *d* distance, *mph* mobile phase. Adopted from Nyiredy (2001)

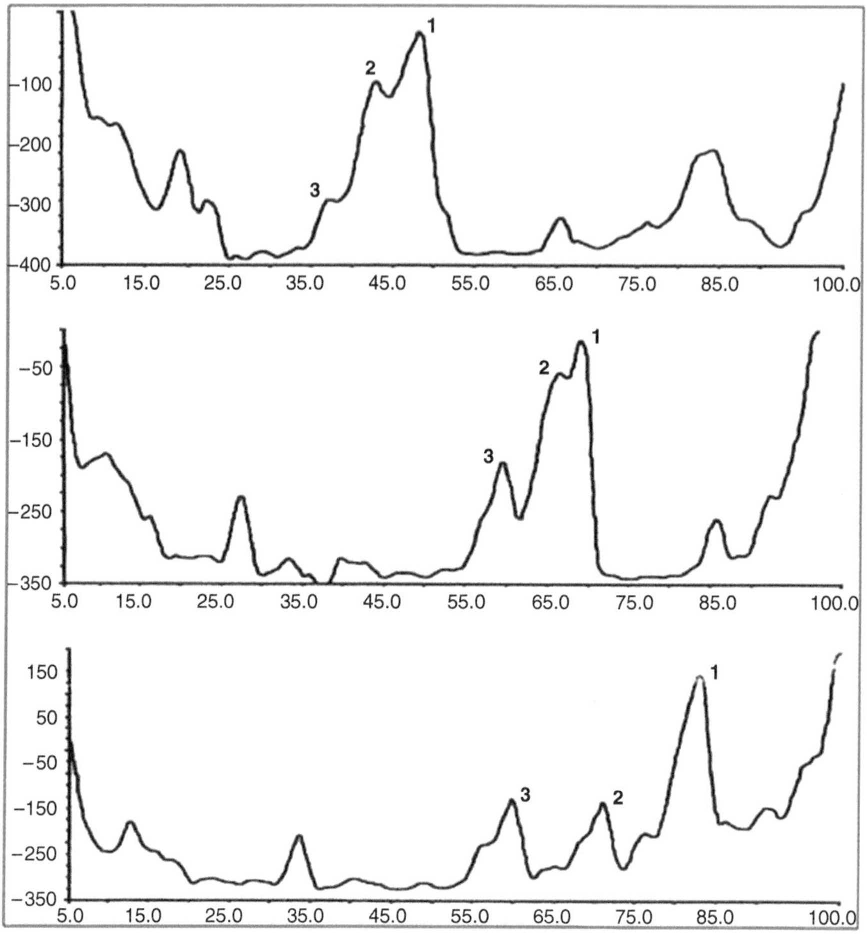

Fig. 5.3 *Fumaria officinalis* isoquinoline alkaloid separation obtained after one-, two-, and threefold development. With permission from Jóźwiak et al. (2000)

resolution markedly increased. The application of UMD in case of *F. officinalis* resulted in the complete resolution of the main alkaloids present in the extract (Fig. 5.3). UMD has also been used hand in hand with other techniques for the resolution of complex mixtures.

Although the application of the UMD technique may bring considerable improvement in the analysis of natural mixtures, there are several cases when it may lead to erroneous results. First of all, the use of multiple development, on silica, should be avoided in case of very labile constituents, which may be oxidized on the surface of polar adsorbents, e.g., carotenoids (Poole and Poole 1995). The European Pharmacopoeia method for the identification of *Angelica sinensis* and its common adulteration *Levisticum officinale* requires double development (Reich and Schibli 2008). Additional spots appearing after the second development are due to the formation of artifacts. Although the second development seems to bring significant improvement, it really causes the method to be inadequate for fingerprint development. Changing chromatographic conditions also may be more beneficial than applying multiple development techniques. For example, fingerprint construction of lavender oil on TLC plates requires double development, while on HPTLC plates, only one development is needed and the separated zones are sharper when compared to those on TLC plate (Reich and Schibli 2008). Changing the application mode also was proven to produce better results when compared to multiple developments in case of fatty oils analysis. In this case, UMD was really needed to focus the spots on bands (Reich and Schibli 2008).

Incremental Multiple Development

In this method, the development distance is increased during each step, with the use of the mobile phase of constant composition (Fig. 5.4) (Nyiredy 2001). When the increment is constant, R_F value of a single compound can be calculated using the equation:

$$R_{Fn}^{\mathrm{lin}} = 1 - (1 - R_F)\frac{1 - (1 - R_F)^n}{n \bullet R_F}$$

where R_F is the retardation factor of the solute after a single development.

This technique has been successfully applied, for example, for the fractionation of alkaloids in extract of *F. officinalis* (Waksmundzka-Hajnos and Jóźwiak 2008), as well as separation of furanocoumarins from *Heracleum sphondylium* fruit extract (Wawrzynowicz et al. 1998). In the first case, the extract was developed on silica plate with the use of eluent: PrOH/CH_3OH/CH_2Cl_2 (4:1:5), 20 mm increment was applied. Seven constituents were separated with the use of the aforementioned conditions (Fig. 5.5). The use of IMD enabled the resolution of three structurally similar furanocuomarins: xanthotoxin, bergapten, and phellopterin.

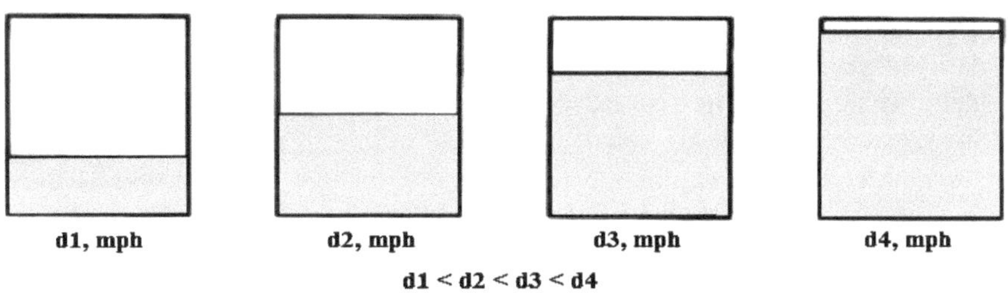

Fig. 5.4 Schematic representation of IMD technique. Symbols: *d* distance, *mph* mobile phase. Adopted from Nyiredy (2001)

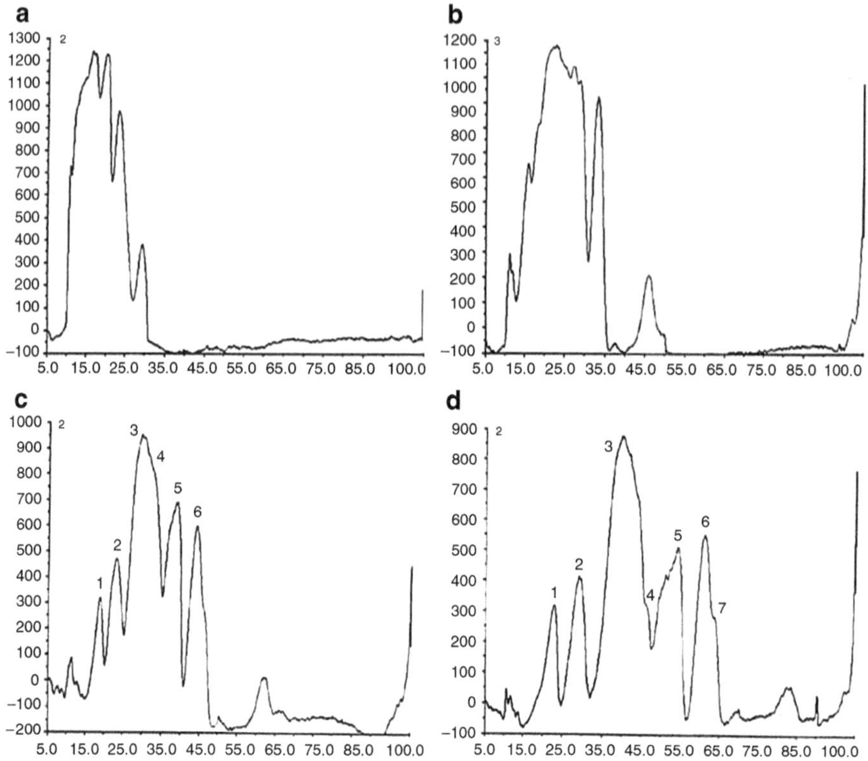

Fig. 5.5 *Fumaria officinalis* isoquinoline alkaloid fraction developed with the use of IMD technique with an increment of 20 mm. With permission from Waksmundzka-Hajnos and Jóźwiak (2008)

Gradient Multiple Development

The definition of GMD is as following: it is a method in which each stage of rechromatography is performed with a mobile phase of increasing strength, while the development distance remains constant (Fig. 5.6) (Nyiredy 2001). This method is especially beneficial in case of samples containing substances spanning a wide polarity range. The difficulty in the separation of samples containing components of widely different polarities is difficult because of the "general elution problem" (Matysik and Soczewiński 1988a). In TLC separation of plant extracts, gradient elution markedly improves the separation of spots owing to stronger displacement effects under condition of numerous adsorptions – desorption processes. According to the above definition, in GMD technique, first the apolar constituents are chromatographed over the entire plate with the use of the weakest mobile phase. This step is followed by detection, and after that, the plate is developed for the second time, for complete resolution of more polar compounds. In order to do that, mobile

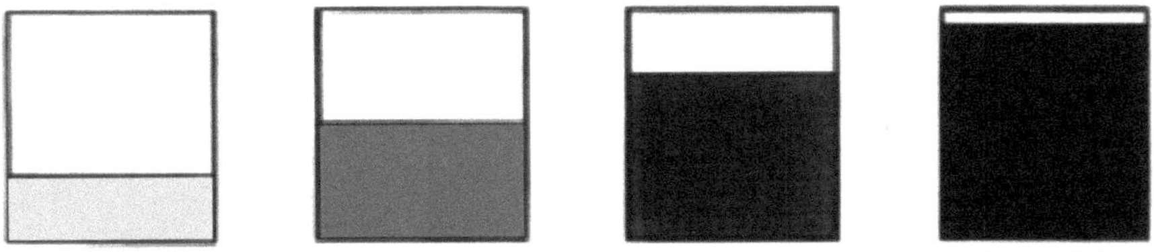

Fig. 5.6 Schematic representation of IMD technique. Symbols: *d* distance, *mph* mobile phase. Adopted from Nyiredy (2001)

phase characterized by the higher eluent strength is used. After the densitometrical evaluation, the plate is developed with eluent of the highest eluent strength. Despite the ability to resolve compounds differing in polarity, this method is quite lengthy as the plate is developed several times over the entire length, and detection is repeated as many times as the number of developments, which can also be considered as an advantage. The method has been successfully applied for the complete resolution of natural complex mixture containing furanocoumarins (apolar compounds), flavonoid aglycones (medium polar), and flavonoid glycosides (polar compounds) (Nyiredy 2001).

Although in its classical version, the definition of GMD includes only methods in which the analyzed sample is developed on the entire plate length, a method proposed by Matysik and Soczewiński can also be described as a multiple gradient method (Matysik and Soczewiński 1988b). In this technique, chromatogram is developed stepwise with the use of eluent of increasing polarity, on such distances that the constituents resolved in earlier steps do not migrate with the solvent front during the consecutive step and remain resolved until the last development. The number of gradient steps depends on the sample character as well as the polarity of its constituents and eluent polarity. The actual position of each band, after each development, should be checked under UV or densitometrically.

The increase of eluent strength of the mobile phase passing through the partly separated starting spot causes that consecutive sample components reach the optimal range in order to increase R_F values. Because the lower edge of the spot is overtaken earlier than the upper edge, by the front of the mobile phase, the spot becomes narrower than the starting band.

Gradient developments can be realized in two modes: stepwise and semicontinuous (Matysik and Soczewiński 1988c). In the first case, mobile phase fractions, of increasing strength, are poured into eluent reservoir in horizontal DS chamber, after the previous eluent has been completely exhausted. More sophisticated equipment is needed for the realization of semicontinuous gradient. Such equipment was introduced by Soczewiński and Matysik: portions of increasing eluent strength are placed in a pipe – they can be divided from each other by air bulbs and then slowly pressed to the eluent distributor (Matysik and Soczewiński 1988c).

This technique was applied for the resolution of a variety of samples, e.g., constituents present in the extract of *Radix rhei* (Fig. 5.7) and *Cortex frangulae*

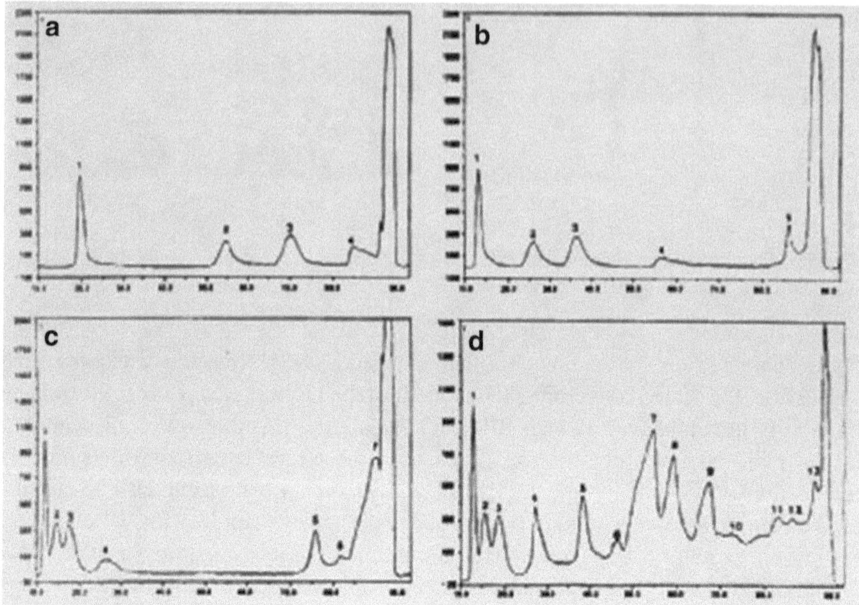

Fig. 5.7 Multiple gradient development of *Radix Rhei* extract. The extract was developed in the following mode: (a) 10% AcOEt (ethyl acetate) + $CHCl_3$, 9 cm, (b) 50% AcOEt + $CHCl_3$, 9 cm, (c) 100% AcOEt, 8 cm, (d) 15% MeOH (methanol) + AcOEt, 5 cm. With permission from Matysik and Soczewiński (1996)

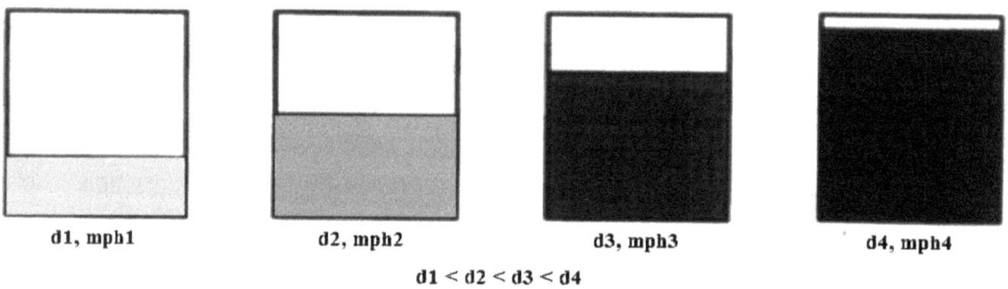

Fig. 5.8 Schematic representation of IMD technique. Symbols: *d* distance, *mph* mobile phase. Adopted from Nyiredy (2001)

(Matysik and Soczewiński 1996), *Herba euphrasiae* (Matysik and Toczołowski 1997), *Herba convallariae majalis* (Matysik et al. 1996), and others. Gradient elution was proven to cause considerable improvement in separation in comparison to isocratic elution. In case of this technique, elution can be performed with the use of gradients, which cannot be applied in case of an automated technique – AMD.

Bivariant Multiple Development and its Automated Version: Automated Multiple Development

BMD involves the step-by-step changing of both the development distance and mobile phase composition. The development distance is increased and the eluent strength of the mobile phase reduced for consecutive steps (Fig. 5.8) (Nyiredy 2001). It is a much simpler technique, when compared to classical GMD, as the separation is detected as a single chromatogram. However, it is not suitable for the analysis of very complex samples, due to its limited spot capacity, which is the result of very short developing distance. This technique was proven to be suitable, for example, for resolution of furanocoumarins present in *H. sphondylium* fruit extract (Wawrzynowicz et al. 1998). Mixtures of ethyl acetate with *n*-heptane, of decreasing eluent strength, were used for the resolution of structural analogs present in the extract.

AMD can be considered as an improved, automated version of BMD. AMD is a technique that uses repeated development of HPTLC plates with decreasing solvent strength on the increasing distance. After each development, the plate is carefully dried by vacuum. The development starts with the most polar solvent (for the shortest development distance) and concludes with the least polar solvent (for the longest migration distance) (Poole et al. 1989). Gradient development with linear eluotropic profile leads to a band reconcentration improving the separation. A successful separation depends mainly on the choice of the solvent components, optimization of the shape of the gradient, the stepwise movement of the elution front, and the repeated developments (Pothier and Galand 2005). AMD is highly recommended in case of samples containing substances of wide polarity or those being structural analogs. For the best resolution of constituents spanning wide polarity range, steep gradient is especially beneficial, while shallow gradient with small increases of developing distance provides good results in case of the analogs (Reich and Schibli 2008). AMD provides a more certain approach to optimizing a gradient separation when compared to other nonautomated TLC gradient methods (Poole et al. 1989). In case of nonautomated gradient elution, the formation of multiple zones of different solvent strength in the direction of chromatography can be observed as a result of solvent demixing (Poole et al. 1989). When compared to manual methods, AMD provides a high degree of gradient reproducibility. One of the disadvantages of AMD is the possibility of losing volatile as well as less volatile constituents present in the analyzed samples, during repetitive drying under vacuum (Reich and Schibli 2008). As far as the phytochemical analysis is concerned, AMD has been applied, for example, for separation of phenolic compounds in a solvent extract from an acidified aqueous suspension of the herb chamomile (Menziani et al. 1990), opiate alkaloids (Pothier and Galand 2005), and calystegines (Scholl et al. 2001). Pothier and Galand give several examples of plant secondary metabolites that have been analyzed with the use of AMD (Pothier and Galand 2005). Gocan et al. describe a simultaneous AMD separation and a comparison with isocratic chromatography for ten plant extracts that contain different classes of compounds: alkaloids (*Cinchona succirubra*, *Berberis vulgaris*), saponins (*Paullinia cupana*, *Aesculus hippocastanum*), terpenes (*Thuja occidentalis*, *Echinacea angustifolia*), flavonoids (*Carduus marianus*, *Lycopus europaeous*, *Baptisia tinctoria*), and coumarins (*Artemisia abrotanum*) (Gocan et al. 1996). Galand et al. compare traditional TLC separations with those obtained with the use of OPLC and AMD, for different natural substance classes (e.g., essential oils or alkaloids) (Galand et al. 2002).

Multidimensional Thin-Layer Chromatography

Theoretically, the analyzed sample can be subjected to unlimited number of separation steps; however, the constructional constrains and difficulties in finding more than two methods characterized by orthogonal mechanisms cause the majority of multidimensional separations to be two-dimensional. 2D-TLC is defined as chromatographic development in one direction followed by a second development in a direction perpendicular to the first. Planar chromatography gives the possibility to conduct two-dimensional separations with the use of the same stationary phase with different eluent systems or by using stationary phase gradient (Nyiredy 2001).

Comprehensive 2D Chromatography on One Adsorbent

In this method, the analyzed sample is spotted at the corner of the stationary phase and developed in the first dimension. Then, after careful drying, the plate is rotated at 90° and redeveloped in the second direction (Fig. 5.9) (Zakaria et al. 1983). This 2D development can be realized with the use of the same eluent in both directions, or by applying two different solvent mixtures, characterized by various selectivity and strength. In the first mode, the increase in resolution is due to the increase of

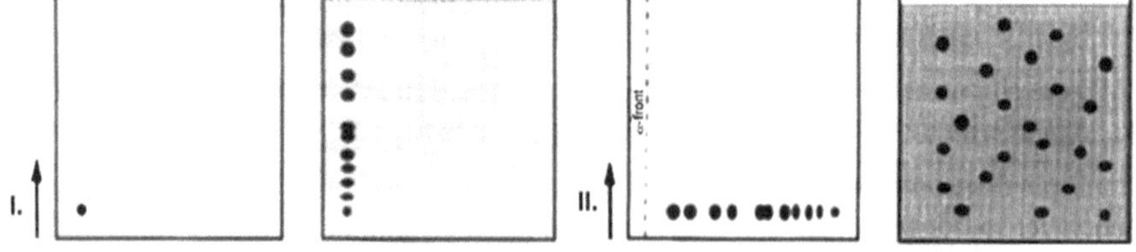

Fig. 5.9 Schematic representation of comprehensive two-dimensional chromatography: (**a**) spotting a sample at the corner of a chromatographic plate, (**b**) development of the plate in the first direction, (**c**) drying of the plate, (**d**) development of the plate in the second direction perpendicular to the first one with the second eluent. With permission from Nyiredy (2001)

migration distance by $\sqrt{2}$. This technique is commonly applied as one of the most important prevalidation tests during method development. It is applied to check whether the analyzed sample undergoes degradation during chromatography or if chemisorptions appear. All spots should be lying on the diagonal and any artifact will be located aside of the line. The spot capacity of such two-dimensional separations is only slightly higher when compared to one-dimensional separations. Thus, this method is not applied for resolving complex natural mixtures, however, remaining an important tool for the sample stability analysis (Reich and Schibli 2008).

The use of a single sorbent layer and two solvent systems of different selectivity is the most popular method of performing two-dimensional separations in medicinal plant research (Cieśla and Waksmundzka-Hajnos 2009b). The greatest spot capacity is obtained when two systems orthogonal in properties are applied in perpendicular directions. This condition is difficult to be fulfilled in case of silica gel – adsorbent being usually the first choice in case of using thin-layer chromatography. This is also the case for other polar adsorbents, e.g., cellulose, polyamide, alumina, magnesium oxide, etc. Several methods have been proposed for the identification of orthogonal systems, one already mentioned, developed by Daszykowski et al. (2008). A computer-aided technique was proposed by Gonnord et al. in which two equations were given for judging the quality of two-dimensional separations (Gonnord et al. 1983). This method has been modified by Steinbrunner et al. (1986). Planar response function (PRF) is another equation-based method for identifying optimal eluents for two-dimensional separations (Nurok et al. 1987). However, the most widely applied method is neither of all mentioned above. The selection of optimal mobile phases is very often based on the correlation coefficients between R_F and R_M values obtained after one-dimensional development, in two different solvent systems, introduced by De Spiegeleer et al. (1987). It appeared to be useful in the analysis of multicomponent mixtures of pesticides (Tuzimski and Soczewiński 2002, 2003; Tuzimski 2004, 2005), coumarins (Härmälä et al. 1990), or polyphenolic compounds (Soczewiński et al. 2001).

Two-Dimensional Separations on Polar Adsorbents: Silica, Cellulose, Polyamide

Although finding two orthogonal systems for the resolution of complex mixtures on the silica surface is rather problematic, several papers have been published presenting such systems in the analysis of plant-derived samples. In case of cellulose and polyamide, finding orthogonal systems is much easier, as it was proved for the resolution of phenolic acids on cellulose (Smolarz and Waksmundzka-Hajnos 1993). The analyzed mixture of polyphenolic compounds was first chromatographed with nonaqueous eluent (adsorption mode) and redeveloped with aqueous solvent mixture perpendicularly (partition mode). The application of this method enabled complete resolution of a 14 compound mixture containing isomeric forms

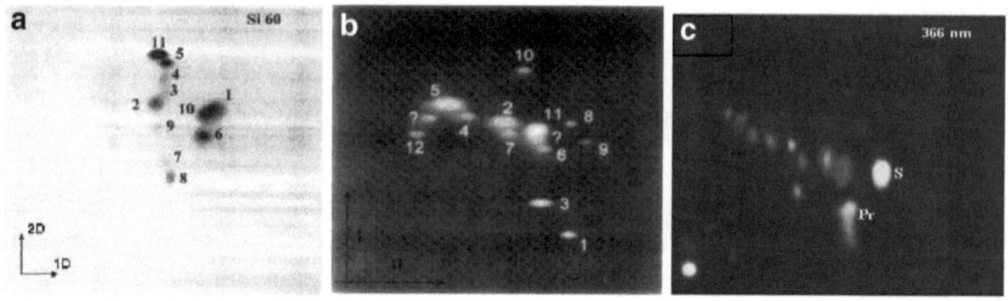

Fig. 5.10 Examples of the comprehensive two-dimensional thin-layer chromatography use for the analysis of phytochemical samples: (**a**) separation of phenolic acid standards, I direction eluent: toluene/dioxan/HCOOH (7:2:1, v/v/v), II direction eluent: mobile phase gradient, for details and symbols, see Glensk et al. (2002); (**b**) separation of an extract from *Flos Sambuci* on CN-silica, I direction eluent: 60% acetone in hexane, II direction eluent: 50% (v/v) methanol in water, for details and symbols, see Hawrył et al. (2002); (**c**) separation of isoquinoline alkaloids from *Fumaria officinalis* on CN-silica plate, I direction eluent: 60% (v/v) methanol in water + 2% ammonia, II direction eluent: 10% (v/v) methanol in isopropanol + 2% ammonia, for details, see Petruczynik et al. (2005)

of different naturally occurring organic acids. Despite its easily noticeable disadvantage – poor spot shape, the proposed system is ideally suited for the resolution of *cis* and *trans* isomers of cinnamic acid derivatives. It has been widely used for the identification of phenolic acids within different plant material, e.g., in the roots and fruits of *Peucedanum verticillare* (Kozyra et al. 2003), in the leaves of *Polygonum amphibium* (Smolarz and Waksmundzka-Hajnos 1993), in the leaves of *Lavatera trimestris* (Głowniak et al. 2005), in the petrioles of *Rheum* spp. (Smolarz et al. 2005), and still others. Better efficiency is observed on silica gel plates, which were also used for two-dimensional separations of phenolic acids or more complex mixtures containing also other polyphenolic derivatives, e.g., flavonoids. Cisowski et al. proposed the resolution of the acid mixture first with the nonaqueous eluent followed by the application of gradient elution mode in the perpendicular direction (Fig. 5.10a) (Glensk et al. 2002). Although spot shape is far better from that obtained on cellulose, some spots remain unresolved. Both aforementioned methods can be used for the resolution of acids' mixture only after the isolation of these compounds from the sample. They are unsuitable for more complex mixtures with more complex matrix. Several authors proposed chromatographic systems that can be used for the analysis of samples containing phenolic acids, only after very simple clean-up procedures. Medič-Šarič et al. introduced an interesting method of quantification of selected flavonoids and phenolic acids in propolis samples (Medić-Šarić et al. 2004). The authors were able to establish the amount of seven flavonoids and two phenoilc acids. Hawrył and Soczewiński applied two pairs of solvent systems, characterized by the lowest correlation coefficients, for the resolution of polyphenolic compounds present in *Betula* leaves (Hawrył and Soczewiński 2001).

Silica has been widely applied for the resolution of different alkaloid classes. Usually, nonaqueous eluents incorporating different additives (ammonia, amines) are applied in the analysis of this group of natural compounds (Mulvena and Slaytor 1982; Chowdhury et al. 1987). The use of silica with aqueous buffered mobile phases in the so-called pseudo-reversed-phase systems has also been reported, e.g., in the analysis of iso quinoline alkaloids present in *F. officinalis* herb extract (Petruczynik et al. 2005). The application of silica gel for the resolution of structural analogues is really not very frequent, due to its poor resolving power for such a group of substances. Thus, there are only few papers describing the application of this adsorbent in the analysis of structural analogs. A good example is the resolution of 16 coumarins from the genus *Angelica* (Härmälä et al. 1990). However, in this case, the resolution was possible due to the application of overpressure in both directions. Other examples of the use of silica in the analysis of naturally occurring substances can be found in the review article by Cieśla and Waksmundzka-Hajnos (2009b).

Two-Dimensional Separations on Polar Chemically-Bonded Stationary Phases: CN-Silica, DIOL-Silica, and NH_2-Silica

The greatest selectivity differences are obtained by combining normal-phase and reversed-phase systems in the perpendicular directions (Nyiredy 2001). The application of such systems is possible with the use of surface-modified adsorbents, especially polar stationary phases bonded to silica matrix. There is no problem of using aqueous and nonaqueous eluents in case of plates covered with the bonded phases. What is more, in case of aminopropyl-bonded layers, they may be used in ion-exchange systems with the application of acidic mobile phases.

The use of polar-bonded stationary phases brings also several disadvantages, which should be taken into account during method development. First of all, the eluents used in the first direction may modify the stationary phase. If the applied solvents cannot be removed from the adsorbent or modification is not reproducible, the use of such adsorbents may lead to irreproducible results. Thus it is recommended to use aqueous systems in the second direction, as water usually modifies the stationary phase, it also causes that the drying step becomes much longer, that extends the method development (Cieśla and Waksmundzka-Hajnos 2009b). Mobile phase additives, such as amines, acids, or ion-pair reagents, should rather be used also in the second development. When the application of basic modifiers (e.g., in the analysis of alkaloids) is essential for obtaining good results, in both directions, the use of ammonia instead of diethylamine is recommended, as it is easier to be removed from the layer. CN-silica plates are the most often choice to perform comprehensive two-dimensional separations on one adsorbent. The mechanism of solute retention on CN-silica plates has been investigated and well-defined (Soczewiński et al. 2001), as well as a rational approach has been proposed for method development for 2D-TLC of multicomponent mixtures on this adsorbent (Tuzimski 2004). They have been used for example for the separation of coumarin fractions present in fruits of different plant species belonging to the *Apiaceae* family (Waksmundzka-Hajnos et al. 2006). Orthogonal systems were also applied for the resolution of this class of compounds on DIOL-silica plates.

DIOL-silica is similar in its properties to deactivated silica; thus, applying aqueous eluents may lead to band tailing. This can be reduced by incorporating organic acids into the mobile phase (e.g., acetic, formic or trifluoroacetic acid), and it was the case in the analysis of coumarins on this type of adsorbent.

The use of CN-silica plates turned out to be beneficial in the resolution of polyphenolic compounds: phenolic acids and flavonoids, very common in almost all plant samples. Hawrył et al. managed to completely separate the extract of *Flos Sambuci* by use of the systems most orthogonal in properties, on CN-silica plates (Fig. 5.10b) (Hawrył et al. 2002). The aforementioned plates were also used in the analysis of extracts containing alkaloids (Petruczynik et al. 2007). Petruczynik et al. investigated different mobile phase additives on the retention, system selectivity, and efficiency for isoquinoline alkaloids. The best systems were ideally suited for the analysis of alkaloid fractions present in different plant species belonging to the *Papaveraceae* family (Fig. 5.10c). Greater selectivity values are obtained with the application of two adsorbents enabling the use of two truly orthogonal chromatographic systems; this can be realized with the use of bilayer plates and in coupled layer 2D TLC.

2D-TLC on Bilayer Plates and Coupled-Layer 2D-TLC

In case of bilayer plates, the sample is first developed on a narrow strip of the first adsorbent; then, after being rotated at 90°, it is redeveloped in the second direction (on the adsorbent adjacent to the first one) (Tuzimski 2007a). The commercially available bilayer plates consist of a narrow silica gel strip (or octadecylsilica gel) coated side by side to C18 (or silica) layer. However, other adsorbent combinations are also possible but not used as often as those mentioned above, e.g., silica gel + alumina, cellulose + silica gel, and others.

The first analytical aim of using bilayer plates was the need to improve performance in case of some one-dimensional separations. The narrow strip of adsorbent was used as the preconcentrating zone. This effect is still used, when bilayer plates are applied. The analyzed substances, when transferred from the first to adjacent adsorbent, become narrower, band-shaped, and thus better resolved even before introducing eluent used in the second direction (Cieśla et al. 2008a). Narrow adsorbent strips may play the same role precolumns play in case of HPLC. They simply are used to remove interfering substances from plant extracts. Two-dimensional separations on bilayer plates can be achieved in two modes: with the use of the same eluent or by applying two different solvent mixtures (Nyiredy 2001). The first one is the opposite of the classical 2D TLC on monolayer, with two different eluents, and is rarely performed. When bilayer plates are applied, the analyzed constituents are separated on silica gel according to differences in polarity, while on octadecyl silica, structural analogs differing with nonpolar fragment are separated.

In case of the analysis of phytochemical samples, the reproducibility of performing two-dimensional separations on bilayer plates may be a problem (Cieśla

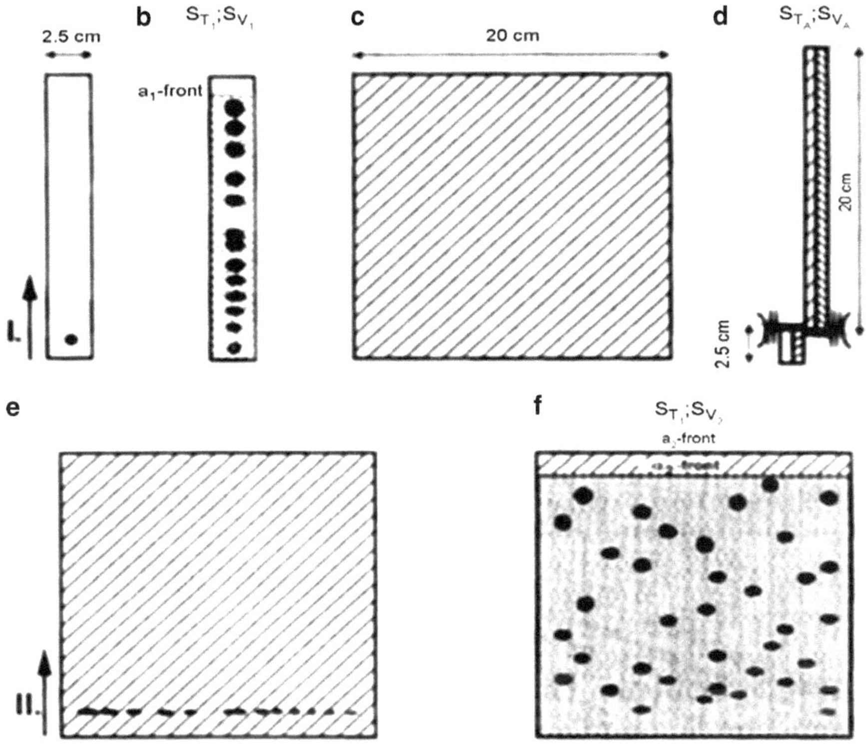

Fig. 5.11 Schematic representation of coupled-layer planar chromatography (graft-TLC): (**a**) sample application to the first plate; (**b**) development of the plate in the first direction; (**c**, **d**) clamping another plate with the narrow strip and transferring partly separated compounds from the first layer; (**e**) drying of the plate; (**f**) development of the second plate in the second direction perpendicular to the first one with the second eluent. With permission from Nyiredy (2001)

and Waksmundzka-Hajnos 2009b). It is due to the fact that water applied in the first direction modifies the silica layer and may lead to difficulties in performing reproducible analysis. In case of any problems of handling bilayer plates, the use of two separate plates coated with two different stationary phases may be recommended. Two different adsorbents may be used when the technique called "graft thin-layer chromatography" is applied. In this method, constituents partially separated, in the first direction, are transferred to other adsorbent and redeveloped in orthogonal chromatographic system (Fig. 5.11). The use of this technique can be somewhat problematic, especially the process of transferring the compounds is the most difficult to perform. Usually, two adsorbents are turned face to face and pressed together, and the layers should overlap. Such way of joining two plates can be realized by putting the plates between two glass plates and joining with the use of clips. However, a device that eases the transfer of partly separated mixtures from one stationary phase to another has also been presented (Łuczkiewicz et al. 2004). The most crucial in performing graft TLC analysis is the selection of proper adsorbents to be used in the first and perpendicular direction, as well as the transferring liquid. Adsorbents, on which the analyzed compounds may be strongly retained, should be avoided. So, for example, apolar substances, such as furanocoumarins or hydrocarbon derivatives, should be first developed on the polar adsorbent, e.g., silica gel, while polar constituents, e.g., alkaloids or flavonoid glycosides, should be first chromatographed on RP-18W plates (Cieśla and Waksmundzka-Hajnos 2009b).

Usually, silica gel and octadecyl silica are applied in graft-TLC as they produce greatest selectivity differences, but other combinations of adsorbents have also been reported: CN-silica + silica (Petruczynik et al. 2008) or DIOL-silica + RP-18W plates (Łuczkiewicz et al. 2004). As far as the transferring solvents are considered, methanol is usually applied; however, there are reports on the use of aqueous methanol (Glensk et al. 2002) or the eluent applied in the second direction as transferring liquids (Łuczkiewicz et al. 2004). In case of the analysis of secondary plant metabolites, bilayer plates have been rarely used due to the problems already mentioned. RP-18W + silica bilayer plates have been applied for the resolution of coumarin standards as well as coumarin fraction from selected *Apiaceae* fruit samples (Fig. 5.12) (Waksmundzka-Hajnos et al. 2006). Multi-K SC5 TLC plates containing a 3 cm × 20 cm strip of silica and a 17 cm × 20 cm layer of C18 were applied for the analysis of tropane alkaloids (Gadzikowska et al. 2005). The authors applied methanol, as in graft-TLC procedure, to transfer the alkaloids to the C18 layer; then the silica strip was cut off to ensure that the resolution is maximized and the separation time is minimized.

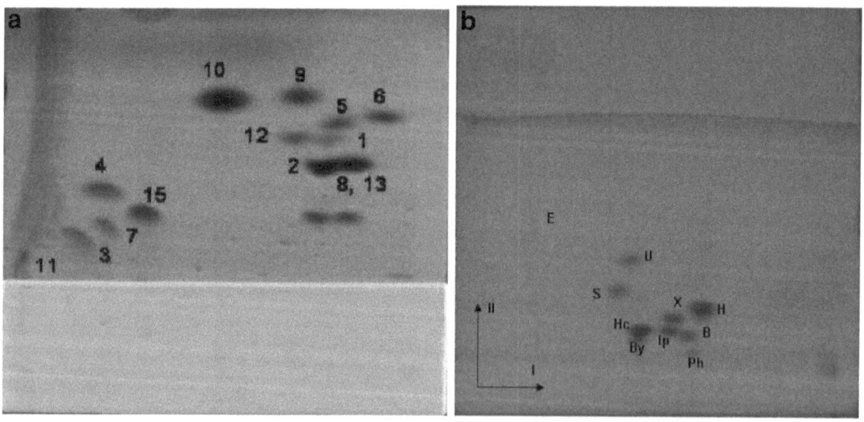

Fig. 5.12 Examples of the use of (**a**) bilayer plates and (**b**) graft-TLC for the analysis of phytochemical samples: (**a**) Separation of coumarin standards on bilayer plate Multi-K CS5 (3 cm zone of C18 parallel to silica layer) I direction eluent: 55% (v/v) methanol in water, II direction eluent: 35% (v/v) ethyl acetate in *n*-heptane, for details, see Waksmundzka-Hajnos et al. (2006); (**b**) separation of furanocoumarins with the use of graft-TLC procedure, the analyzed sample was first chromatographed on silica gel plate with the use of eluent: 35% (v/v) ethyl acetate in *n*-heptane (triple developed), after transfer with methanol, it was redeveloped on RP-18W plate with 55% (v/v) methanol in water, for details, see Cieśla et al. (2008c)

Graft-TLC has been chosen more often in comparison to the analysis of bilayer plates. It was successfully applied for the resolution of a phenolic acids mixture (Glensk et al. 2002). Complete resolution of eleven phenolic acids has been obtained after the application of reversed-phase system in the first direction and normal-phase one orthogonally. Phenolic compounds present in the extracts of *Polygonum* and *Verbascum* species were also resolved with the use of graft-TLC (Hawrył and Waksmundzka-Hajnos 2006). Krauze-Baranowska et al. reported the use of coupled-layer thin-layer chromatography for the analysis of flavonol and truxinic esters from *Pseudotsuga menziesii* (Krauze-Baranowska et al. 2002). Graft-TLC has been used in the analysis of quinolizidine (Łuczkiewicz et al. 2004) and isoquinoline alkaloids (Petruczynik et al. 2008). In the first case, the analyzed samples were first chromatographed in normal phase system on DIOL-silica and, after the transfer to RP-18 plate, were developed with aqueous eluent. Although the chromatographic system applied in the first direction is characterized by rather poor efficiency, the use of graft-TLC enabled complete separation of the investigated alkaloids. In case of the isoquinoline alkaloids, cyanopropyl-silica was used as the first direction adsorbent in normal-phase system, with the addition of ammonia. Then the alkaloids were rechromatographed on RP-18W plates or on silica. The systems characterized by the best selectivity and efficiency were used for the analysis of alkaloid fractions from the herbs of *C. majus*, *F. officinalis*, and *Glaucium flavum*. Coupled layer TLC was also used in the analysis of ginsenosides (Glensk et al. 2001). The pharmaceutical formulation containing *Ginseng* extract was first developed on RP-18 plate and rechromatographed on silica layer. Similar combination of chromatographic systems was used also in the analysis of saponins present in *Silene vulgaris* (Glensk and Cisowski 2000). Due to great selectivity differences obtained with the use of graft-TLC, it is a technique that may bring several other successful separations of constituents from a variety of plant samples.

Combination of MD-PC Methods: Hyphenated Techniques

In combined MD-PC methods, at least two different modes of development are combined in both directions. Usually, multiple development techniques, in one direction, are used together with two-dimensional procedures. There are several possibilities of performing such separations, which was well summarized by Nyiredy (2001). One of the possibilities is the application of BMD technique in the first direction – the analyzed substances is separated according to differences in their polarities. Then orthogonally, another multiple development technique may be used, for example, UMD or IMD. Nyiredy proposed a method in which in the first direction, the components are developed with solvent mixture of decreasing strength, at constant selectivity. In orthogonal direction, selectivity of the mobile phase is varied, while solvent strength remains constant.

In case of the analysis of phytochemical samples, only few real MD-PC methods have been already applied. The use of UMD with comprehensive two-dimensional chromatography and graft-TLC has been reported (Cieśla et al. 2008a, b, c). Such combination of multidimensional planar chromatographic methods was used for the resolution of coumarins. The analyzed compounds were first chromatographed on silica or CN-silica plates and then redeveloped on RP-18W or silica gel plates, respectively. The use of UMD caused the bands to be wider before the introduction of the second eluent, which led to better resolution. What is more, well shaped spots, obtained after the application of this combined technique, enabled performing quantitative analysis, after two-dimensional separation (Cieśla et al. 2008b). This technique turned out to be useful for fingerprint construction of closely related plant species, varieties, and forms from the Apiaceae family (Cieśla et al. 2008).

Matysik et al. proposed a method of joining multiple gradient development technique with two-dimensional separation on one adsorbent. The author managed to completely separate anthraquinone aglycones and glycosides on silica gel plate (Matysik et al. 2008).

Tuzimski proposed a new MD-PC method in the analysis of very complex samples (Tuzimski 2007b). A technique applying different modes of multidimensional thin-layer chromatography, as far as phytochemistry is considered, has been used for the analysis of methanolic extracts of different *Peucedanum* species (Cieśla et al. 2009).

Hyphenated techniques, in which TLC is combined with other complementary techniques, are another means of realizing multidimensional separations. Thin-layer chromatography can be used as the first or second dimension method (Poole and Poole 1995). The first case is rather problematic due to difficulties in obtaining reproducible results, as in this method, scrapping off the adsorbent is required, which often leads to analysis errors. When TLC is used as the second dimension technique, it is usually applied along with HPLC, GC, supercritical fluid chromatography, capillary electrophoresis, counter-current chromatography, and others. Hyphenated techniques have been rarely used in the analysis of phytochemical samples. The most common is the use of combination of HPLC and TLC in off-line mode.

A hyphenated HPLC–TLC procedure for separation of coumarins has been proposed by Hawrył et al. A mixture of 12 coumarins from *Archangelica officinalis* was completely separated as a result of the different selectivity of the two combined chromatographic techniques, RP-HPLC and NP-TLC (Hawrył et al. 2000). After separation on the HPLC column, all fractions were collected, evaporated, and developed on silica gel. The combination of these methods gave successful results, although both methods, if used separately, failed to give good resolution. RP-HPLC and NP-TLC off-line method was also proposed for the separation of a mixture of ten flavonoids (Hawrył and Soczewiński 2000). This method should rather be used for micropreparative purposes. The use of off-line methods, for the analysis of natural compounds, rather than on-line modes is caused by the frequent practice of using mobile phase additives, such as organic acids, amines, and buffers, which are difficult to be handled in on-line modes.

Notes

Multidimensional and multimodal TLC techniques are commonly applied in the phytochemical analysis due to high complexity of plant samples. Its potential to produce greater spot capacity, when compared to one-dimensional chromatography, can be used as one of the first laboratory methods for examining botanical material. They are especially of value in the analysis of samples containing substances spanning a wide polarity range or being structural analogs. However, it should be always checked whether the application of such methods is really needed, as some changes in chromatographic conditions may bring better results. The amount of multidimensional TLC methods applied in the field of phytochemistry is still growing and, in the future, will remain an important tool in the analysis of samples of natural origin, along with HPLC, HPLC–MS, HPLC–NMR, GC, and others.

References

Chowdhury BK, Rai PP and Bhattacharayya P. Separation of carbazole alkaloids by two-dimensional thin-layer chromatography. Chromatographia, 1987, 23, 205–206.

Cieśla Ł and Waksmundzka-Hajnos M. A comparison of TLC methods for constructing medicinal plant fingerprints. Sep Sci 2009a; 6, 22–27.

Cieśla Ł and Waksmundzka-Hajnos M. Two-dimensional thin-layer chromatography in the analysis of secondary plant metabolites. J Chromatogr A, 2009b, 1216, 1035–1052.

Cieśla Ł, Bogucka-Kocka A, Hajnos M, Petruczynik A and Waksmundzka-Hjanos M. Two-dimensional thin-layer chromatography with adsorbent gradient as a method of chromatographic fingerprinting of furanocoumarins for distinguishing selected varieties and form of *Heracleum* spp. J Chromatogr A, 2008, 1207, 160–168.

Cieśla Ł, Petruczynik A, Hajnos M, Bogucka-Kocka A and Waksmundzka-Hajnos M. Two-dimensional thin-layer chromatography of structural analogs. Part I: graft TLC of selected coumarins. J Planar Chromatogr, 2008a, 21, 237–241.

Cieśla Ł, Petruczynik A, Hajnos M, Bogucka-Kocka A and Waksmundzka-Hajnos M. Two-dimensional thin-layer chromatography of structural analogs. Part II: Method for quantitative analysis of selected coumarins in plant material. J Planar Chromatogr, 2008b, 21, 447–452.

Cieśla Ł, Skalicka-Woźniak K, Hajnos M, Hawrył M and Waksmundzka-Hajnos M. New chromatographic procedure for separation of complex natural mixtures spanning a wide polarity range, application for fingerprint construction and investigation of systematic relationships within the Peucedanum genus. Acta Chromatogr 2009 641–657.

Daszykowski M, Hawrył M, Waksmundzka-Hajnos M and Walczak B. Identification of similar and orthogonal chromatographic thin-layer systems for two-dimensional separations of flavonoids and their analogues. Acta Chromatogr, 2008, 20, (3), 283–307.

De Spiegeleer B, van der Bossche W, de Moerlose P and Massart D. A strategy for two-dimensional, high-performance thin-layer chromatography, applied to local anesthetics. Chromatographia, 1987, 23, 407–411.

Gadzikowska M, Petruczynik A, Waksmundzka-Hajnos M, Hawrył M and Jóźwiak G. Two-dimensional planar chromatography of tropane alkaloids from *Datura innoxia* Mill. J Planar Chromatogr, 2005, 18, 127–131.

Galand N, Pothier J, Dollet J and Viel C. OPLC and AMD, recent techniques of planar chromatography: their interest for separation and characterization of extractive and synthetic compounds. Fitoterapia, 2002, 73, 121–134.

Giddings JC (1990). In Cortes HJ (ed.): Multidimensional Chromatography, 1st edition, published by Marcel Dekker, New York.

Glensk M and Cisowski W. Two-dimensional TLC with adsorbent gradient for analysis of saponins in *Silene vulgaris* Garcke. J Planar Chromatogr, 2000, 13, 9–11.

Glensk M, Czekalska M and Cisowski W. Resolution of saponins from a *Ginseng* preparation by 2D TLC with an adsorbent gradient. J Planar Chromatogr, 2001, 14, 454–456.

Glensk M, Sawicka U, Mażol I and Cisowski W. 2D TLC – graft planar chromatography in the analysis of a mixture of phenolic acids. J Planar Chromatogr, 2002, 15, 463–465.

Głowniak K, Skalicka K, Ludwiczuk A and Jop K. Phenolic compounds in the flowers of *Lavatera trimestris* L. (Malvaceae). J Planar Chromatogr, 2005, 18, 264–268.

Gocan S, Cimpan G and Muresan L. Automated multiple development thin layer chromatography of some plant extracts. J Pharm Biomed Anal, 1996, 14, 1221–1227.

Gonnord MF, Levi F and Guiochon G. Computer assistance in the selection of the optimum combination of systems for two-dimensional chromatography. J Chromatogr, 1983, 264, 1–6.

Härmälä P, Botz L, Sticher O and Hiltunen R. Two-dimensional planar chromatographic separation of a complex mixture of closely related coumarins from the genus *Angelica*. J Planar Chromatogr, 1990, 3, 515–520.

Hawrył M and Soczewiński E. Separation of some flavonoids using RP–HPLC–NP–TLC off-line coupled system. Chromatographia, 2000, 62, 175–178.

Hawrył MA and Soczewiński E. Normal phase 2D TLC separation of flavonoids and phenolic acids from *Betula* sp. leaves. J Planar Chromatogr, 2001, 14, 415–421.

Hawrył MA and Waksmundzka-Hajnos M. Separation of phenolic compounds by NP and RP two-dimensional thin layer chromatography on connected plates. J Planar Chromatogr, 2006, 19, 92–97.

Hawrył MA, Soczewiński E and Dzido T. Separation of coumarins from *Archangelica officinalis* in high-performance liquid chromatography and thin-layer chromatography systems. J Chromatogr A, 2000, 886, 75–81.

Hawrył MA, Hawrył A and Soczewiński E. Application of normal- and reversed-phase 2D TLC on a cyanopropyl-bonded polar stationary phase for separation of phenolic compounds from the flowers of *Sambucus nigra* L. J Planar Chromatogr, 2002, 15, 4–10.

Jóźwiak G, Wawrzynowicz T and Waksmundzka-Hajnos M. Optimization of the preparative chromatography of an extract of alkaloids from *Fumaria officinalis*. J Planar Chromatogr, 2000, 13, 447–451.

Kozyra M, Głowniak K and Zadubiec A. Phenolic acids in *Peucedanum verticillare* L. Koch ex. J Planar Chromatogr, 2003, 16, 421–424.

Krauze-Baranowska M, Malinowska I and Skwierawska J. TLC of flavonol truxinic esters from *Pseudotsuga menziesii*. J Planar Chromatogr, 2002, 15, 437–441.

Łuczkiewicz M, Migas P, Kokotkiewicz A, Walijewska M and Cisowski W. Two-dimensional TLC with adsorbent gradient for separation of quinolizidine alkaloids in the herb and in-vitro cultures of several *Genista* species. J Planar Chromatogr, 2004, 17, 89–94.

Matysik G and Soczewiński E. Stepwise gradient development in thin-layer chromatography. J Chromatogr, 1988a, 446, 275–282.

Matysik G and Soczewiński E. Gradient thin-layer chromatography of plant extracts. Chromatographia, 1988b, 26, 178–180.

Matysik G and Soczewiński E. A miniaturized device for the generation of eluent composition gradients for sandwich thin-layer chromatography. Chem Anal, 1988c, 33, 363–369.

Matysik G and Soczewiński E. Modified programmed multiple development (MGD) in the analysis of complex plant extracts. Chromatographia, 1996, 43, 39–43.

Matysik G and Toczołowski J. TLC densitometric investigation of aucubin penetration from *Herba Euphrasiae* extract to the aqueous humor of the eye. Chem Anal, 1997, 42, 221–226.

Matysik G, Kowalski J and Staszewska H. Quantitative determination of convallatoxine from *Herba Convallariae*. Chromatographia, 1996, 43, 559–562.

Matysik G, Skalska-Kamińska A, Stefańczyk B, Wójciak-Kosior M and Rapa D. Application of a new technique in two-dimensional TLC of multicomponent mixtures. J Planar Chromatogr, 2008, 21, 233–236.

Medić-Šarić M, Jasprica I, Mornar A, Smolčić-Bubalo A and Golja P. Quantitative analysis of flavonoids and phenolic acids in propolis by two-dimensional thin layer chromatography. J Planar Chromatogr, 2004, 17, 459–463.

Menziani E, Tosi B, Bonora A, Reschiglian P and Lodi G. Automated multiple development high-performance thin-layer chromatographic analysis of natural phenolic compounds. J Chromatogr, 1990, 511, 396–401.

Mulvena D and Slaytor M. Separation of tryptophan derivatives in *Phalaris aquatica* by thin-layer chromatography. J Chromatogr, 1982, 245, 155–157.

Nurok D, Habibi-Goudarzi S and Kleyle R, Statistical approach to solvent selection as applied to two-dimensional thin-layer chromatography. Anal Chem, 1987, 59, 2424–2428.

Nyiredy Sz (2001). Planar Chromatography – A Retrospective View for the Third Millennium, 1st edition, published by Springer, Budapest.

Petruczynik A, Waksmundzka-Hajnos M and Hajnos M. The effect of chromatographic conditions on the separation of selected alkaloids on silica layers. J Planar Chromatogr, 2005, 18, 78–84.

Petruczynik A, Waksmundzka-Hajnos M, Michniowski T, Plech T, Tuzimski T, Hajnos ML, Gadzikowska M and Jóźwiak G. Thin-layer chromatography of alkaloids on cyanopropyl bonded stationary phases. Part I. J Sep Sci, 2007, 45, 447–454.

Petruczynik A, Waksmundzka-Hajnos M, Plech T, Tuzimski T, Hajnos MŁ, Jóźwiak G, Gadzikowska M and Rompała A. TLC of alkaloids on cyanopropyl bonded stationary phases. Part II. Connection with RP-18 and silica plates. J Sep Sci, 2008, 46, 291–297.

Poole CF and Poole SK. Multidimensionality in planar chromatography. J Chromatogr A, 1995, 703, 573–612.

Poole CF, Poole SK, Fernando WPN, Dean TA, Achmed HD and Berndt JA. Multidimensional and multimodal thin layer chromatography: pathway to the future. J Planar Chromatogr, 1989, 2, 336–345.

Poole CF, Poole SK and Belay MT. Insights into the mechanism and applications of unidimensional multiple development in thin layer chromatography, J Planar Chromatogr, 1993, 6, 438–445.

Pothier J and Galand N. Automated multiple development thin-layer chromatography for separation of opiate alkaloids and derivatives. J Chromatogr A, 2005, 1080, 186–191.

Reich E and Schibli A (2008). High-Performance Thin-Layer Chromatography for the Analysis of Medicinal Plants, 1st edition, published by Thieme, New York.

Scholl Y, Asano N and Dräger B. Automated multiple development thin layer chromatography for calystegines and their biosynthetic precursors. J Chromatogr A, 2001, 928, 217–224.

Smolarz HD and Waksmundzka-Hajnos M. Two-dimensional TLC of phenolic acids on cellulose. J Planar Chromatogr, 1993, 6, 278–281.

Smolarz HD, Medyńska E and Matysik G. Determination of emodin and phenolic acids in the petioles of *Rheum undulatum* and *Rheum rhaponticum*. J Planar Chromatogr, 2005, 18, 319–322.

Soczewiński E, Hawrył MA and Hawrył A. Retention behavior of some flavonoids in 2D-TLC systems on cyano bonded polar phases. Chromatographia, 2001, 54, 789–794.

Steinbrunner JE, Johnson EK, Habibi-Goudarzi S and Nurok D 1986. In: Kaiser RE (ed.), Planar Chromatography, vol. 1, Hüthig, Heidelberg.

Tuzimski T. Separation of a mixture of eighteen pesticides by two-dimensional thin-layer chromatography on a cyanopropyl-bonded polar stationary phase. J Planar Chromatogr, 2004, 17, 328–334.

Tuzimski T. Two-dimensional TLC with adsorbent gradients of the type silica-octadecyl silica and silica-cyanopropyl for separation of mixtures of pesticides. J Planar Chromatogr, 2005, 18, 349–357.

Tuzimski T. Separation of multicomponent mixtures of pesticides by graft thin-layer chromatography on connected silica and octadecyl silica layers. J Planar Chromatogr, 2007a, 20, 13–18.

Tuzimski T. A new procedure for separation of complex mixtures of pesticides by multidimensional planar chromatography, J Sep Sci, 2007b, 30, 964–970.

Tuzimski T and Soczewiński E. Correlation of retention parameters of pesticides in normal- and reversed-phase systems and their utilization for the separation of a mixture of 14 triazines and urea herbicides by means of two-dimensional thin-layer chromatography. J Chromatogr A, 2002, 961, 277–283.

Tuzimski T and Soczewiński E. Correlation of retention data of pesticides in normal- and reversed-phase systems and utilization of the data for separation of a mixture of ten urea herbicides by two-dimensional thin-layer chromatography on cyanopropyl-bonded polar stationary phase and on a two-adsorbent-layer Multi-K SC5 plate. J Planar Chromatogr, 2003, 16, 263–267.

Waksmundzka-Hajnos M and Cieśla Ł. (2009) Special techniques of development – tool in separation of plant extracts by thin-layer chromatography. Medicinal Plants in Poland and in the World, 17, 24–30.

Waksmundzka-Hajnos M and Jóźwiak G. Special modes of development in preparative-layer chromatography of extracts of alkaloids from *Fumaria officinalis*. J Planar Chromatogr, 2008, 21, 61–66.

Waksmundzka-Hajnos M and Wawrzynowicz T. On-line purification and micropreparative separation of contaminated fruit extracts by TLC in equilibrium sandwich chambers. J Planar Chromatogr, 1992, 5, 169–174.

Waksmundzka-Hajnos M, Gadzikowska M and Hajnos M. Strategy for preparative separation of quaternary alkaloids from *Chelidonium majus* L. by thin-layer chromatography. J Planar Chromatogr, 2002, 15, 289–293.

Waksmundzka-Hajnos M, Petruczynik A, Hajnos M, Tuzimski T, Hawrył A and Bogucka-Kocka A. Two-dimensional thin-layer chromatography of selected coumarins. J Sep Sci, 2006, 44, 510–517.

Wawrzynowicz T, Czapińska K and Markowski W. The use of special development techniques for micropreparative TLC separation of closely related compounds. J Planar Chromatogr, 1998, 11, 388–393.

Zakaria M, Gonnord MF and Guiochon G. Applications of two-dimensional thin-layer chromatography. J Chromatogr, 1983, 271, 127–192.

Chapter 6
Stability-Indicating HPTLC Determination of Imatinib Mesylate in Bulk Drug and Pharmaceutical Dosage

P. Musmade, N. Vadera, and G. Subramanian

Abstract A simple, selective, precise and stability-indicating high-performance thin-layer chromatographic method of analysis of imatinib mesylate both as a bulk drug and in formulations was developed and validated. The method employed HPTLC aluminum plates precoated with silica gel 60F-254 as the stationary phase. The solvent system consisted of chloroform:methanol (6:4, v/v). The system was found to give compact spot for imatinib mesylate (R_f value of 0.53 ± 0.02). Densitometric analysis of imatinib mesylate was carried out in the absorbance mode at 276 nm. The linear regression analysis data for the calibration plots showed good linear relationship with $r^2 = 0.9966 \pm 0.0013$ with respect to peak area in the concentration range 100–1,000 ng per spot. The mean value ± SD of slope and intercept were 164.85 ± 0.72 and 1168.3 ± 8.26, respectively, with respect to peak area. The method was validated for precision, recovery, and robustness. The limits of detection and quantitation were 10 and 30 ng per spot, respectively. Imatinib mesylate was subjected to acid and alkali hydrolysis, and oxidation and thermal degradation. The drug undergoes degradation under acidic, basic, oxidation, and heat conditions. This indicates that the drug is susceptible to acid, base hydrolysis, oxidation, and heat. Statistical analysis proves that the method is repeatable, selective, and accurate for the estimation of the said drug. The proposed developed HPTLC method can be applied for identification and quantitative determination of imatinib mesylate in bulk drug and dosage forms.

Imatinib mesylate, designated chemically as 4-[(4-methyl-1-piperazinyl) methyl]-*N*-[4-methyl-3-[[4-(3-pyridinyl)-2-pyrimidinyl]amino]-phenyl]benzamide methanesulfonate, is a white to off-white to brownish or yellowish tinged crystalline powder (Maki 2004). It is a protein kinase inhibitor (PTK) that potently inhibits the Abelson tyrosine kinase. PTKs are enzymes that can transfer the terminal

P. Musmade (✉), N. Vadera, and G. Subramanian
Department of Pharmaceutical Quality Assurance, Manipal College of Pharmaceutical Sciences, Manipal, Karnataka 576104, India
e-mail: prashant.pqa@manipal.edu

MM. Srivastava (ed.), *High-Performance Thin-Layer Chromatography (HPTLC)*,
DOI 10.1007/978-3-642-14025-9_6, © Springer-Verlag Berlin Heidelberg 2011

phosphate of an adenosine triphosphate molecule to a tyrosine residue of cytoplasmic protein substrate. PTKs are key modulators of cellular signal transduction pathways. If for any reason, these signaling proteins are subjected to oncogenic mutation(s), a cellular deregulation may occur, yielding an imbalance between cell proliferation, cell growth, and cell death (apoptosis). Hence, PTKs have emerged as important therapeutic targets for intervention in cancer (Adcock et al. 2006). Various methods are available for the analysis of imatinib mesylate in literature like HPLC (Velapandian et al. 2004, 2003), LC–MS–MS (Bakhtiar et al. 2002; Parise et al. 2003). But there is no analytical method for the estimation of imatinib mesylate in bulk drug and dosage form by HPTLC. Moreover, none of them is stability-indicating method. The International Conference on Harmonization (ICH) guideline entitled "stability testing of new drug substances and products" requires the stress testing to be carried out to elucidate the inherent stability characteristics of the active substance (International Conference 1993). Susceptibility to oxidation is one of the required tests. Also, the hydrolytic and thermal stability are required. An ideal stability-indicating method is one that quantifies the drug per se and also resolves its degradation products. Nowadays, HPTLC is becoming a routine analytical technique because of its advantages (Sethi 1996; Kulkarni and Amin 2000; Thoppil et al. 2001; Makhija and Vavia 2001; Ivanovic et al. 2004). The major advantage of HPTLC is that several samples can be run simultaneously using a small quantity of mobile phase unlike HPLC, thus lowering analysis time and cost per analysis. Mobile phase having pH 8 and above can be employed. Suspensions, dirty, or turbid samples can be directly applied. It facilitates automated application and scanning in situ. HPTLC facilitates repeated detection (scanning) of the chromatogram with the same or different parameters. Simultaneous assay of several components in a multicomponent formulation is possible. The aim of this work is to develop an accurate, specific, repeatable, and stability-indicating method for the determination of imatinib mesylate in the presence of its degradation products as per ICH guidelines (A and B 1994; Bakshi and Singh 2002).

Imatinib mesylate was a gift sample from Natco Pharmaceuticals, Hyderabad, India. All chemicals and reagents used were of analytical grade and purchased from Qualigens Fine Chemicals, Mumbai, India.

The samples were spotted in the form of bands of width 6 mm with a Camag microliter syringe on precoated silica gel aluminum Plate 60F-254 (20 cm $\times$ 10 cm with 0.2 mm thickness, E. Merck, Germany) using a Camag Linomat IV (Switzerland). A constant application rate of 100 nl/s was employed and space between two bands was 8 mm. The slit dimension was kept 5 mm $\times$ 0.45 mm micro and 5 mm/s scanning speed was employed. The mobile phase consisted of chloroform: methanol (6:4, v/v). Linear ascending development was carried out in twin-trough glass chamber saturated with mobile phase. The optimized chamber saturation time for mobile phase was 30 min at room temperature. The length of chromatogram run was approximately 70 mm. Subsequent to the development, TLC plate was dried in a current of air with the help of an air-dryer. Densitometric scanning was performed on Camag TLC scanner III in the absorbance mode at 276 nm. The source of radiation utilized was deuterium lamp.

Method Validation

Precision

Repeatability of sample application and measurement of peak area were carried out using six replicates of the same spot (500 ng per spot of imatinib mesylate). The intra- and inter-day variation for the determination of imatinib mesylate was carried out at three different concentration levels of 300, 500, and 700 ng per spot.

Robustness

By introducing small changes in the mobile-phase composition, the effects on the results were examined. Mobile phases having different composition of chloroform: methanol (5.5:4.5 and 6.5:3.5, v/v) were tried and chromatograms were run. The amount of mobile phase, temperature, and relative humidity was varied in the range of ±5%. The plates were prewashed by methanol and activated at 60 ± 5°C for 2, 5, and 7 min prior to chromatography. Time from spotting to chromatography and from chromatography to scanning was varied from 0, 20, 40, and 60 min. Robustness of the method was done at three different concentration levels: 300, 500, and 700 ng per spot.

Detection and Limit of Quantification

In order to determine detection and quantification limit, imatinib mesylate concentrations in the lower part of the linear range of the calibration curve were used. Imatinib mesylate solutions of 10, 12, 14, 16, 18, and 20 μg/ml were prepared and applied in triplicate (10 μl each). The amount of imatinib mesylate by spot versus average response (peak area) was graphed and the equation for this was determined. The standard deviations (SD) of responses were calculated. The average of standard deviations was calculated (ASD). Detection limit was calculated by $(3.3 \times \text{ASD})/b$ and quantification limit was calculated by $(10 \times \text{ASD})/b$, where "b" corresponds to the slope obtained in the linearity study of method.

Specificity

The specificity of the method was ascertained by analyzing standard drug and sample. The spot for imatinib mesylate in sample was confirmed by comparing the R_f values and spectra of the spot with that of standard. The peak purity of imatinib mesylate was assessed by comparing the spectra at three different levels, i.e., peak start (S), peak apex (M), and peak end (E) positions of the spot.

Recovery Studies

The analyzed samples were spiked with extra 50, 100, and 150% of the standard imatinib mesylate and the mixture were analyzed by the proposed method. At each level of the amount, six determinations were performed. This was done to check the recovery of the drug at different levels in the formulations.

Analysis of Imatinib Mesylate

To determine the concentration of imatinib mesylate in capsules (labeled claim: 100 mg per capsule), the contents of 20 capsules were weighed, their mean weight determined, and they were finely powdered. The powder equivalent to 10 mg of imatinib mesylate was weighed. The drug from the powder was extracted with methanol. To ensure complete extraction of the drug, it was sonicated for 30 min and the volume was made up to 10 ml. The resulting solution was centrifuged at 3,000 rpm for 5 min and supernatant was analyzed for drug content. From the supernatant, 0.5 ml was taken and made up to 10 ml by methanol to obtain 50 g/ml solution. The above solution (500 ng per spot) was applied on TLC plate followed by development and then scanned (Fig. 6.1). The analysis was repeated in triplicate. The possibility of excipient interferences in the analysis was studied.

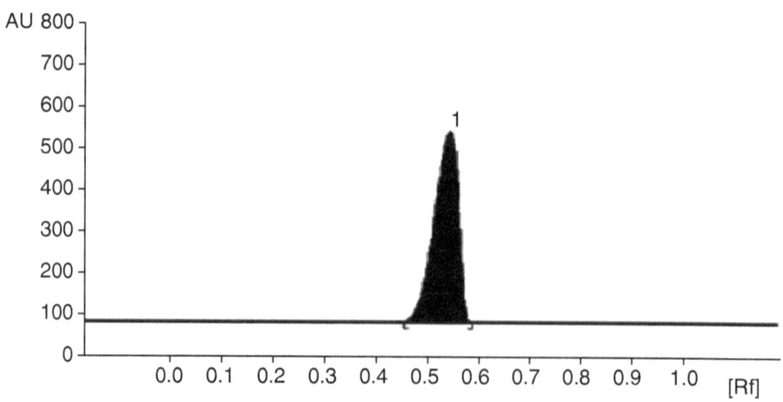

Fig. 6.1 A typical HPTLC chromatogram of imatinib mesylate ($R_f = 0.53$)

Forced Degradation of Imatinib Mesylate

Acid- and Base-Induced Degradation

The 10 mg of imatinib mesylate was separately dissolved in 10 ml of methanolic solution of 0.1 M HCl and 1 M NaOH. These solutions were kept for 8 h at room temperature in the dark in order to exclude the possible degradative effect of light. The 1 ml of above solutions was taken and neutralized, then diluted up to 10 ml with methanol. The resultant solution were applied on TLC plate in triplicate (10 μl each, i.e., 1,000 ng per spot) (Figs. 6.2 and 6.3).

Hydrogen Peroxide-Induced Degradation

Separately, 10 mg of imatinib mesylate was dissolved in10 ml of methanolic solution of hydrogen peroxide (3.0%, v/v). The solution was kept for 8 h at room

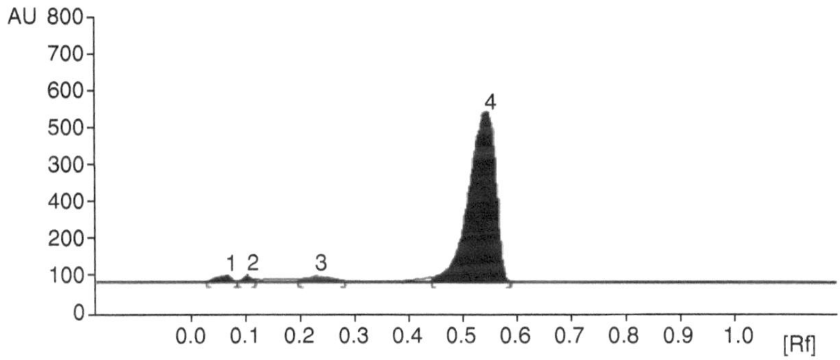

Fig. 6.2 HPTLC chromatogram of acid-degraded imatinib mesylate

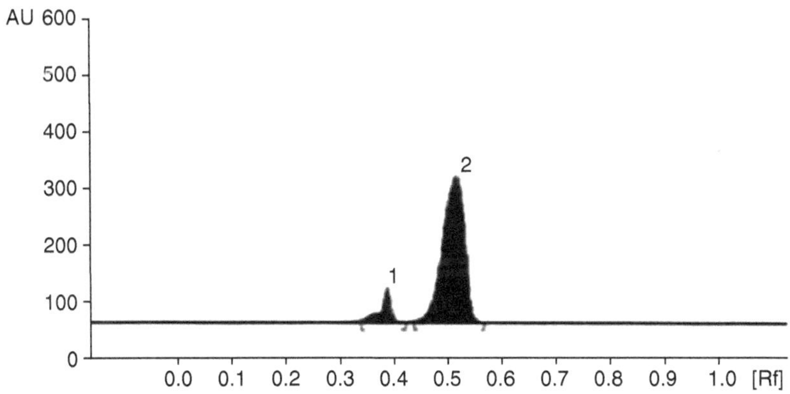

Fig. 6.3 HPTLC chromatogram of base-degraded imatinib mesylate

temperature in the dark in order to exclude the possible degradative effect of light. One ml of above solution was taken and diluted up to 10 ml with methanol. The resultant solution was applied on TLC plate in triplicate (10 μl each, i.e., 1,000 ng per spot). The chromatograms were run; shown in Fig. 6.4.

Dry Heat Degradation Product

The powdered drug was stored at 55°C for 3 h under dry hot conditions and showed significant degradation. The degraded products were resolved from the standard. In all degradation studies, the average peak areas of imatinib mesylate after application (1,000 ng per spot) of three replicates were obtained (Fig. 6.5).

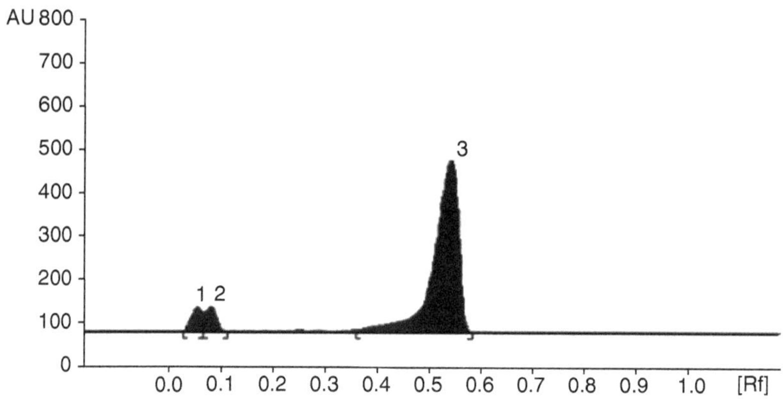

Fig. 6.4 HPTLC chromatogram of hydrogen peroxide-degraded imatinib mesylate

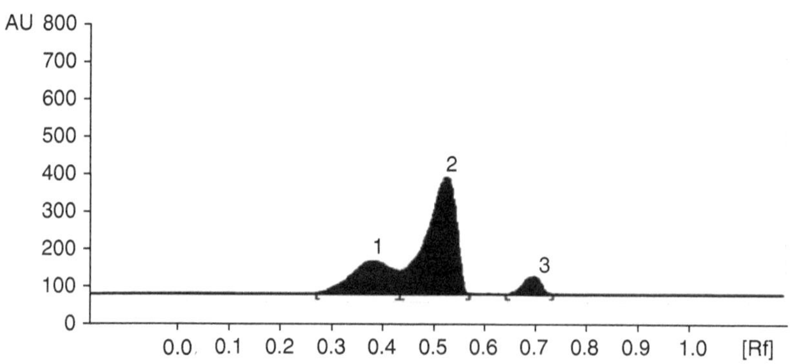

Fig. 6.5 HPTLC chromatogram of heat-degraded imatinib mesylate

Optimum Mobile Phase

TLC procedure was optimized with a view to developing a stability-indicating assay method. Initially, chloroform: methanol (7:3, v/v) gave good resolution with R_f value of 0.53 for imatinib mesylate but typical peak nature was missing. Finally, the mobile phase consisting of chloroform:methanol (6:4, v/v) gave a sharp and well-defined peak at R_f value of 0.53 (Fig. 6.1). Well-defined spots were obtained when the chamber was saturated with the mobile phase for 30 min at room temperature.

Calibration Curves

A stock solution of imatinib mesylate (1,000 µg/ml) was prepared in methanol. Different volumes of stock solution, 0.1, 0.2, 0.3, 0.4, 0.5, 0.6, 0.7, 0.8, 0.9, and 1.0 ml were taken and volume made up to 10 ml by methanol, to obtain 10, 20, 30, 40, 50, 60, 70, 80, 90, and 100 µg/ml solution, respectively. From each of the above solutions,10 µl was spotted in replicate on TLC plates to obtain concentration of 100, 200, 300, 400, 500, 600, 700, 800, 900, and 1,000 ng per spot of imatinib mesylate, respectively. The data of peak area versus drug concentration were treated by linear least-square regression.

The linear regression data for the calibration curves ($n = 3$) as shown in Table 6.1 showed a good linear relationship over the concentration range 100–1,000 ng per spot with respect to peak area. No significant difference was observed in the slopes of standard curves (ANOVA, $P > 0.05$)

Validation

Precision

The repeatability of sample application and measurement of peak area were expressed in terms of % RSD and results are depicted in Table 6.2, which revealed intra- and inter-day variation of imatinib mesylate at three different concentration levels of 300, 500, and 700 ng per spot.

Table 6.1 Linear regression data for the calibration curves[a]

Linearity range (ng per spot)	100–1,000
$r^2 \pm$ SD	0.9966 ± 0.0013
Slope ± SD	164.85 ± 0.72
Confidence limit of slope[b]	164.03–165.66
Intercept ± SD	1168.3 ± 8.26
Confidence limit of intercept[b]	1158.95–1177.64

[a] $n = 3$
[b] 95% confidence limit

Table 6.2 Intra- and inter-day precision of HPTLC method[a]

Amount (ng/spot)	Intra-day precision				Inter-day precision			
	Mean area	SD	%RSD	S.E.[b]	Mean area	SD	%RSD	S.E.b
300	5960.18	35.21	0.54	14.37	5882.36	54.12	0.92	22.09
500	9522.25	42.56	0.45	17.37	9488.92	62.22	0.66	25.40
700	13310.50	54.24	0.41	22.14	13,194	84.81	0.64	34.62

[a]$n = 6$
[b]Standard error

Table 6.3 Robustness of the method[a]

Parameter	SD[b] of peak area	%RSD[b]
Mobile-phase composition	1.23	0.98
Amount of mobile phase	1.06	0.85
Temperature	0.92	0.68
Relative humidity	0.88	0.56
Plate pretreatment	0.59	0.41
Time from spotting to chromatography	0.41	0.36
Time from chromatography to scanning	0.36	0.31

[a]$n = 6$
[b]Average of three concentrations: 300, 500, and 700 ng per spot

Robustness of the Method

The standard deviation of peak areas was calculated for each parameter and %RSD was found to be less than 2%. The low values of %RSD values, shown in Table 6.3, indicated robustness of the method.

LOD and LOQ

Calibration curve in this study was obtained by a plot between amount of analyte versus average response (peak area) and the regression equation was obtained ($Y = 154.27X + 1313.9$) with a regression coefficient of 0.9964. Detection limit and quantification limit were calculated by the method as described in Section 2.4.3 and found to be 6.51 and 19.72 ng, respectively. However, by experiment, we got LOD and LOQ was 10 and 30 ng, respectively. This indicates the adequate sensitivity of the method.

Specificity

The peak purity of imatinib mesylate was assessed by comparing the spectra at peak start, peak apex, and peak end positions of the spot, i.e., r^2 (S, M) = 0.9998 and r^2 (M, E) = 0.9988. Good correlation ($r^2 = 0.9989$) was also obtained between standard and sample spectra of imatinib mesylate.

Table 6.4 Recovery studies[a]

Excess drug added to analyte (%)[b]	Amount recovered (mg)	Recovery (%)	% RSD	S.E.
0	101.21	101.21	0.42	0.34
50	150.93	100.62	0.68	0.55
100	198.68	99.34	0.72	0.57
150	247.15	98.86	0.48	0.38

[a] $n = 6$
[b] Matrix containing 100 mg drug

Recovery Studies

The proposed method when used for extraction and subsequent estimation of imatinib mesylate from pharmaceutical dosage forms after spiking with 50, 100, and 150% of additional drug afforded recovery of 98–102% as listed in Table 6.4. The data of summary of validation parameters were listed in Table 6.4.

Analysis of Prepared Formulation

A single spot of R_f 0.53 was observed in chromatogram of the imatinib mesylate samples extracted from capsules. There was no interference from the excipients commonly present in the capsules. The imatinib mesylate content was found to be 99.2% with a %RSD of 0.63. It may therefore be inferred that degradation of imatinib mesylate had not occurred in the formulation that were analyzed by this method. The low %RSD value indicated the suitability of this method for routine analysis of imatinib mesylate in pharmaceutical dosage forms.

Stability-Indicating Property

The chromatogram of samples degraded with acid, base, hydrogen peroxide, and heat showed well-separated spots of pure imatinib mesylate as well as some additional peaks at different R_f values. The spots of degraded product were well resolved from the drug spot as shown in Figs. 6.2–6.5. The number of degradation product with their R_f values, content of imatinib mesylate remained, and percentage recovery were calculated and listed in Tables 6.5 and 6.6.

Notes

The developed HPTLC technique is precise, specific, accurate, and stability-indicating. The developed method was validated based on ICH guidelines (Ivanovic et al. 2004). Statistical analysis proves that the method is repeatable and selective

Table 6.5 Summary of validation parameters

Parameter	Data
Linearity range (ng per spot)	100–1,000
Correlation coefficient	0.9966 ± 0.0013
Limit of detection (ng per spot)	10
Limit of quantitation (ng per spot)	30
Recovery ($n = 6$)	100.01 ± 1.32
Precision (%RSD)	
Repeatability of application ($n = 6$)	0.35
Repeatability of measurement ($n = 6$)	0.16
Inter-day ($n = 6$)	0.74
Intra-day ($n = 6$)	0.48
Robustness	Robust
Specificity	Specific

Table 6.6 Forced degradation of imatinib mesylate

S. No.	Sample exposure condition	Number of degradation products (R_f value)	Imatinib mesylate remained (ng/1,000 ng) (±SD, $n = 3$)	S.E.	(%) Recovery
1	0.1 M HCl, 8 h, RT	3 (0.8, 0.11, 0.24)	920.21 (±1.52)	0.88	92.02
2	1 M NaOH,8 h, RT	1 (0.38)	721.86 (±4.26)	2.46	72.19
3	3% H_2O_2, 8 h, RT	2 (0.5, 0.8)	811.46 (±3.91)	2.26	81.15
4	Heat, 3 h, 55°C	2 (0.38, 0.7)	775.26 (±5.98)	3.45	77.53

for the analysis of imatinib mesylate as bulk drug and in pharmaceutical formulations. The method can be used to determine the purity of the drug available from the various sources by detecting the related impurities. It may be extended to study the degradation kinetics of imatinib mesylate and for its estimation in plasma and other biological fluids. As the method separates the drug from its degradation products, it can be employed as a stability indicating one.

References

A and B. International Conference on Harmonization guideline Q2. Validation of Analytical Procedures: Text and Methodology, Geneva, October, 1994.

Adcock IM, Chung KF, Caramori G and Ito K. Kinase inhibitors and airway inflammation. Eur J Pharmacol, 2006, 533, 118–132.

Bakhtiar R, Khemani L, Hayes M, Bedman T and Tse F. Quantification of the anti-leukemia drug STI571 (Gleevec) and its metabolite (CGP 74588) in monkey plasma using a semi-automated solid phase extraction procedure and liquid chromatography–tandem mass spectrometry. J Pharm Biomed Anal, 2002, 28, 1183–1194.

Bakshi M and Singh SJ. Development of validated stability-indicating assay methods-critical review. J Pharm Biomed Anal, 2002, 28, 1011–1040.

International Conference on Harmonization guideline Q1A, 'stability testing of new drug substances and products', Geneva, October, 1993.

Ivanovic D, Medenica M, Jancic B and Malenovic A. Reversed-phase liquid chromatography analysis of imatinib mesylate and impurity product in Glivec capsules. J Chromatogr B, 2004, 800, 253–258.

Kulkarni SP and Amin PD. Stability indicating HPTLC determination of timolol maleate as bulk drug and in pharmaceutical preparations. J Pharm Biomed Anal, 2000, 23, (6), 983–987.

Makhija SN and Vavia PR. Stability indicating HPTLC method for the simultaneous determination of pseudoephedrine and cetirizine in pharmaceutical formulations. J Pharm Biomed Anal, 2001, 25, 663–667.

Maki RG. Gastrointestinal stromal tumors respond to tyrosine kinase-targeted therapy. Curr Treat Opt Gastroenterol, 2004, 7, 13–17.

Parise RA, Ramanathan RK, Hayes MJ and Egorin MJ. Liquid chromatographic-mass spectrometric assay for quantitation of imatinib and its main metabolite (CGP 74588) in plasma. J Chromatogr B Anal Technol Biomed Life Sci, 2003, 79, 39–44.

Sethi PD (1996), High Performance Thin Layer Chromatography, Quantitative Analysis of Pharmaceutical Formulations, published by CBS Publishers, New Delhi.

Thoppil SO, Cardoza RM and Amin PD. Stability indicating HPTLC determination of trimetazidine as bulk drug and in pharmaceutical formulations. J Pharm Biomed Anal, 2001, 25, (1), 15–20.

Velapandian T, Mathur R, Agarwal NK, Arora B, Kumar L and Gupta SK. Development and validation of a simple liquid chromatographic method with ultraviolet detection for the determination of imatinib in biological samples. J Chromatogr B, 2004, 80, 4431–4434.

Vivekanand VV, Rao S, Vaidyanathan G, Sekher NM, Kelkar SA and Puranik PR. A validated LC method for imatinib mesylate. J Pharm Biomed Anal, 2003, 33, 879–889.

Chapter 7
HPTLC Fingerprint Analysis: A Quality Control for Authentication of Herbal Phytochemicals

Mauji Ram, M.Z. Abdin, M.A. Khan, and Prabhakar Jha

Abstract Authentication and consistent quality are the basic requirement for Indian traditional medicine (TIM), Chinese traditional herbal medicine (TCHM), and their commercial products, regardless of the kind of research conducted to modernize the TIM and TCHM. The complexities of TIM and TCHM challenge the current official quality control mode, for which only a few biochemical markers were selected for identification and quantitative assay. Referring too many unknown factors existed in TIM and TCHM, it is impossible and unnecessary to pinpoint qualitatively and quantitatively every single component contained in the herbal drug. Chromatographic fingerprint is a rational option to meet the need for more effective and powerful quality assessment to TIM and TCHM. The optimized chromatographic fingerprint is not only an alternative analytical tool for authentication, but also an approach to express the various pattern of chemical ingredients distribution in the herbal drugs and preserve such "database" for further multifaced sustainable studies. Analytical separation techniques, for example, high-performance liquid chromatography (HPLC), gas chromatography (GC) and mass spectrometry (MS) were among the most popular methods of choice used for quality control of raw material and finished herbal product. Fingerprint analysis approach using high-performance thin-layer chromatography (HPTLC) has become the most potent tool for quality control of herbal medicines because of its simplicity and reliability. It can serve as a tool for identification, authentication, and quality control of herbal drugs. In this chapter, attempts are being made to expand the use of HPTLC and at the same time create interest among prospective researcher in

M. Ram (✉), M.Z. Abdin, and M.A. Khan
Department of Biotechnology, Faculty of Science, Centre for Transgenic Plant Development, Jamia Hamdard, Hamdard Nagar, New Delhi 110062, India
e-mail: tyagi_mjt@rediffmail.com

P. Jha
Department of Botany, Faculty of Science, Jamia Hamdard, Hamdard Nagar, New Delhi 110062, India

MM. Srivastava (ed.), *High-Performance Thin-Layer Chromatography (HPTLC)*,
DOI 10.1007/978-3-642-14025-9_7, © Springer-Verlag Berlin Heidelberg 2011

herbal analysis. The developed method can be used as a quality control tool for rapid authentication from a wide variety of herbal samples. Some examples demonstrated the role of fingerprinting in quality control and assessment.

The paradigms of traditional Chinese (TCHM) and Indian (TIM) herbal medicines are featured as holistic system; the common clinical use of Indian and Chinese medicines require the complex recipes and formulae derived from historical and anecdotal evidences of Indian and Chinese medicinal practitioners. On the basis of the ancient Chinese philosophy, a typical therapeutic formula is symbolized as an active "cabinet" consisting of "Monarch" drug, "ministers" drug, "assistants" drug and "messengers", or "servants" drug, which work together harmoniously and serve various functions respectively, to adjust, balance, and restore the body's functions guided by Chinese ancient philosophy and culture. Consequently, no single active constituent is responsible for the overall efficacy of the whole formula, even single herbal drug, which contain numerous chemical compounds with holistic efficacy rather than a single active compound. It is definitely different from the Western single-chemical drug, and is also distinguished from Western herbal/botanical medicines because TIM and TCHM are the carriers loading the information of the philosophy and culture of traditional medicine. Hence, we should keep in mind that it will be nothing different from Western herbal drug, once the TCM information is unloaded from TCHM matrices. On the new wave of modernization of TIM and TCHM under the economic globalization environment, there is trend towards the seeking of TIM and TCHM as a source for discovery of new chemical drugs without considering the synergic effect of all the ingredients in the herbal drugs. That means such an approach is mostly concerned with only the interested single target and ignores the total quality of the herbal drug itself. The consequence would lead to loss of ancestors' wisdom (the culture and philosophy of TIM and TCHM) and exclusive clinical experiences accumulated the rough generations. Referring to the complexity and difficulty in merging ancient Indian, Chinese culture, and modern Western science, the first step in the process of TIM and TCHM modernization, we should preserve the total information loaded in the TIM and TCHM as much as possible in order to avoid only rash affirmation on the pros and cons of TCHM without discreet study. From the viewpoint of chemistry and biology, the total chemical ingredients pattern in every entity of TIM and TCHM should be preferably expressed in the appropriate chromatographic fingerprint consisting of detectable ingredients. The optimized fingerprint can serve as "chemical signatures" of the TIM and TCHM for consecutive multifaceted research. The following examples illustrate the role of chromatographic fingerprints in TIM and TCHM.

The construction of chromatographic fingerprints plays an important role in the quality control of complex herbal medicines (Gong et al. 2003). Chemical fingerprints obtained by chromatographic techniques are strongly recommended for the purpose of quality control of herbal medicines, since they might represent appropriately the "chemical integrities" of the herbal medicines and therefore be used for authentication and identification of the herbal products. On the basis of the concept of phytoequivalence, the chromatographic fingerprints of herbal medicines could be utilized for addressing the problem of quality control of herbal medicines (Liang et al. 2004). By definition, a chromatographic fingerprint of a herbal medicine is, in practice, a chromatographic pattern of pharmacologically active and or chemically characteristic constituents present in the extract (State Drug Administration of China 2000; Ong 2002; Xie 2001). This chromatographic profile should be featured by the fundamental attributions of "integrity" and "fuzziness" or "sameness" and "differences" so as to chemically represent the herbal medicines investigated. This suggest that chromatographic fingerprint can successfully demonstrate both "sameness" and "differences" between various samples and the authentication and identification of herbal medicines can be accurately conducted even if the number and/or concentration of chemically characteristic constituents are not very similar in different samples of herbal medicine. Thus, chromatographic fingerprint should be considered to evaluate the quality of herbal medicines globally considering multiple constituents present in the herbal medicines.

Phytochemical analysis of medicinal herbs remains challenging issues for analytical chemists, as herbs are a complicated system of mixtures. Analytical separation techniques, for example, high-performance liquid chromatography (HPLC), high-performance thin-layer chromatography (HPTLC), gas chromatography (GC), and mass spectrometry (MS) are among the most popular methods of choice used for quality control of raw material and finished herbal product. The application of infrared (IR) spectroscopy in herbal analysis is still very limited compared to its applications in other areas (food and beverage industry, microbiology, pharmaceutical, etc.). In this chapter, attempts have been made to expand the use of HPTLC method and at the same time creating interest among prospective researcher in herbal analysis. The developed method can be used as a quality control tool for rapid authentication from a wide variety of herbal samples. It is also found useful in phytochemical studies as a "fingerprinting" device, for comparing a natural with commercial sample.

Chromatographic Fingerprints

Herbal medicines have a long therapeutic history and are still serving many of the health needs of a large population of the world. But the quality control and quality assurance still remains a challenge because of the high variability of chemical components involved. Herbal drugs, singularly and in combinations, contain a myriad of compounds in complex matrices in which no single active constituent is responsible for the overall efficacy. This creates a challenge in establishing quality control standards for raw materials and standardization of finished herbal drugs (World Health Organization 2001). Traditionally, only a few markers of pharmacologically active constituents were employed to assess the quality and authenticity of complex herbal medicines. However, the therapeutic effects of herbal medicines are based on the complex interaction of numerous ingredients in combination, which are totally different from those of chemical drugs. Thus, many kinds of chemical fingerprint analysis methods to control the quality of herbal drugs have gradually come into being, such as thin-layer chromatography, GC, HPLC, etc. chromatographic fingerprint analysis of herbal drugs represents a comprehensive qualitative approach for the purpose of species authentication, evaluation of quality, and ensuring the consistency and stability of herbal drugs and their related products. The entire pattern of compounds can then be evaluated to determine not only the presence or absence of desired markers or active constituents but the complete set of ratios of all detectable analytes. The chemical fingerprints obtained by chromatographic and electrophoretic techniques, especially by hyphenated chromatographies, are strongly recommended for the purpose of quality control of herbal medicines, since they might represent appropriately the "chemical integrities" of herbal medicines and therefore be used for authentication and identification of the herbal products.

Techniques Used for Fingerprint Analysis of Herbal Medicines

Thin-Layer Chromatography

Thin layer chromatography (TLC) is frequently used for the analysis of herbal medicines since various pharmacopeias such as American Herbal Pharmacopoeia (AHP), Chinese Drug Monographs and analysis, Pharmacopoeia of People's Republic of China, etc., still use TLC to provide first characteristic fingerprints of herbs. Rather, TLC is used as an easier method of initial screening with a semiquantitative evaluation together with other chromatographic techniques. High-performance TLC has the advantages of manifold possibilities of detection in analyzing herbal medicines.

In addition, HPTLC analysis is rather simple and can be employed for multiple sample analysis. With the help of CAMAG video store system (CAMAG Switzerland) and TLCQA-UV methods, it is possible to get useful qualitative and quantitative information from the developed TLC plate (Chau et al. 1998). The advantages of using TLC to construct the fingerprints of herbal medicines are its simplicity, versatility, high velocity, specific sensitivity, and simple sample preparation. Thus, TLC is a convenient method of determining the quality and possible adulteration of herbal products. It summarized the progress in forced-flow-planer chromatography (FFPC) and demonstrated the importance of the different techniques like rotation planar chromatography (RPC), overpressured-layer chromatography (OPLC), and electroplanar chromatography (EPC) (Nyiredy 2003).

Gas Chromatography

GC is usually used for fingerprint analysis when pharmacologically active components in herbal medicines are volatile chemical compounds. The advantage of GC clearly lies in its high sensitivity of detection for almost all the volatile chemical compounds. This is especially true for the usual FID detection and GC-MS. Furthermore, the high selectivity of capillary columns enables separation of many volatile compounds simultaneously within comparatively short times. However, the most serious disadvantage of GC is that it is not convenient for the analysis of samples of planer and nonvolatile compounds (Majlat 1982; Angerosa et al. 1996; Ylinen et al. 1986).

High-Performance Liquid Chromatography

HPLC is a popular method for the analysis of herbal medicines because it is easy to learn and use and is not limited by the volatility or stability of the sample compound. In general, HPLC can be used to analyze almost all the compounds in the herbal medicines. Reversed-phase (RP) columns may be the most popular columns used in the analytical separation of herbal medicines. It is necessary to notice that the optimal separation condition for the HPLC involves many factors, such as the different compositions of the mobile phases, their pH adjustment, pump pressures, etc. Thus, a good experimental design for the optimal separation seems, in general, necessary (Sanyal and Bhattacharya 2003; Thanawiroon and Linhardt 2003; Lin and Chen 2003). In order to obtain better separation, some new techniques have been recently developed in research field of liquid chromatography. These are micellar electrokinetic capillary chromatography (MECC) (Albert et al. 1997), high-speed counter-current chromatography (HSCCC), low-pressure size-exclusion chromatography (SEC) (Pervin et al. 1995), reversed-phase ion-pairing HPLC (RP-IPC-HPLC) (Longanathan et al. 1990; Parmanos et al. 1997), and strong anion-exchange HPLC (SAX-HPLC) (Rice and Linhardt 1989). They will provide new opportunities for good separation for some specific extracts of some herbal medicines.

Hyphenated Procedures

In the past two decades, combining a chromatographic separation system on-line with a spectroscopic separation system on-line with a spectroscopic detector in order to obtain structural information on the analytes present in a sample has become the most important approach for the identification and/or confirmation of the identity of target and unknown chemical compounds. For most (trace-level) analytical problems in research field of herbal medicines, the combination of column liquid chromatography or capillary GC with a UV–vis or a mass spectrophotometer become the preferred approach for the analysis of herbal medicines. Various hyphenated procedures used for the analysis of herbal drugs are HPLC-DAD, CE-DAD, GC-MS, LC-MS, HPLC-MS, HPLC-DAD-MS and LC-DAD-MS. The data obtained from such hyphenated instruments are the so-called two-way data; say one way for chromatogram and the other way for spectrum, which could provide much more information than the classic one-way chromatography. A "total analysis device" has been recently demonstrated in the case of on-line HPLC-UV (DAD)-FTIR-NMR-MS analyses (Liang et al. 2004).

Traditional Medicines and Chromatographic Fingerprints

Owing to low toxicity and known pharmacological activity, traditional medicine, containing mainly of herbal drugs, has been popularly and extensively used for many centuries. However, it is not easy to conduct quality control and assurance of herbal medicines because of high variability of chemical components involved. Thus, as pointed out by World Health Organization (WHO), since the quality and quantity of safety and efficacy data on traditional medicine are far from sufficient to meet the criteria needed to support its use worldwide and there is still a lack of adequate or accepted research methodology for evaluating traditional medicine up to now, traditional medicine has not been officially recognized in most countries. Under this situation, the U.S. IODA does not definitely emphasize the developments of very clear pictures of all chemical components from herbal medicines, and thus, the fingerprinting approach has been recently recognized and accepted for quality assessment of traditional medicines. The concept of phytoequivalence was developed in Germany in order to ensure consistency of herbal products. According to this concept, a chemical profile, such as chromatographic fingerprint, for herbal product should be constructed and compared with the profile of a clinically proven reference product. Chinese State Food and Drug Administration have said to regulate the compositions of liquid injections with herbal ingredients using stringent quality procedures such as chemical assay and standardization. Fingerprints of herbal medicinal liquid injections are compulsorily carried out for this purpose. In addition, among the various experimental techniques, chromatographic methods are highly recommended for finding out fingerprints of herbal products because of the high separation ability of chromatography (Table 7.1).

Authentication of the Species Prone to Confusion

Ginseng (root of *Panax ginseng*) and American ginseng (root of *Panax quinquefolium*) are two close species containing very similar chemical ingredients. The functions of the two species are different according to TCHM in clinical use. It is difficult to distinguish them by only selecting single ginsenosides, but the HPTLC images with digital scanning profiles as a whole can easily be differentiated between them (Xie 2005a) (Fig. 7.1).

Table 7.1 List of spices and potential adulterants as well as the suggested method

Product/herbs	Adulterant	Recommended method
R. graveolens	*E. dracuveuloids*	HPTLC profiling
C. angustifolia	*C. tora*	HPTLC profiling
C. reflexa	*C. chinesis*	Microscopic
P. nigrum	*C. papaya*	Microscopic
G. glabra	*A. precatorius*	TLC
Z. jujube	*Z. mauritiana*	Microscopic
Capsicums	Sudan red and related dyes, mono or disaccharides	HPTLC profiling
Oregano	Noncompliant herbs, i.e., Savory, thyme, marjoram	Microscopic
Saffron	Added artificial color	TLC
Cinnamon	Coffee husks	Microscopic
Nutmeg	Coffee husks	Microscopic
Cloves	Essential oils may have been removed	TLC
Chilli powder	Saw dust and color may be added	HPTLC
Coffee	Chicory	HPTLC
Cinnamon	Cassia bark which resembles cinnamon in taste and odor	HPTLC

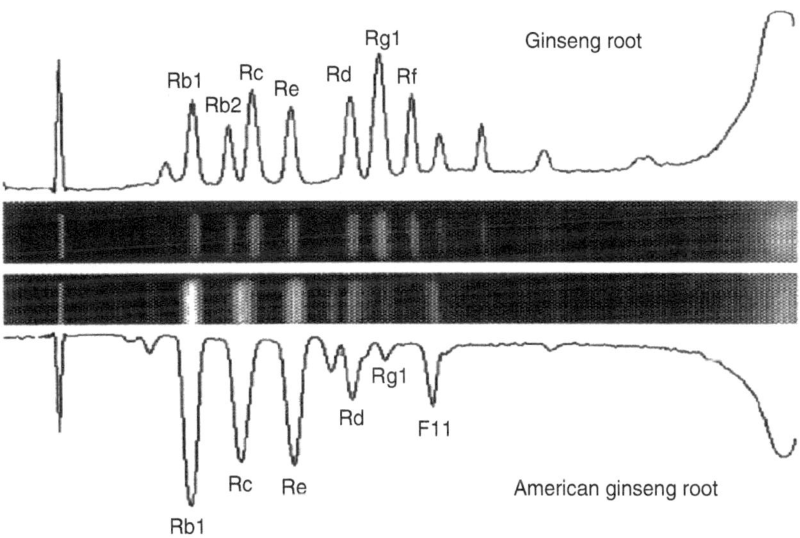

Fig. 7.1 HPTLC fingerprint of Ginseng and American ginseng

Monitoring the Dynamic Change Due to Interaction of Mixed Herbal Drugs during Extraction

The empirical formula of Sheng Mai Yin (SMY) consisted of ginseng root (Renshen), Ophiopogon root (Maidong) and Schizandra fruits (*Wuweizi*). A quality survey of commercial SMY products in the market by means of fingerprinting revealed that the ginsenosides, the main active constituent in ginseng had been destroyed; in some, none of the primary ginsenosides were even detected. The reason was obviously that ginsenosides in ginseng were hydrolyzed uncontrollably by the organic acid in *Wuweizi* when the mixture was subjected to extended heating in water during manufacture. Analysis showed that the primary ginsenosides were hydrolyzed rapidly when the mixed ginseng and *Wuweizi* was boiled with water under rigorous heating for 20 min and destroyed afterwards (Fig. 7.2, left). However, such hydrolysation behavior would stabilize under gentle heating for 120 min, similar to that undergone by home-made detection. The ginsenosides Rb1, Re and Rg1 were hydrolysed into ginsenoside Rg3 and Rh with a rather consistent state of the hydrolysed ginsenoside pattern (Xie 2005b) (Fig. 7.2, right). It is well known that ginsenoside Rg3 and Rh are active components for the cardiovascular system. The conventional home-made SMY detection probably generated a "hidden" added positive value for preventing and curing diseases.

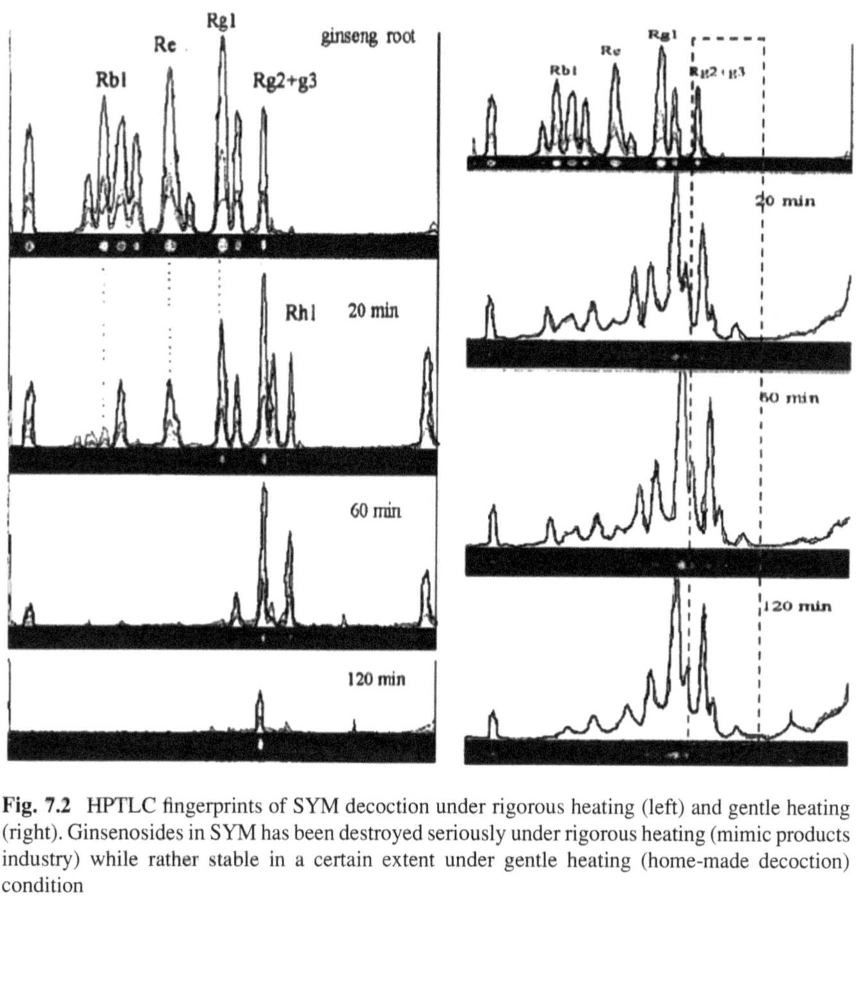

Fig. 7.2 HPTLC fingerprints of SYM decoction under rigorous heating (left) and gentle heating (right). Ginsenosides in SYM has been destroyed seriously under rigorous heating (mimic products industry) while rather stable in a certain extent under gentle heating (home-made decoction) condition

Quality Evaluation of the Crude Drugs

There are two species of *Ge Gen* (Kudzu root), Chinese Pharmacopoeia *Ye Ge* (*Radix Pueraria lobatae*) and *Gan Ge* (*Radix P. thomsonii*), which were used as the same herbal drug for a long time (Xie 2005b). But the content determination of main isoflavonoid, Puerarin and HPTLC fingerprint analysis showed the great disparity of the content of puerarin, and the total chemical pattern expressed by the fingerprint revealed that the puerarin content and the chemical components' concentration distribution in the *Gan Ge* fingerprint was 8–15 times lower than that of *Ye Ge* (Fig. 7.3); thus, it is impossible that both species are bio-equivalent. Hence, *Ye Ge* should be the appropriate candidate for *Ge Gen* (Kudzu root) in the prescription by TCM practitioners.

Adulterants and the Authentic Sample

The general practice of quality assessment of extracts of *Emblica officinalis* fruit and *Cassia angustifolia* leaves with their adulterants, *Impoea batata* and *Cassia tora*, respectively, were distinguished by the help of HPTLC fingerprint analysis without the provision of other detailed quality information (Fig. 7.4a–d). The HPTLC fingerprint of total compounds disclosed the distribution pattern with the detectable peak intensities and peak-to-peak ratio, which expresses the inherent quality. Any significant change in the pattern appearance will hint at the quality fluctuation of the products.

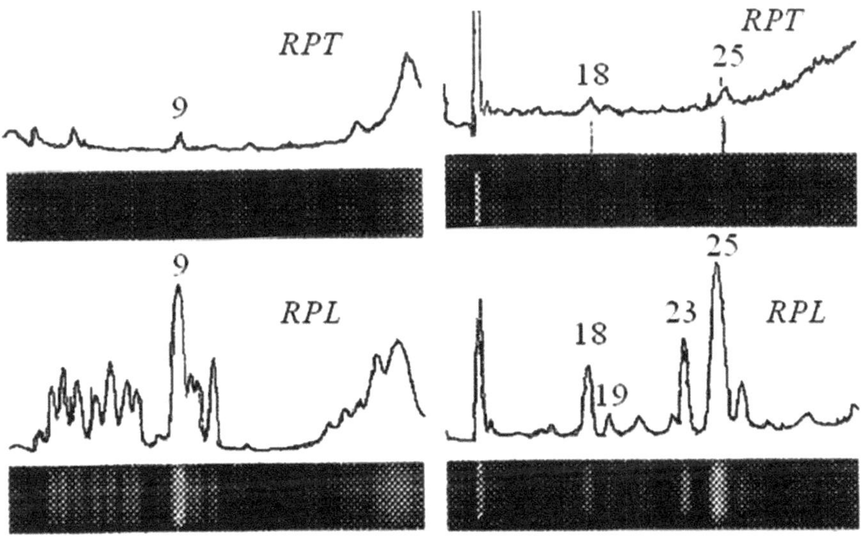

Fig. 7.3 HPTLC fingerprints of Ye Ge (root of *Puerariae lobatae*) and Gan Ge (root of *Puerarae thomsonii*). Left: Isoflavonoides, Right: Aglycones, PRL: Root of *Pueraria lobata*, PRT: Root of *Pueraria thomsonii*

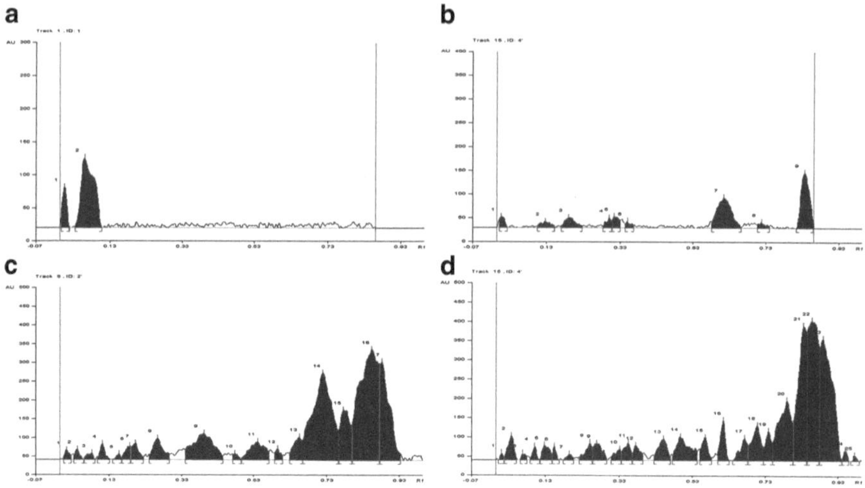

Fig. 7.4 HPTLC fingerprints of medicinal drugs with their adulterants (**a**) *Emblica officinalis* (**b**) Adulterant (*Impoea batata*) (**c**) *Cassia angustifolia* (**d**) Adulterant (*Cassia tora*)

Notes

The problem of quality assurance of herbal medicines has been solved to a great extent with the help of chromatographic fingerprint analysis. The variation determination of common peaks/regions in a set of chromatographic fingerprints could provide useful qualitative and quantitative information on the characteristic components of herbal medicines investigated. On the other hand, whether the real samples were identified as the herbs with the same quality grade could be determined successfully by way of comparing the chromatographic fingerprints with the similarity index and linear correlation analysis. Furthermore, pattern recognition can be used to discriminate different kinds of samples of herbal medicines investigated. Thus, chromatographic fingerprint analysis serves as a promising quality control tool for herbal medicines.

Authentication and consistent quality are the basic requirements for Indian Traditional Medicine (TIM), Chinese Traditional Medicine (TCHM) and its commercial products, regardless of the kind of research conducted to modernize the TIM and TCHM. Chromatography fingerprinting is a rational option to meet the need for more effective and powerful quality assessment to TIM. The optimize chromatographic fingerprint is not only an alternative analytical tool for identification but also approach to various patterns of chemical ingredients distribution in the herbal drugs and to preserve such wholeness target "Database" for further multiphase studies.

The reported HPTLC method was found to be rapid, simple, and accurate for authentication of the herbal plant materials. The described method is suitable for routine use and authenticates the herbal plant materials. The processing of samples and standards together at the same time (in-system calibration) leads to improved reproducibility and accuracy. Fingerprint developed by HPTLC can be used for routine authentication of various herbal plant materials purchased from market. Biochemical markers also can be developed by this method to authenticate the herbal plant materials.

References

Albert K, Lee W, Chan WF, Liang YZ and Fang KT. An example of a sequential design: Application in capilllary electrophoresis. Chemom Intell Lab Syst 1997; 39 (1), 11.

Angerosa F, Alessandro ND, Corana F, Mellerio G. (1996) Characterization of phenolic and secoiridoid aglycons present in virigin olive oil by gas chromatography-chemical ionization mass spectrometry. J Chromatogr A; 736 (1–2), 195.

Chau FT, Chan TP and Wang J. TLCQA: quantitative study of thin-layer chromatography. Bioinformatics 1998; 14, 540–541.

Gong F, Liang YZ, Xie PS and Chau FT. Information theory applied to chromatographic fingerprint of herbal medicine for quality control. J Chromatogr A 2003; 1002 (1), 25–40.

Liang YZ, Xie P and Chan K. Quality control of herbal medicines. J Chromatogr B 2004; 812 (1–2), 53–70.

Lin CH and Chen BH. Determination of carotenoids in tomato juice by liquid chromatography. J Chromatogr 2003; 1012, 103.

Longanathan D, Wang HM, Mallis LM and Linhardt RJ. Structural variation in the antithrombin III binding site region and its occurrence in heparin from different sources. Biochemistry 1990; 29 (18), 4362.

Majlat P. Gas chromatographic assay of atropine and phenobarbital in pharmaceutical preparations containing Valeriana Liquid extract. J Chromatogr 1982; 241, 399.

Nyiredy S. Progress in forced-flow planar chromatography. J Chromatogr 2003; 1000, 985.

Ong ES. Chemical assay of glycyrrhizin in medicinal plants by pressurized liquid extraction (PLE) with capillary zone electrophoresis (CZE). J Sep Sci 2002; 25 (13), 825.

Parmanos NK, Vanky P, Tzanakakis GN, Tsegenidis T and Hjerpe A. Ion-pair high performance liquid chromatography for determining disaccharide composition in heparin and heparan sulphate. J Chromatogr 1997; 765 (2), 169.

Pervin A, Gallo C, Jandik KA, Han XJ and Linhardt RJ. Preparation and structural characterization of large heparin-derived oligosaccharides. Glycobiology 1995; 5 (1), 83–95.

Rice KG and Linhardt RJ. Study of structurally defined oligosaccharide substrate of heparin and heparan monosulphate lysates. Carbohydr Res 1989; 190 (2), 219–233.

Sanyal U, Bhattacharya S, Patra A and.Hazra B. Liquid chromatographic separation of derivatives of diospyrin, a bioactive bisnaphthoquinonoid plant-product and analogous naphthyl compounds. J Chromatogr 2003; 1017 (1–2), 225–232.

State Drug Administration of China, Technical requirements of the fingerprint in injection of Chinese Materia Medica. Chin Tradit Pat Med 2000; 22, 825.

Thanawiroon C and Linhardt RJ. Separation of complex mixture of heparin-derived oligosaccharides using reversed-phase high-performance liquid chromatography. J Chromatogr 2003; 1014 (1–2), 215.

World Health Organization (WHO), WPR/RC52/7: A Draft Regional Strategy for Traditional Medicine in Western Pacific. WHO Regional Committee, 52nd Session Darussalam, 10–14 Sept. 2001.

Xie PS. A feasible strategy for applying chromatography fingerprint to assess quality of Chinese herbal medicine. Tradit Chin Drug Res Clin Pharm 2001; 12(3), 141–169.

Xie PS. (2005a) The Chromatographic fingerprint of Chinese Herbal Medicine published by People's Health Publishing House, Tianjin.

Xie PS (2005b) The Chromatographic fingerprint of Chinese Herbal Medicine. People's Health Publishing House published by Tianjin.

Ylinen M, Naaranlahti T, Lapinjoki S, Huhtiakangas A, Salonen ML, Simola LK and Lounasmaa M. Trapane alkaloids from atropa belladona; Part I. capillary gas Chromatography analysis. Planta Med 1986; 52 (2), 85–87.

Chapter 8
HPTLC in Herbal Drug Quantification

Devanand B. Shinde, Machindra J. Chavan, and Pravin S. Wakte

Abstract For the past few decades, compounds from natural sources have been gaining importance because of the vast chemical diversity they offer. This has led to phenomenal increase in the demand for herbal medicines in the last two decades and need has been felt for ensuring the quality, safety, and efficacy of herbal drugs. Phytochemical evaluation is one of the tools for the quality assessment, which include preliminary phytochemical screening, chemoprofiling, and marker compound analysis using modern analytical techniques. High-performance thin-layer chromatography (HPTLC) has been emerged as an important tool for the qualitative, semiquantitative, and quantitative phytochemical analysis of the herbal drugs and formulations. This includes developing TLC fingerprinting profiles and estimation of biomarkers. This review has an attempt to focus on the theoretical considerations of HPTLC and some examples of herbal drugs and formulations analyzed by HPTLC.

Standardized manufacturing procedures and suitable analytical tools are required to establish the necessary framework for quality control in herbals. Among those tools, high-performance liquid chromatography (HPLC), high-performance thin-layer chromatography (HPTLC), and capillary electrophoresis are the most widely used to establish reference fingerprints of herbs, against which raw materials can be evaluated and finished products can be assayed. High-performance thin-layer chromatography, also known as planar chromatography, is a modern technique with high separation power and reproducibility superior to classical TLC. Main

D.B. Shinde (✉) and P.S. Wakte
Department of Chemical Technology, Dr. Babasaheb Ambedkar Marathwada University, Aurangabad 431 001, Maharashtra, India
e-maill: dbshinde97@rediffmail.com

M.J. Chavan
Department of Pharmacognosy, Amrutvahini College of Pharmacy, Sangamner, S.K. Dist, Ahmednagar, Maharashtra 422 605, India

MM. Srivastava (ed.), *High-Performance Thin-Layer Chromatography (HPTLC)*,
DOI 10.1007/978-3-642-14025-9_8, © Springer-Verlag Berlin Heidelberg 2011

difference of HPTLC and TLC is particle and pore size of sorbents. Special features of HPTLC are as follows:

- Simultaneous processing of sample and standard – better analytical precision and accuracy and less need for internal standard
- Several analysts can work simultaneously
- Lower analysis time and less cost per analysis
- Low maintenance cost
- Simple sample preparation – can handle samples of divergent nature
- No prior treatment for solvents like filtration and degassing
- Low mobile-phase consumption per sample
- No interference from previous analysis – fresh stationary and mobile phases for each analysis, no contamination
- Visual detection possible – open system
- Non-UV-absorbing compounds detected by postchromatographic derivatization

Protocol Used for Drug Analysis

Selection of Chromatographic Layer

The precoated plates with different supporting materials (glass, aluminum, plastic) and with different sorbent layers are available in different format and thickness. Usually plates with solvent thickness of 100–250 μm are used for quantitative and qualitative analysis. Commonly available precoated plates are:

Silica gel 60F: More than 80% of analysis is done on these plates.

Aluminum oxide: Basic substances, alkaloids, steroids.

Microcrystalline cellulose: Amino acids, sugars, antibiotics.

RP-2, RP-8and RP-18: These are the chemically modified silica gel plates commonly used for analysis of fatty acids, carotenoids, steroids, and cholesterol and its esters.

Sample and Standard Preparation

To avoid interference from impurities, low signal-to-noise ratio, straight base line, and improvement of limit of detection (LOD) are employed. An increase in signal-to-noise (*S/N*) ration can be resulted if noise contribution from the electrical and optical noise can be lowered. Below 250 nm, the main cause of signal variation is due to the particulate nature of the layer. Reduction of layer thickness from 0.25 to 0.15 mm is an alternative and preferred method of maintaining the light flux for a favorable *S/N* ratio. Solvents used for sample preparations are methanol, chloroform:methanol (1:1), ethyl acetate:methanol (1:1), chloroform:methanol:ammonia (90:10:1), methylene chloride:methanol (1:1), 1% ammonia or 1% aceticacid.

Activation of Precoated Plates

Freshly open box of plates do not require activation. Plates which are exposed to high humidity or kept on hand for long time requires activated. Activation of the plates can be done by placing the plates in an oven at 110–120°C for 30′ prior to spotting. Aluminum sheets should be placed in between two glass plates and kept in oven at 110–120°C for 15 min.

Application of Sample and Standard

Usual concentration range is 0.1–1 μg/μl; ranges above this lead to poor separation. Sample and standard can be applied by using automatic sample applicator with nitrogen gas sprays on TLC plates as bands or spot. Band-wise application has an advantage of better separation and high response to densitometer

Selection of Mobile Phase

Poor grade of solvent used in preparing mobile phases have been found to decrease resolution, spot definition and R_f reproducibility. Mobile phase commonly called solvent system is traditionally selected by controlled process of trial and error and also based on one's own experience and literature. Mobile phase should be chosen taking into consideration chemical properties of the analytes and sorbent layer. Use of mobile phase containing more than three or four components should normally be avoided as it is often difficult to get the reproducible ratios of different components. If normal stationary phase is polar and mobile phase selected is nonpolar, then nonpolar compounds are eluted first because of lower affinity with stationary phase and polar compounds retained because of higher affinity with the stationary phase and vice versa.

Preconditioning (Chamber Saturation)

Chamber saturation has pronounced influence on the separation profile. Unsaturated chamber causes high R_f values. Saturated chamber by lining with filter paper for 30 min prior to development leads to uniform distribution of solvent vapors and requires less solvent for the sample to travel.

Chromatographic Development and Drying

A glance at the apparatus available for TLC reveals quite clearly that significant advances have been made with scanners but that chromatography itself is carried out in miniaturized paper chromatography troughs. It is, therefore, to be welcomed that the AMD system (Automated multiple developments) is now available for completely automatic development. The development is always made ascending with fixed plate positioning, without chamber saturation in classical sense and with only 8 ml mobile phase on each run. The distance of the development is determined by time control. The chamber is not opened until all chromatographic steps have been completed. The plate is dried between the various developments and freed from solvent on completion of development.

Detection and Visualization

Detection under UV light is first choice and is nondestructive. Spots of fluorescent compounds can be seen at 254 nm (short wave length) or at 366 nm (long wave length). Spots of nonfluorescent compounds can be seen by using fluorescent stationary phase like silica gel GF. Non-UV-absorbing compounds are visualized by dipping the plates in 0.1% iodine solution. When individual component does not respond to UV, then derivatization is required with suitable visualizing agent.

Quantification

Sample and standard should be chromatographed on same plate and after development chromatogram is scanned. Densitometry is a simple way of quantifying the desired sample components or amount directly applied on the plate. The resolution of compounds to be separated on the chromoplate is followed by measuring the optical density of the separated spots directly on the plate. The sample amounts are determined by comparing them to a standard curve from reference materials chromatographed simultaneously under the same condition. The original data evaluation using the conventional methods of scanning was done by measuring the optical density of the transmitted light as a function of the concentration of the sample or standard delivered on stationary phase. The scanning densitometer is a more advanced workstation for evaluation of TLC/HPTLC and objects by measuring the absorbance (optical density) of fluorescence. This advanced workstation for evaluation of scanning of the chromatogram is done in reflectance or in transmittance mode by absorbance or fluorescent and is termed a scanner.

Validation of Analytical Method

All validation parameters such as precision, accuracy, LOD, LOQ, ruggedness, and robustness can be performed. HPTLC plays a major role in the characterization of the phytoconstituents for the development of the standardization parameters of different plant extracts. To strengthen this comment some examples has been cited in the following section with some potent plant species from the reference:

Simultaneous Estimation of Androgrpholide and Wadelolactone

Andrographolide and wedalolactone are active components of *Andrographis paniculata* and *Eclipta alba* respectively. The extracts of these plants are used in many traditional hepatoprotective formulations. An attempt has been made to develop an accurate, precise, and specific HPTLC method to quantify simultaneously both these chemical markers of diversified chemical structure in different dosage forms such as tablet and syrups.

Precoated Silica gel 60 F254 plates with toluene:acetone:formic acid (9:6:1) as a mobile phase and detection wavelength of 254 nm were used. The method was validated in terms of linearity, accuracy, precision, and specificity. The calibration curve was found to be linear between 200 and 400 ng/spot for andrographolide and 100–200 ng/spot for wedalolactone. The LOD and the limit of quantification for andrographolide were 26.16 and 79.28 ng/spot, respectively, and for wedalolactone 5.06 and 15.32 ng/spot, respectively (Patel et al. 2008).

Quantification of Eugenol, Luteoline, Ursolic Acid and Oleanolic Acid in Black and Green Varieties of Ocimum Sanctum

O. sanctum (Lamiaceae) is a reputed drug of Ayurveda, commonly known as Tulsi or Holy Basil. In traditional medicine, the plant is used in cardiopathy, blood disorders, leucoderma, asthma, bronchitis, genitourinary disorders, and skin diseases.

Preliminary experiments showed that, of the four compounds, eugenol and ursolic acid were in free form, whereas luteolin and olenolic acid were detected only after hydrolysis. Consequently, eugenol and ursolic acid were quantified from methanolic extracts and the samples were hydrolyzed to obtain the aglycons of luteolin and oleanolic acid. The optimized mobile phase resolved all the marker compounds with the following R_f values: eugenol, 0.77; luteolin, 0.27; ursolic acid, 0.50: and oleanolic acid 0.56. Other compounds in the same extracts did not interfere. Simultaneous quantification of all four markers was not possible even though they were resolved in the same solvent system because of the following reasons.

- Eugenol and ursolic acid are present in free form, but eugenol is detected under UV light (λ_{max} 280 nm) without derivatization, whereas ursolic acid can be detected only after derivatization with anisaldehyde–sulphuric acid reagent (λ_{max} 530 nm).
- Luteoline and oleanolic acid are present in bound form, but luteolin is detected under UV light (λ_{max} 350 nm) without derivatization, whereas ursolic acid can be detected only after derivatization with anisaldehyde–sulphuric acid reagent (λ_{max} 530 nm).
- The plates were scanned at the respective wavelength of the four markers for quantification. All the four are visible after derivatization, and this feature can be used for TLC-fingerprinting purpose, where the sample extracts can be cochromatographed with markers and visualized after derivatization.

This method is helpful in distinguishing the two varieties of *O. Sanctum*, Black and green (Anandjiwala et al. 2006).

Determination of Phyllanthin and Hypophyllanthin

HPTLC was performed using commercially prepared preactivated silica gel 60 F 254 plates. The application parameters were identical for all the analysis. Each TLC plates was developed to a height of about 10 cm with a mobile phase of hexane: acetone:ethyl acetate (74:12:8) under laboratory conditions. During the HPTLC study of the variation of Phyllanthin and hypophyllanthin (Fig. 8.1) in samples of phyllanthus species collected from different geographical source, none of the sample was found to contain the higher concentrations of hypophyllanthin as compared to phyllanthin. This is because the reported higher concentration of hypophyllanthin was due to other lignans present at the same R_f as that of hypophyllanthin. Therefore, a number of mobile phases of different composition were tried. The phase finally chosen, namely, hexane:acetone:ethyl acetate (24:12:8) gave good resolution of Phyllanthin and hypophyllanthin from other closely related lignans. The reagent used for color development gave intense blue color at R_f values

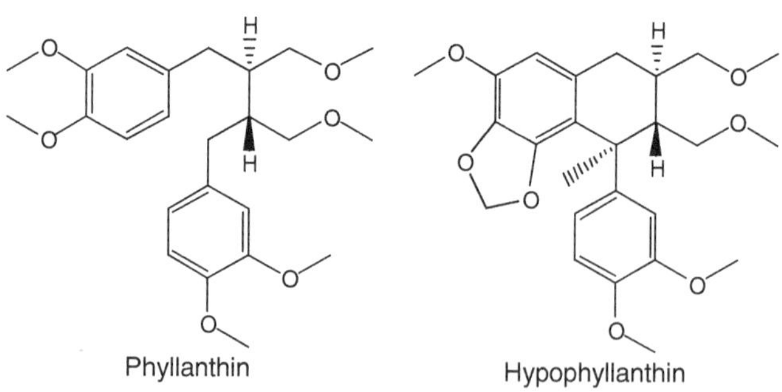

Fig. 8.1 Structures of phyllanthin and hypophyllanthin

of 0.24 and 0.29, respectively (Fig. 8.2). The scanning of the TLC plates was performed at 580 nm in the absorbance reflectance mode (Tripathy et al. 2006).

Determination of Cucuminoids

The rhizomes of the plant in the genus curcuma have a traditionally important role as a coloring agent in food, cosmetics, and textiles. *Curcuma amada* Roxb., commonly known as mango ginger, is one of the species with rhizomes having the characteristic odor of raw mango. Simple HPTLC method for rapid analysis of major curcuminoids (Fig. 8.3) in *Curcuma longa* and *C. amada* was developed. The method was found suitable for rapid screening of the plant material for their genotypic assessment and can be performed without any special sample pretreatment. Chromatography has been performed on preactivated silica gel plates 60 F 254. Samples and standards were applied to the plates as 6-mm wide bands. The TLC plates were developed with mobile-phase chloroform:methanol (95:5) to a height of about 8 cm. Peaks corresponding to curcumin (A), demethoxy curcumin

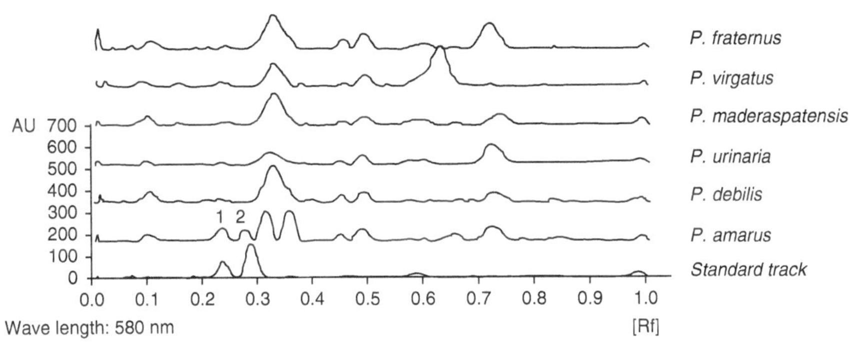

Fig. 8.2 The HPTLC separation of phyllanthin (1) and hypophyllanthin (2) in extracts of *Phyllanthus* species and in the standard track

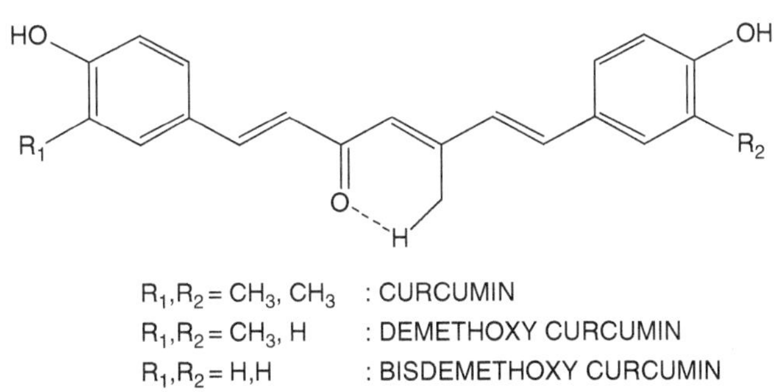

Fig. 8.3 Structures of curcuminoids

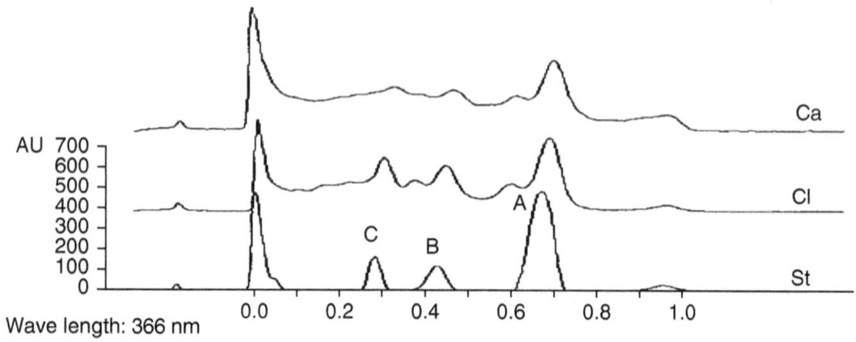

Fig. 8.4 HPTLC separation of curcumin (A), demethoxy curcumin (B), and bisdemethoxy curcumin (C) in standard (St), *Curcuma longa* (Cl) and *Curcuma amada* (Ca) tracks at 366 nm UV absorption/reflection mode

(B), and bis-demethoxy curcumin (C) were at R_f: 0.69, 0.44, and 0.29, respectively (Fig. 8.4). Resolution by UV scanning was better than that observed by visible light as some of the compounds were not fully responding in the sample track when the visible mode was applied (Gupta et al. 1999).

Micro Analytical Technique for Determination of Podophyllotoxin by RP-HPTLC

Podophyllum hexandrum (Berberidaceae) is a herbaceous, rhizomatous species of great medicinal importance. The rhizomes of *P. hexandrum* yielded cytotoxic lignan podophyllotoxin and resin due to which podophyllum possesses antitumor activity. Moreover, important drugs used in the treatment of testicular and small cell lung cancer, namely etoposide and teniposide, are produced by semisynthesis from the plant derived lignan podophyllotoxin.

HPTLC analysis was carried out by applying bands of width 6 mm on RP18 F254 TLC plates and developed using mobile-phase acetonitrile:water (50:50). Densitometric scanning was performed in the absorbance–reflectance mode at 217 nm. The spot for podophyllotoxin was ascertained by comparing the R_f values and spectra of sample with those of the standard podophyllotoxin (Fig. 8.5) (Mishra et al. 2005).

HPTLC in Morphological and Geographical Variations in the Herbal Raw Material

Asteracantha longifolia (Acantheceae) is a common weed growing in marshy and waterlogged area. The plant is used commercially as an ingredient of some over-the-counter (OTC) formulations used in liver disorders and those prescribed as general

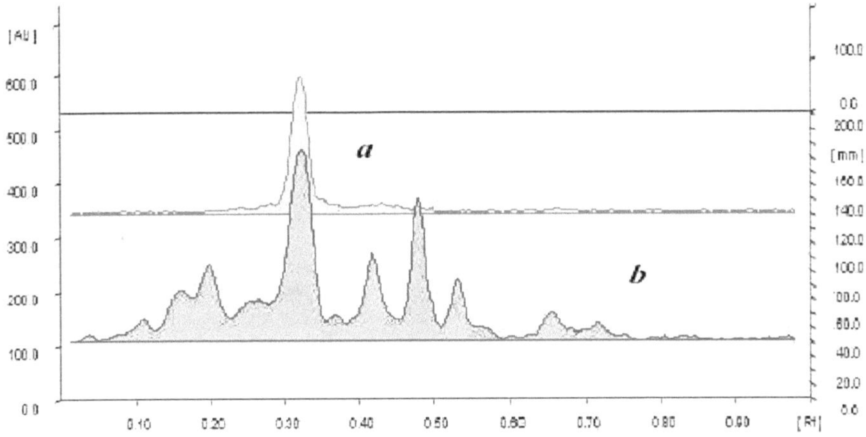

Fig. 8.5 HPTLC 3D overlay chromatogram of (a) standard track and (b) resin sample of roots of *Podophyllum hexandrum*

tonic. Since different parts of *A. longifolia* finds use for different clinical indications, it is important to establish quality of the plant raw material for its constituent plant part composition. HPTLC method can help distinctively to identify the source of the plant-powder raw material and also its constituent's plant part.

Chromatography was performed by using mobile-phase toluene:ethyl acetate: methanol (15:3:1.5). The plates were developed to a distance of 80 mm. After development, the plates were derivatized in Liebermann–Burchard reagent. Densitometric evaluation of the plates was performed at 366 nm. Presence of lupeol and β-sitosterol standards was detected and quantified. Developed chromatograms have a distinct phytochemical variation in morphological parts as well as in geographical regions (Figs. 8.6 and 8.7) (Sunita and Abhishek 2008).

HPTLC Studies of Various Indian Herbal Formulations

Sennoside Content

Several polyherbal formulations containing senna (*Cassia angustifolia*) leaves are available in Indian market for the treatment of constipation. The purgative effect of senna is due to presence of two unique hydroxyl anthracene glycosides, sennosides A and B. The HPTLC method for the quantification of sennosides A and B present in the formulation has been developed. Each 4 g of different branded formulations was sonicated with 70% methanol for about 45 min. Concentrated methanol extract was redissolved and finally reconstituted in 20 ml methanol prior to HPTLC analysis. Chromatograph was developed with mobile-phase 2-propanol:ethyl acetate:water:formic acid (17:19:12:2). The sennosides A and B were quantified at

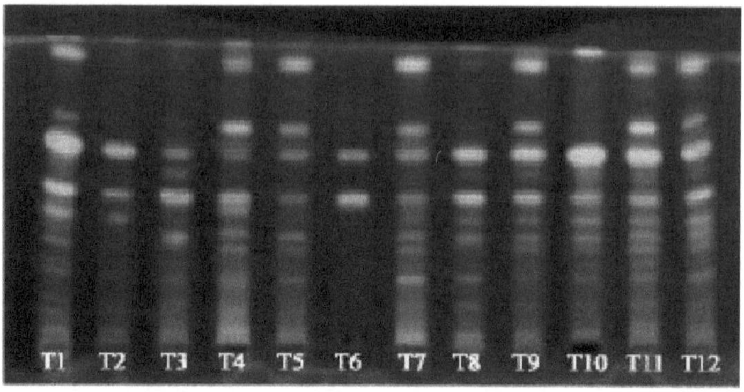

Fig. 8.6 The chromatographic plate of standard lupeol and β-sitosterol with *Asterecantha longifolia* Nees. Methanolic extracts of T1: *A. longifolia* (root); T2: *A. longifolia* (Stem); T3: *A. longifolia* (Leaf); T4: *A. longifolia* (Seed); T5: *A. longifolia* (Thane); T6: Standard Lupeol and β-sitosterol; T7: *A. longifolia* (Patalganaga); T8: *A. longifolia* (Mahad); T9: *A. longifolia* (Karjat); T10: *A. longifolia* (Kolhapur); T11: *A. longifolia* (Deharadun); T12: *A. longifolia* (Nazimabad)

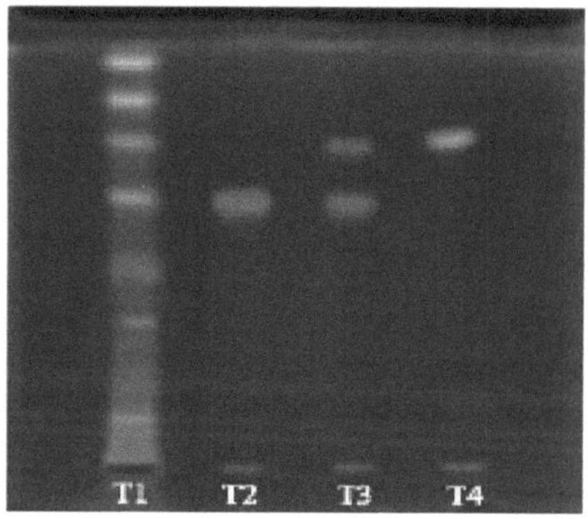

Fig. 8.7 The chromatographic plate of standard lupeol and β-sitosterol with *Asterecantha longifolia* Nees. methanolic extracts. T1: methanolic extract of whole plant powder of *A. longifolia* T2: β-sitosterol, T3: β-sitosterol + Lupeol T4: Lupeol

350 nm. The identification of sennoside A and B in the formulation was confirmed by superimposing the UV spectra of sample and standards within the same R_f window (Figs. 8.8 and 8.9) (Aktar et al. 2008).

Andrographolide in Extracts and Formulations

Andrographolide is an antiinflammatory, antipyretic, antiviral, immunostimulatory, antihyperglycemic, and antioxidant agent. HPTLC method was applied for the estimation of andrographolide (Fig. 8.10) in four different *A. paniculata* herb samples, three different extracts, and three different dosage forms: two tablets and one capsule. The interference studies revealed that the presence of commonly

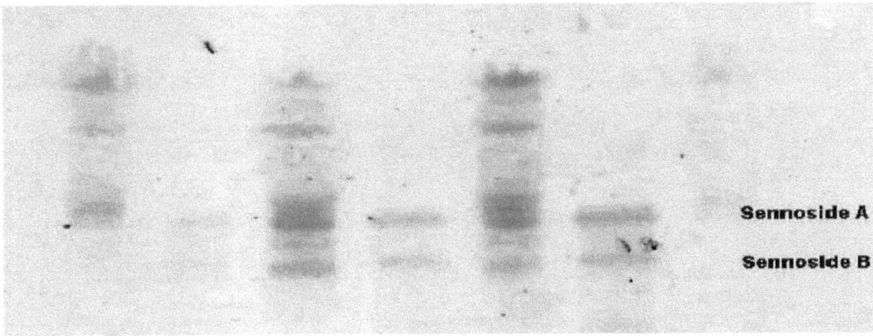

Fig. 8.8 TLC plate showing the distinct separation of sennoside A and B after development in twin-trough chamber

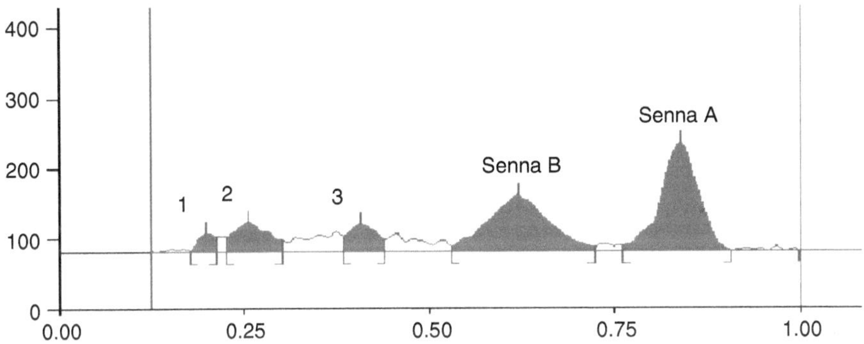

Fig. 8.9 Scan (at 350 nm) showing the separation of sennoside A and B in the extract of a laxative formulation

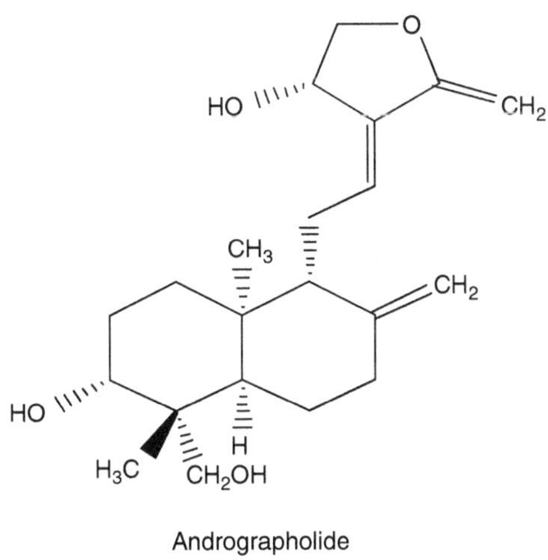

Fig. 8.10 Structure of andrographolide

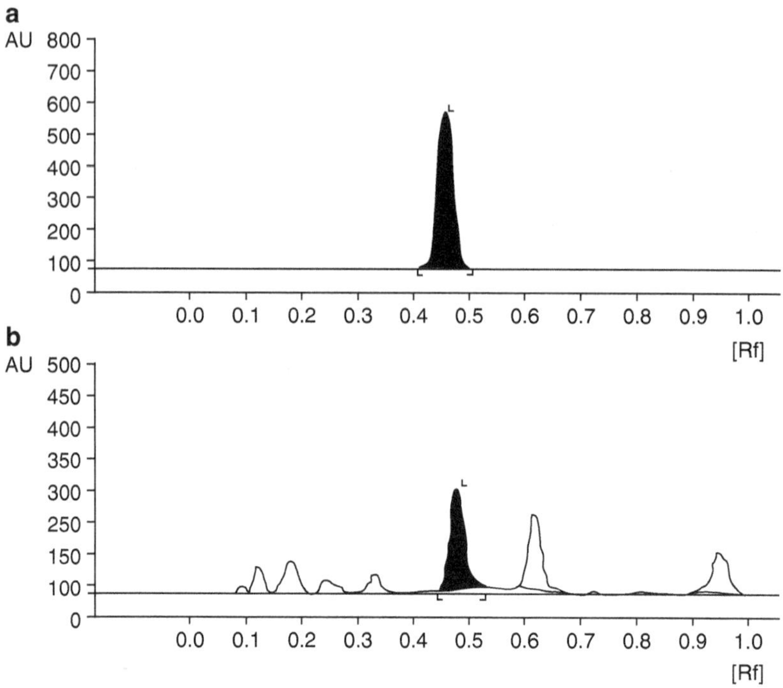

Fig. 8.11 HPTLC chromatograms of (**a**) andrographolide standard (**b**) *Andrographis paniculata* plant extract

used excipients like starch, talc, gelatin, malto dextrin, colloidal silicon dioxide, magnesium stearate, and plant extracts, triphala, *Tinospora cordifolia*, *E. alba*, *Boerhavia diffusa* does not interfere in the estimation of andrograoholide (Fig. 8.11) (Vijaykumar et al. 2007).

Determination of Hyperforin in Hypericum perforatum

Hypericum perforatum (St. John's Wort) has been widely used as an antiinflammatory and healing agent in traditional medicine. Hypericin, flavonoids, and hyperforin have been the constituents contributing to the antidepressant activity of the plant. Accurately weighed 6 mg extract of *H. perforatum* extracted with methanol by vortexing. The extract was concentrated and final volume was made to 10 ml with methanol. Two milligram standard hyperforin was dissolved in 10 ml of methanol and calibration curve from 0.2 to 2 μg was prepared and checked for reproducibility, linearity and validating the method. Silica gel 60 F 254 plates were used with petroleum ether:ethyl acetate (90:10) as a mobile phase. The plates were scanned at 290 nm for quantification (Figs. 8.12 and 8.13) (Tiwari et al. 2008).

Determination of Corosolic Acid

Lagerstroemia speciosa (Lythraceae), popularly known as Banaba, is an ornamental plant and has been traditionally used for the prevention and treatment of diabetes. Corosolic acid, the active marker compound of Banaba extracts displayed potential antidiabetic activity. To estimate the content of corosolic acid in *L. speciosa* leaves extracts, aliquots of 10 μl were subjected to HPTLC and plates were developed to a distance of 8 cm in a chloroform:methanol (9:1) mobile phase. The plates were scanned at 20 nm (Fig. 8.14) (Vijaykumar et al. 2006).

Evaluation and Validation of a Polyherbal Formulation

Raktadosh Nasak Vati is a polyherbal formulation, consisting of 13 ingredients of plant origin and it is widely used as antioxidant, antiseptic, and antiallergic. The major active components in the oleo gum resin of *Balsamodendron mukul* are Eguggulsterone and Z-guggulsterone that are responsible for its several activities.

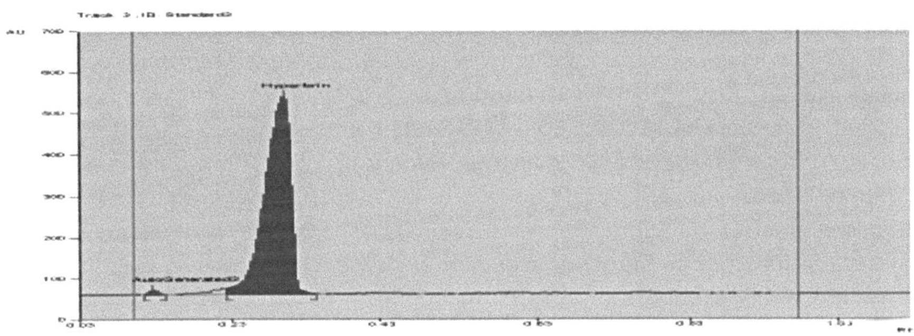

Fig. 8.12 TLC chromatogram of working standard of Hyperforin

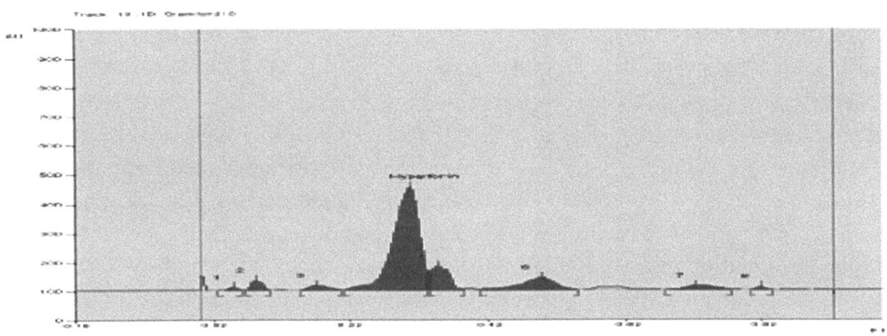

Fig. 8.13 TLC chromatogram test sample (*Hypericum perforatum* extract -IE-1)

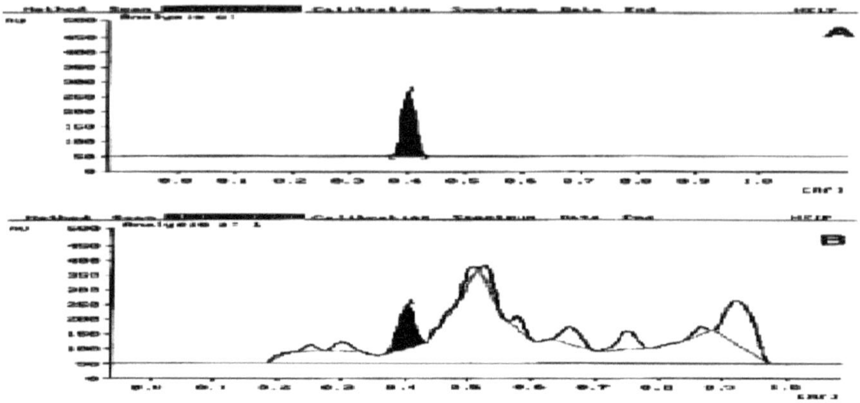

Fig. 8.14 HPTLC chromatograms of (**a**) corosolic acid standard (**b**) *Lagerstroemia speciosa* leaf extract

Guggul is the major ingredient of the formulation and responsible for the activity of the preparation. E-guggulsterone and Z-guggulsterone content reflects the quality of guggul used in the formulation. Berberine from *berberis* species (18), curcumin from *C. longa*, gallic acid, and ellagic acid from *Triphala* were obtained. Gallic acid and ellagic acid are present in many ingredients including the fruit pulp of *Terminalia chebula*, *Terminalia belerica,* and *Emblica officinalis*. TLC densitometric methods were developed using HPTLC for the quantification of six marker compounds from the polyherbal formulation *Raktadosh Nasak Vati*. Solvent systems were optimized to achieve best resolution of the marker compounds from the other components of the sample extracts. Of the various solvent systems tried, the one containing toluene:ethyl acetate:methanol:formic acid (3:3:0.2:0.8) gave the best resolution of berberine ($R_f = 0.27$), curcumin ($R_f = 0.74$), gallic acid ($R_f = 0.63$), and ellagic acid ($R_f = 0.51$) in the presence of other compounds in the sample extract and enabled the quantification of marker compounds. It was not possible to resolve Zguggulsterone and E-guggulsterone from methanolic extract of sample formulation because of the interference of other compounds. Hence, to resolve these two compounds sample preparation was modified. Solvent system of petroleum ether:ethyl acetate:methanol (6:2:0.5) gave the best resolution of E-guggulsterone ($R_f = 0.64$) and Z-guggulsterone ($R_f = 0.71$) (Fig. 8.15). In these two solvent systems, the marker compounds were well resolved from the other components of the formulation. The identity of the bands in the sample extracts were confirmed by comparing the R_f and the absorption spectra by overlaying their UV absorption spectra with those of their respective standard using TLC Scanner. The purity of the bands due to E-guggulsterone, Z-guggulsterone, berberine, curcumin, gallic acid, and ellagic acid bands in the sample extract was confirmed by overlaying the absorption spectra recorded at start, middle, and end position of the band in the sample tracks. The methods were validated in terms of precision, repeatability, and accuracy (Dhalwal et al. 2008).

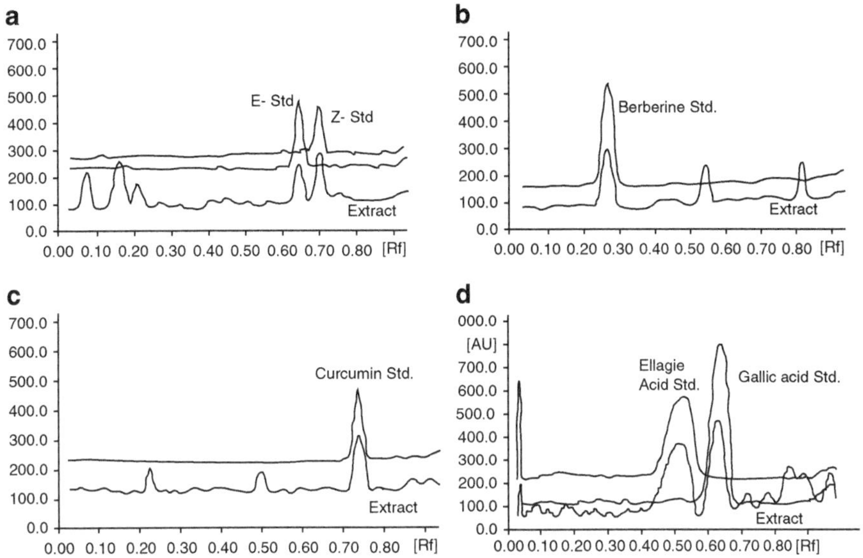

Fig. 8.15 TLC densitograms of sample solutions of *Raktadosh Nasak Vati* with their respective standard scanned at different wavelength. A–E-Guggulsterone and Z-Guggulsterone scanned at 254, B– Berberine at 366, C– Curcumin at 429 and D–Gallic acid and Ellagic acid at 280

Ingredients	Weight per tablet (mg)
Rubia cordifolia	20
Berbaris species	20
Curcuma aromatica	20
Tinospora cordifolia	20
Tribulus terrestris	20
Shuddha Gandhak	20
Azadirachta indica	30
Hemidesmus indicus	30
Balsamodendron mukul	30
Triphala	120

*Quantitative Analysis of Alkaloids in Hardin Grass (***Phalaris aquatical***)*

Silica gel 60 F254 preconcentration zone HPTLC glass was used. After the screening of several mobile phase solvent systems, a preferred mobile phase solvent system was developed that consisted of ethyl acetate/chloroform/7 N NH_4OH in methanol (8:2:1, v/v/v). Plates were developed until the mobile-phase solvent front was within 2 cm of the top of the plate. For double-development, the dried plates were allowed to be placed for a second time in mobile-phase solvent and documented by photography under UV 254 and UV 365 irradiation, after spraying with acidified anisaldehyde reagent consisting of ethanol/H_2SO_4/acetic acid/anisaldehyde (36:2:0.4:2, v/v/v/v). The quantification method was evaluated via linear regression analysis of a series of known standards on a HPTLC plate. Alkaloid standards gramine, 5-methoxydimethyltryptamine, methyltyramine, and hordenine (sulfate salt), the four major alkaloids in harding grass were mixed. One microliter each of the mixed dilutions was applied to the HPTLC plate. The plate was developed, dried, and scanned (Fig. 8.16) (Lili et al. 2006).

Densitometric Determination of Hecogenin from Agave americana *Leaf*

A simple TLC densitometric method for the quantification of hecogenin from the leaves of *A. americana* was developed using HPTLC. The method was validated for precision, repeatability, and accuracy. The method was found to be precise with RSD of 0.78 (intraday) and 0.82 (interday) for different concentrations of hecogenin. The content of hecogenin in different samples was estimated by the proposed method and was found to be in the range of 0.05–0.14% w/w in the samples analyzed. Accuracy of the method was checked by conducting recovery study at three different levels for hecogenin and the average percentage recovery was 98.98, 101.92, and 103.33%, respectively. The TLC densitometry method developed for the quantification of hecogenin was found to be simple, precise, specific, sensitive, and accurate and can be used in routine quality control (Fig. 8.17) (Ghogari and Rajani 2006).

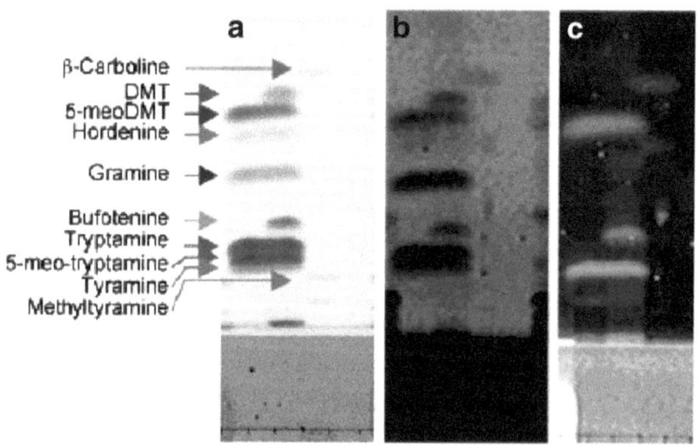

Fig. 8.16 Chromatography for the separation of 10 standards in HPTLC 20′ 10 preconcentration zone plate: (**a**) anisaldehyde reagent spray; (**b**) UV 254 *l*m imaging; (**c**) UV 365 *l*m imaging. (lane 1) six mixed standards; (lane 2) eight mixed standards; (lane 3) standard for ,carboline and ethyltyramine. Double-development was performed using a mobile phase consisting of ethyl acetate/chloroform/7 N NH_4OH in methanol (8:2:1). DMT, dimethyltryptamine; 5-meoDMT, 5-methoxydimethyltryptamine; 5-meo-tryptamine, 5-methoxytryptamine

Determination of Triterpenoid Acids (Arjungenin, Arjunolic Acid) from **Terminalia arjuna** *Stem Bark*

The stem bark of *T. arjuna* Linn. (fam: Combretaceae), commonly known as Arjuna in Indian systems of medicine, is a reputed drug used for various cardiac disorders. *T. arjuna* stem bark is reported to contain different groups of chemical constituents including phenolics, tannins, saponins, and triterpenoid acids. With tannin containing herbal drugs, tannins interfere in the extraction of certain compounds and hence in their quantification. A sample preparation is method to overcome the interference of the tannins by adsorbing them with carboxy methyl cellulose, which facilitates the efficient extraction of the triterpenoid acids. Further the TLC densitometric methods for the quantification of two of the triterpenoid acids of *T. arjuna* stem bark viz., arjungenin and arjunolic acid was developed using HPTLC. The methods were validated in terms of accuracy, precision, and repeatability. The percentage of arjungenin and arjunolic acid were found to be 0.324 and 0.524% w/w, respectively, in the stem bark by this modified method of extraction, which was many times higher than when compared to that using the extraction method without CMC (0.018 and 0.049% respectively). The study reiterates the importance of sample preparation in the quantification of nonpolar phytochemicals from herbal raw materials, such that the compounds of interest are extracted efficiently, overcoming the interference of other compounds like tannins in the matrix of plant material (Figs. 8.18 and 8.19) (Kalala and Rajani 2006).

Quantification of Valerenic Acid in **Valeriana jatamansi** *and* **Valeriana officinalis**

A simple, rapid, cost-effective, and accurate high-performance thin-layer chromatographic method has been developed for the quantification of valerenic acid

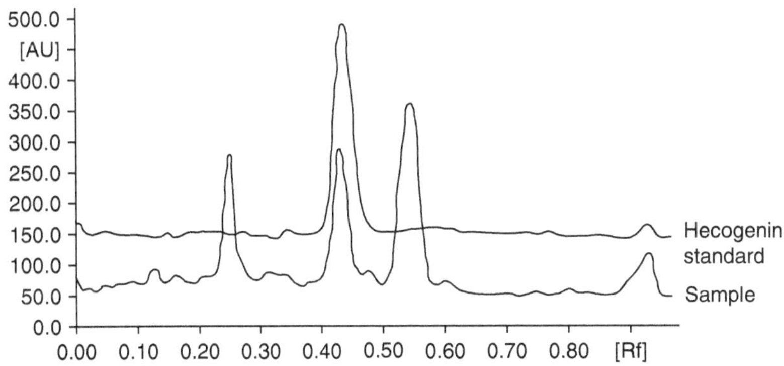

Fig. 8.17 TLC densitometric chromatogram of methanolic extract of *Agave americana* leaf, along with standard hecogenin, scanned at 430 nm

in *V. jatamansi* and *V. officinalis* which is one of the stable compounds of *V. officinalis* and designated as a key marker compound. Valerenic acid makes substantial contribution to the sedative and spasmolytic activity of the essential oil and extract of *V. officinalis*. Separation and quantification was achieved by HPTLC using ternary mobile phase of hexane:ethyl acetate:acetic acid (80:20:0.5 v/v) on precoated silica gel 60F254 aluminum plates and densitometric determination was carried out after derivatization with anisaldehyde–sulphuric acid reagent at 700 nm, in absorption–reflectance mode. The calibration curves were linear in the range 500 ng–2.5 lg (Fig. 8.20) (Singh et al. 2006).

Determination of Rutin in Amaryllis belladonna *L. Flowers*

Development with the mobile phase on the HPTLC silica gel layers produced compact, dark bands of rutin (R_f 0.3) when viewed under a 254 nm UV light. Selective baseline separation was observed between rutin and kaempferol-3-*O*-

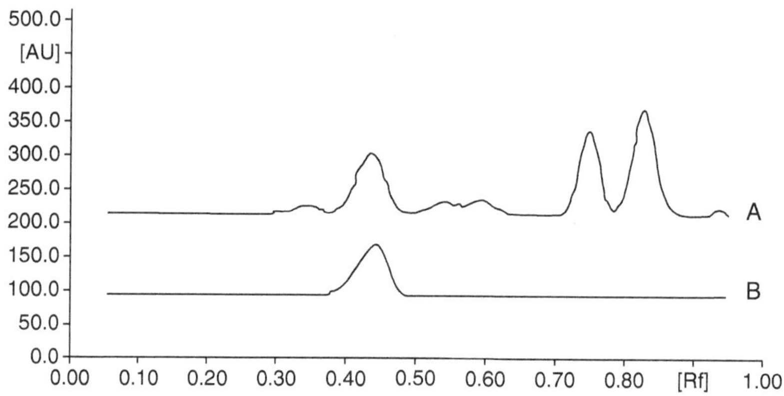

Fig. 8.18 TLC densitometric chromatograms of *Terminalia arjuna* stem bark extract (A) and Arjungenin standard (B)

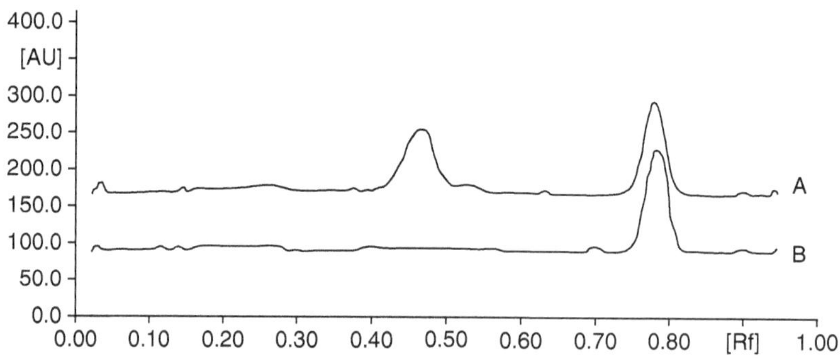

Fig. 8.19 TLC densitometric chromatograms of *Terminalia arjuna* stem bark extract (A) and Arjunolic acid standard (B)

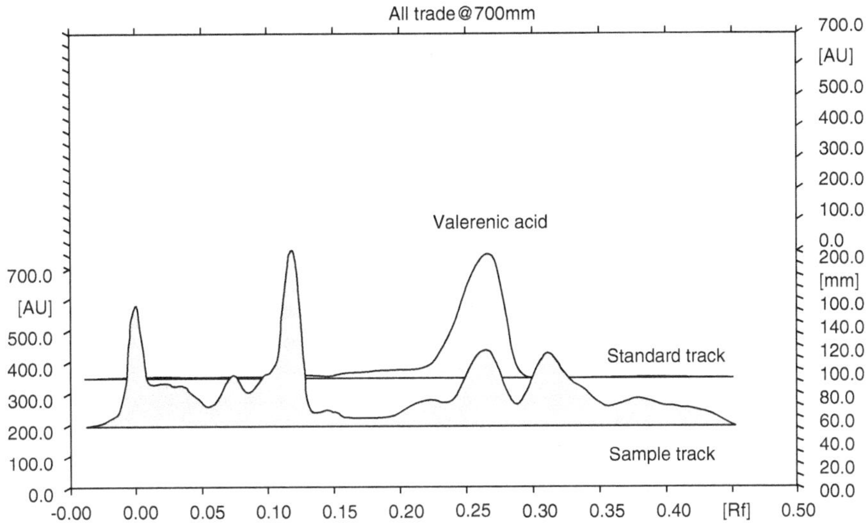

Fig. 8.20 3D overlay chromatogram of standard track and methanolic extract of *Valeriana* sample

rutinoside. Baseline separation was also obtained between rutin and other matrix components. Linear correlation was obtained between the area under the peak and concentration over the range 0.4–2.4 μg (Fig. 8.21) (Abou-Donia et al. 2006).

Determination of Quercetin, Rutin, and Coumaric Acid in Flowers of Rhododendron arboreum

A simple and fast method was developed for simultaneous quantitative determination of three biologically active phenolic compounds i.e., quercetin, rutin, and coumaric acid in flowers of *R. arboreum* using high-performance thin-layer chromatography (HPTLC). The separation was performed on TLC aluminum plates precoated with silica gel RP-18 F254S. Good separation was achieved in the mobile phase of methanol–water–formic acid (40:57:3, v/v/v) and densitometric determination of these compounds was carried out at 280 nm in reflectance/absorbance mode. Accuracy of the method was checked by recovery study conducted at two different levels with the average recovery of 99.90, 99.02, and 99.16% for quercetin, rutin, and coumaric acid, respectively (Fig. 8.22) (Swaroop et al. 2005).

Determination of Rutin in Amaranthus spinosus *Linn*

A simple, precise, and accurate high-performance thin-layer chromatographic method has been established for the determination of rutin in the whole plant

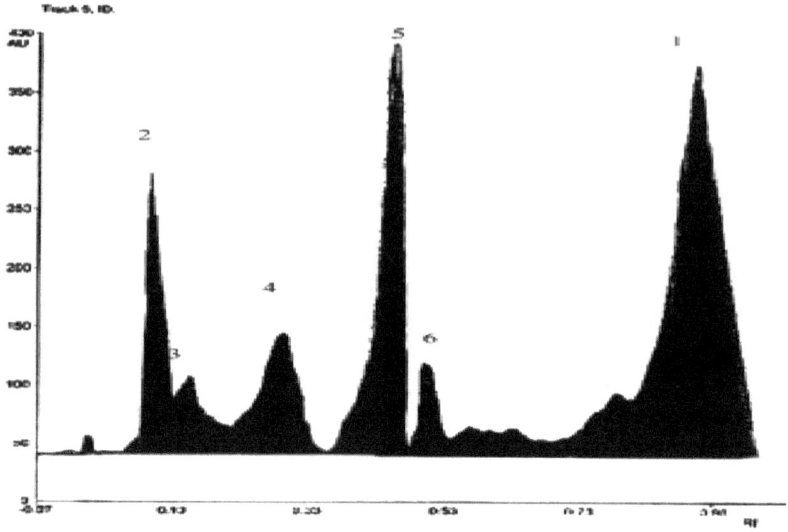

Fig. 8.21 HPTLC scan-densitogram showing the separation of rutin (5) and kaempferol-3-*O*-rutinoside (6) from other matrix components (1, 2, 3, 4)

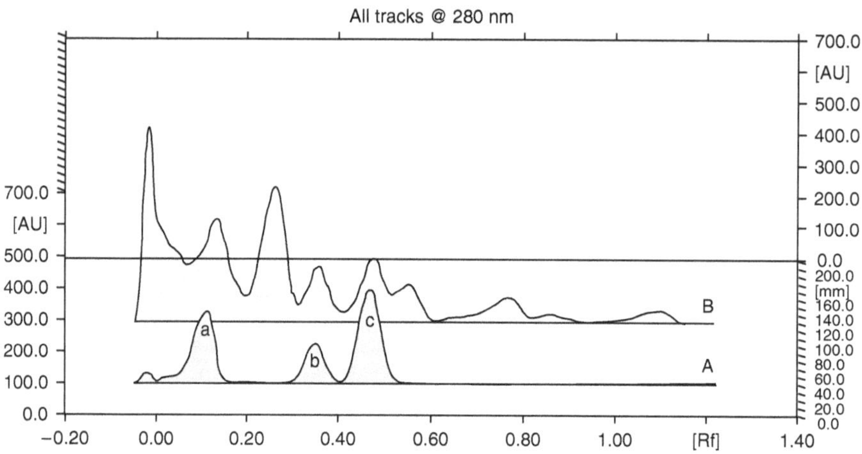

Fig. 8.22 3D overlay chromatogram of standard track (A) and sample track (B). Peaks a, b, and c represents quercetin, rutin, and coumaric acid, respectively

powder of Amaranthus spinosus Linn. Rutin has been reported to have antidiabetic, antithrombotic, antiinflammatory, and anticarcinogenic activity. A methanol extract of the whole plant powder was used. The concentration of rutin in the whole plant powder was found to be 0.15%. Separation was performed on silica gel 60 F254 HPTLC plates with ethyl acetate:formic acid:methanol:distilled water in the proportion 10:0.9:1.1:1.7 (v/v), as mobile phase. The determination was carried out using the densitometric absorbance mode at 363 nm (Fig. 8.23) (Suryavanshi et al. 2007).

Determination of Lycorine in Amaryllidaceae Plants Extracts

Lycorine, the most frequent Amaryllidaceae alkaloid, has proven to have a wide spectrum of biological activities including antiviral, cytotoxic, antimalarial, and antiinflammatory. The mobile phase consisted of chloroform:methanol (9:1). Ascending development of the plates was carried out. Plates were developed to a distance of 7 cm beyond the origin. The development time was 11 min. After development, the plates were air-dried for 5 min. Densitometric scanning was performed

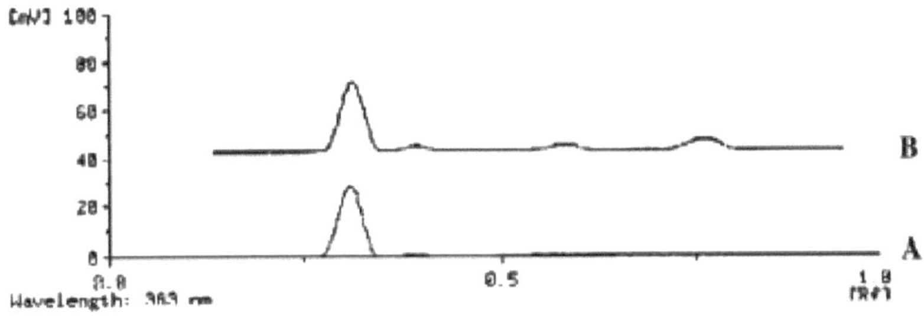

Fig. 8.23 A typical HPTLC chromatogram of standard rutin and plant. A standard rutin; B *Amaranthus spinosus* Linn

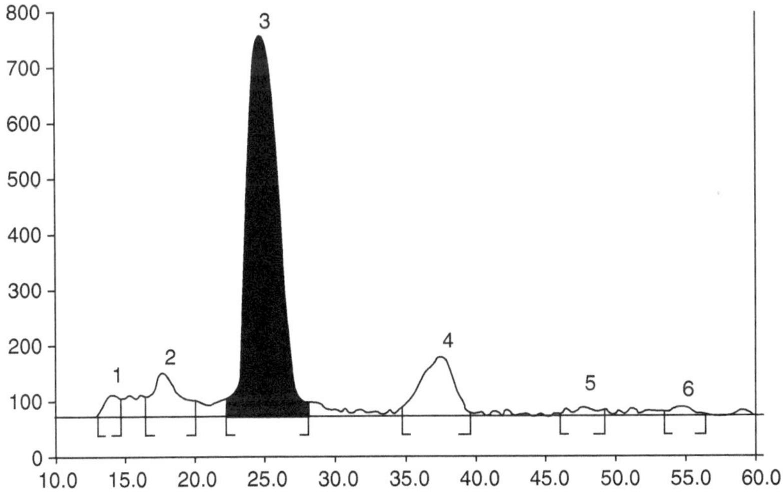

Fig. 8.24 HPTLC scan-densitogram showing the separation of lycorine y from other components in the sample at 368 nm

on TLC scanner in the reflectance–fluorescence mode at 368 nm (Fig. 8.24) (Abou-Donia et al. 2007).

Notes

HPTLC-based quantitative determination of many herbal extracts and herbal formulations have been reported. The present review is an attempt to present the theoretical and technical information to perform the reliable and reproducible HPTLC to establish the identity, purity, and quality of raw materials, extracts, and finished products. The present review has called attention to some examples of quantitative estimation of biomarkers present in herbal extracts and formulations which helps in performing the quantification of herbal drugs.

References

Abou-Donia AH, Toaima SM, Hammoda HM and Shawky E. Determination of rutin in *Amaryllus belledonna* L. flowers by HPTLC and spectrophotometry. Chromatographia, 2006, 64, (1/2), 109–112.

Abou-Donia AH, Toaima SM, Hammoda HM and Shawky E. New rapid validated HPTLC method for determination of Lycorine in Amaryllidaceae plant extracts. Chromatographia, 2007, 65.

Aktar MW, Poi R and Bhattacharyya A. Status of sennosides content in various Indian herbal formulations: method standardization by HPTLC. Bangladesh J Pharmacol, 2008, 3, 64–68.

Anandjiwala S, Kalola J and Rajani M. Quantification of eugenol, luteolin, ursolic acid and oleanolic acid in black and green varieties of *Ocimum sanctum* Linn. Using high performance thin layer chromatography. J AOAC Int, 2006, Nov–Dec, 712–718.

Dhalwal K, Biradar YS, ShindeVM, Mahadik KR, Rajani M. Phytochemical evaluation and validation of polyherbal formulation using HPTLC. Phcog Mag, 2008, 4, (14), 89–95.

Ghogari AM and Rajani M. Densitometric determination of hecogenon from *Agave americana* leaf using HPTLC. Chromatographia, 2006, 64 (1/2), 113–116.

Gupta AP, Gupta MM and Kumar S. Simultaneous determination of curcuminoids in *curcuma* samples using high performance thin layer chromatography. J Liq Chrom Rel Technol, 1999, 22, (10), 1561–1569.

Kalala J and Rajani M. Extraction and TLC Desitometric determination of triterpenoid acids (Arjungenin, Arjunolic Acid) from *Terminalia arjuna* stem bark without interference of tannins. Chromatographia, 2006, 63, (9/10), 475–481.

Lili Z, Andrew AH and Sumner LW. Efficient and sensitive method for quantitative analysis of alkaloids in Hardin grass (*Phalaris aquatica* L.). J Agric Food Chem, 2006, 54, (25), 9287–9291.

Mishra N, Acharya R, Gupta AP, Singh B, Kaul VK and Ahuja PS. A simple microanalytical technique for determination of podophyllotoxin in *Podophyllum hexandrum* roots by quantitative RP-HPLC and RP-HPTLC. Curr Sci, 2005, 88, (9), 1372–1373.

Patel MB, Kadakia VM and Mishra H. Simultaneous estimation of andrographolide and wedelolactone in herbal formulation. Ind J Pharm Sci, 2008, Sep–Oct, 689–693.

Singh H, Gupta AP, Singh B and Kaul VK. Quantification of valerenic acid in *Valeriana jatamansi* and *Valeriana officinalis* by HPTLC. Chromatographia, 2006, 63, (3/4), 209–213.

Sunita S, Abhishek S. A comparative evaluation of phytochemical fingerprints of *Asteracantha longifolia* Nees. using HPTLC. Asian J Plant Sci, 2008, 7, (6), 611–614.

Suryavanshi VL, Sathe PA, Baing MM, Singh BR and Lakshmi SN. Determination of rutin in *Amaranthus spinosus* Linn. Whole plant powder by HPTLC. Chromatographia, 2007, 65, (11/12), 767–769.

Swaroop A, Gupta AP and Sinha AK. Simultaneous determination of quercetin, ritin and coumaric acid in flowers of *Rhododendron arboreum* by HPTLC. Chromatographia, 2005, 62, (11/12), 649–652.

Tiwari N, Menghani H, Hamarapurkar P, Suthar A and Chauhan VS. Developing a new high performance thin layer chromatography method for quantitative estimation of hyperformin in *Hypericum perforatum*. Phcog Mag, 2008, 4, (16), 303–307.

Tripathy AK, Verma RK, Gupta AK and Khanuja SPS. Quantitative determination of phyllantin and hypophyllanthin in *Phyllanthus* species by high performance thin layer chromatography. Phytochem Anal, 2006, 17, 394–397.

Vijaykumar K, Murthy PBS, Kannababu S, Syamsunder B and Subbaraju GV. Quantitative determination of corosolic acid in *Lagerstromia speciosa* leaves, extracts and dosage forms. Int J Appl Sci Eng, 2006, 4, (2), 103–114.

Vijaykumar K and Murthy PBS, Kannababu S, Syamsunder B, Subbaraju GV. Estimation of andrographolide in *Andrographis paniculata* herb, extracts and dosage forms. Int J Appl Sci Eng, 2007, 5, (1), 27–39.

Chapter 9
HPTLC Determination of Artemisinin and Its Derivatives in Bulk and Pharmaceutical Dosage

Suraj P. Agarwal and Shipra Ahuja

Abstract A simple, selective, accurate, and precise high-performance thin-layer chromatographic (HPTLC) method has been established and validated for the analysis of artemisinin and its derivatives (artesunate, artemether, and arteether) in the bulk drugs and formulations. The artemisinin, artesunate, artemether, and arteether were separated on aluminum-backed silica gel 60 F_{254} plates with toluene: ethyl acetate (10:1), toluene: ethyl acetate: acetic acid (2:8:0.2), toluene:butanol (10:1), and toluene:dichloro methane (0.5:10) mobile phase, respectively. The linear detector response for concentrations between 100 and 600 ng/spot showed good linear relationship with r value 0.9967, 0.9989, 0.9981 and 0.9989 for artemisinin, artesunate, artemether, and arteether, respectively. Statistical analysis proves that the method is precise, accurate, and reproducible and hence can be employed for the routine analysis.

Artemisinin and its derivatives (artesunate, artemether, and arteether) are promising and potent antimalarial drugs, which meet the dual challenge, posed by drug-resistant parasites and rapid progression of malarial illness (Thaitong and Beale 1985; Webster and Lehnert 1994). It is a potent antimalarial drug isolated from the plant *Artemisia annua* (Thaitong and Beale 1985). Chemically, it is a sesquiterpene lactone with an endoperoxide bridge (–C–O–O–C–), to which its antimalarial activity is attributed (Webster and Lehnert 1994). The use of artemisinin as antimalarial agent is hampered by its poor solubility in oil and water and its poor efficacy on oral administration. Thus, new structural modifications in artemisinin were essential (Xinyi et al. 1985). A water-soluble derivative of artemisinin, the sodium salt of artesunic acid (succinic acid half-ester derivative of dihydroartemisinin) can be administered by i.v. injection, a property that makes it especially useful in the treatment of advanced and potentially lethal cases of *Plasmodium falciparum* infection (Ferreira et al. 1994).

S.P. Agarwal (✉) and S. Ahuja
Department of Pharmaceutics, Jamia Hamdard University, Hamdard Nagar, New Delhi 110062, India

MM. Srivastava (ed.), *High-Performance Thin-Layer Chromatography (HPTLC)*, DOI 10.1007/978-3-642-14025-9_9, © Springer-Verlag Berlin Heidelberg 2011

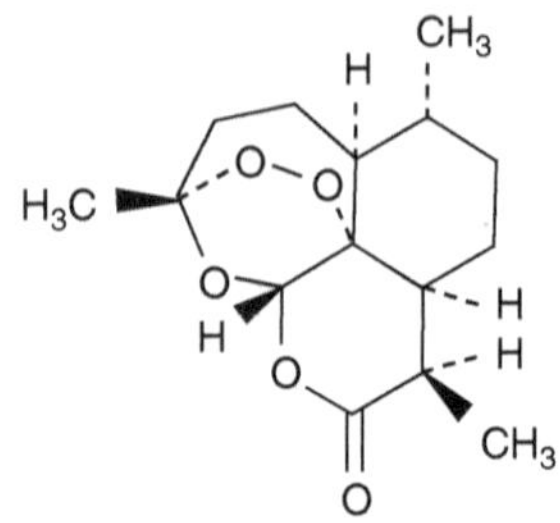

Artemisinin

Artemisinin Derivatives

Artemether: Artemether, a methyl ether derivative of artemisinin possess superior lipid solubility and antimalarial activity comparable to parent artemisinin (Dhingra et al. 2000; Zhao and Song 1990). Dihydroartemisinin, a mixture of epimer is produced by reduction of artemisinin with sodium borohydride. The mixture is treated with methanol and an acid catalyst to provide artemether (Haynes and Vonwiller 1994). It is effective in cerebral malaria by intramuscular injection (Salako et al. 1994).

Arteether: Arteether, a *p*-anomer of the ethyl ether of dihydroartemisinin is synthesized by boronhydride reduction of artemisinin to dihydroartemisinin, etherification in boron trifluoride ethyrate, and finally separation of anomers (Brossia et al. 1988; Li et al. 1981). Arteether is virtually insoluble in water, but is soluble in 50–100% ethanol in water and in a variety of oils. It has been stable in accelerated stability studies at 50°C for several months (Davidson 1994). It is a potent antimalarial and is very effective against drug-sensitive strains of *P. fulciparum* in vitro as well as in vivo in the mouse malaria model (Brossia et al. 1988).

Artesunate: Artesunate is a water-soluble hemisuccinate derivative of artemisinin. It is the only artemisinin analog that can be given intravenously (Batty et al. 1996). However, since it is unstable in water, it should be reconstituted with 5% sodium bicarbonate solution prior to injection.

Analytical Methods

Presently, TLC[3], GC4, HPLC with UV[5]/chemiluminescent (Green et al. 1995)/ electrochemical (Haynes and Vonwiller 1994) detector, RIA (Ferreira and Janick 1996), and ELISA (Jaziri et al. 1993) methods are documented/employed for the determination of artemisinin. But these methods are tedious and time-consuming. Literature survey reveals few analytical methods for the estimation of artesunate in biological fluids and pharmaceutical formulations, which include HPLC with electrochemical detection (Dhingra et al. 2000) and with ion pairing (Green et al. 1995), reversed-phase HPLC (Jastrebova et al. 1998), and colorimetric methods (Ferreira and Janick 1996; Jaziri et al. 1993). Therefore, an attempt has been made to develop an HPTLC method which is specific, accurate, precise, and reproducible (Agarwal et al. 2007, 2008, 2009). The developed method is also utilized to determine purity and quality of the market formulation HPTLC is becoming a routine analytical technique because of advantages (Bartsch et al. 1999; Kotiyan and Vavia 2000; Puthli and Vavia 2000), which include the small amount of mobile phase required, the speed of the method, and the possibility of analysis of several samples simultaneously (i.e., on the same plate), unlike HPLC. It thus reduces analysis time and cost per analysis. HPTLC does not, moreover, suffer from pH restrictions – mobile phases of pH 8 and above can be used, in contrast with HPLC. Cloudy samples and suspensions can also be analyzed directly by HPTLC. Automatic sample application is possible in HPTLC and repeated scanning can be performed on the same plate, so scanning conditions can be changed.

A gift sample of artemisinin was obtained from Saokin Co-operation Ltd., Vietnam. Artesunate bulk drug, arteether bulk drug, tablets (50 mg), and injection (60 mg) were obtained as gift samples from Skymax Laboratories Pvt. Ltd., Gujarat, India. Other samples of artesunate tablets (50 mg) and combination dosage form containing artesunate (100 mg), sulfadoxine (500 mg), and pyramethamine (25 mg) were from Medicamen Biotech Ltd. India. Artemether bulk drug was obtained as a gift sample from IPCA laboratories Ltd., M.P. All other chemicals and reagents were AR/HPLC grade.

Chromatographic Conditions

The instrument used in the present study was Camag-HPTLC system (Switzerland) comprising Camag Linomat V automatic sample applicator, Camag TLC SCANNER III with WINCATS software. The samples were spotted in the form of bands of width 5 mm using a Camag microlitre syringe on precoated silica gel 60 F_{254} TLC precoated aluminum plates (E. Merck), 10 cm × 10 cm size with 200-μm layer thickness using a Camag Linomat V applicator. A constant application rate of 150 nl/s was employed and space between two bands was 5.5 mm. The slit dimension was kept at 4 mm × 0.1 mm and 20 mm/s scanning speed was employed. The chromatogram was developed in a Camag twin-trough glass chamber using a linear ascending technique. The chamber saturation time for mobile phase was optimized to 30 min at room temperature. The length of chromatogram run was 80 mm. The mobile phase was used and densitometric scanning was performed on Camag TLC scanner III as mentioned below (Table 9.1). The source of radiation utilized was deuterium or tungsten lamp.

Table 9.1 HPTLC conditions

Drugs	Mobile phase	UV detection (nm)
Artemisinin	Toluene:ethyl acetate (10:1)	520
Artesunate	Toluene:ethyl acetate:acetic acid (2:8:0.2)	520
Artemether	Toluene:butanol (10:1)	546
Arteether	Toluene:dichloromethane (0.5:10)	546

Calibration Curve

A stock solution of artemisinin, artesunate, artemether, and arteether (100 μg/ml) in methanol were prepared by dissolving 10 mg of drug in 100 ml of methanol. Different volumes of stock solution 1, 2, 3, 4, 5, and 6 μl were spotted in duplicate on TLC plate with the help of automatic sample applicator, to obtain concentrations of 100, 200, 300, 400, 500, and 600 ng/spot. The plates were developed in the presaturated twin-trough chamber and densitometrically scanned at 520 nm for artemisinin and artesunate, and 546 nm for artemether and arteether using opto-mechanical scanning technique. The data of peak area versus drug concentration were treated by linear least-square regression analysis. Linear regression data for the calibration plots for artemisinin and its derivatives (artesunate, artemether, and arteether) ($n = 3$) were indicative of a good linear relationship between peak area response and concentration over the range 100–600 ng per spot (Table 9.2). There was no significant difference between the slopes of the calibration plots.

Validation of the Method as per ICH Guideline

Precision

The precision of the method was evaluated by repeatability of sample application and measurement of peak area using six replicates of the same spot (500 ng/spot) with % RSD of 0.84, 0.88, 0.66, and 0.56 for artemisinin, artesunate, artemether, and arteether, respectively. The intra- and interday precision of the artemisinin was carried out at three different concentration levels of 200, 600, and 3,000 ng/spot. This test was performed to justify the suitability of the proposed method (Table 9.3).

Robustness

The effect of altering various experimental conditions like mobile-phase composition, amount of mobile phase, plate treatment, time from spotting to chromatography, and time from chromatography to scanning on the results were examined.

Table 9.2 Linearity range

Parameter	Artemisinin	Artesunate	Artemether	Arteether
Linearity range (ng)	100–600	100–600	100–600	100–600
Correlation coefficient	0.9967	0.9989	0.9981	0.9989

Table 9.3 Intra- and interday precision

Amount (ng/spot)	Drugs	Intraday precision			Interday precision		
		Mean area	SD	% RSD	Mean area	SD	% RSD
200	Artemisinin	859.00	1.90	0.22	918.96	2.69	0.30
	Artesunate	1473.35	15.61	1.02	1383.54	14.88	1.08
	Artemether	2076.01	2743	1.32	2207.39	23.56	1.06
	Arteether	2165.55	13.34	0.61	2066.26	26.23	1.26
600	Artemisinin	1658.54	3.80	0.21	1718.25	2.91	0.17
	Artesunate	3156.41	29.68	0.94	3054.75	32.10	1.05
	Artemether	4862.54	39.83	0.81	4683.43	28.18	0.60
	Arteether	6662.59	25.62	0.38	6742.59	25.85	0.38
3,000	Artemisinin	7320.65	41.31	0.56	7529.83	32.76	0.61
	Artesunate	7629.21	37.30	0.48	7716.97	0.44	13.92
	Artemether	15100.13	38.52	0.25	15238.02	64.11	0.42
	Arteether	19251.94	31.32	0.16	19376.77	37.24	0.19

It was carried out at three different concentration levels of 200, 600, and 3,000 ng/spot in triplicate. For artemisinin and artesunate, % RSD value of 0.42 and 0.43 were found, respectively (Tables 9.4 and 9.5). However, it was 0.48 each for artemether and arteether(Tables 9.6 and 9.7).

Limit of Detection and Limit of Quantification

In order to estimate the LOD and LOQ, blank methanol was spotted six times on HPTLC plate and then developed, sprayed, and scanned in a way similar to that for calibration curve, and then signal–noise ratio was determined.

The limit of detection may be expressed as

$$\mathrm{LOD} = \frac{3.3\,\sigma}{S}.$$

The limit of quantitation may be expressed as

$$\mathrm{LOQ} = \frac{10\,\sigma}{S}.$$

Table 9.4 Robustness of the method for artemisinin ($n = 3$)

Parameters	Artemisinin			
	% RSD			Mean % RSD
	200 ng/μl	600 ng/μl	3,000 ng/μl	
Mobile-phase composition	0.41	0.28	0.65	0.45
Amount of mobile phase	0.30	0.20	0.58	0.36
Plate treatment	0.35	0.25	0.50	0.37
Time from spotting to chromatography	0.50	0.36	0.68	0.51
Time from chromatography to scanning	0.45	0.33	0.52	0.42

Table 9.5 Robustness of the method for artsunate ($n = 3$)

Parameters	Artesunate			
	% RSD			Mean % RSD
	200 ng/μl	600 ng/μl	3,000 ng/μl	
When plate was activated for 40 min	0.25	0.54	0.49	0.42
When 15 ml of mobile phase was used	0.39	0.29	0.55	0.41
After 15 min of development of plate, scanning was carried out	0.41	0.62	0.26	0.43
After 5 min of spotting, development of plate was carried outs	0.28	0.36	0.44	0.36
Mobile-phase composition	0.40	0.33	0.63	0.55

Table 9.6 Robustness of the method for artemether ($n = 3$)

Parameters	Artemether			
	% RSD			Mean % RSD
	200 ng/μl	600 ng/μl	3,000 ng/μl	
When plate was activated for 40 min	0.23	0.69	0.52	0.45
When 15 ml of mobile phase was used	0.33	0.20	0.65	0.48
After 10 min of development of plate, scanning was carried out	0.35	0.67	0.59	0.39
After 5 min of spotting, development of plate was carried out	0.50	0.38	0.82	0.56
Mobile-phase composition	0.49	0.65	0.23	0.53

Table 9.7 Robustness of the method for arteether ($n = 3$)

Parameters	Arteether			
	% RSD			Mean % RSDS
	200 ng/μl	600 ng/μl	3,000 ng/μl	
When plate was activated for 40 min	0.66	0.25	0.13	0.34
When 15 ml of mobile phase was used	0.35	0.44	0.53	0.44
After 15 min of development of plate, scanning was carried out	0.49	0.62	0.24	0.45
After 5 min of spotting, development of plate was carried out	0.85	0.44	0.60	0.63
Mobile-phase composition	0.74	0.52	0.38	0.54

Where σ = the standard deviation of the response and S = the slope of the calibration curve. The limit of detection and limit of quantitation for artemisinin, artesunate, artemether, and arteether were calculated as follows.

Recovery Studies

The accuracy of the method was evaluated by % recovery (standard addition method) of the drug. Stock solution of marketed preparation of artemisinin, artesunate, artemether, and arteether having concentration of 100, 500, 800, and 750 μg/ml, respectively, were prepared. This solution was first analyzed by the proposed method. In the analyzed sample, an extra 80, 100, and 120% of the drug was spiked and then the mixture was reanalyzed. The experiment was conducted in triplicate. It was done to find out the recovery of drug at different levels in the formulation. The average recovery of artemisinin was found to be 99.60 with average % RSD value of 0.27 (Tables 9.8 and 9.9). However, the average % recovery of artesunate was found to be 99.89 ± 1.006. The average recovery of artemether and arteether were found to be 99.49 ± 0.687 and 99.50 ± 0.590, respectively (Table 9.10).

Table 9.8 LOD/LOQ

Drugs	LOD (ng/spot)	LOQ (ng/spot)
Artemisinin	25	75
Artesunate	30	90
Artemether	25	28
Arteether	75	84

Table 9.9 Recovery studies for artemisinin ($n = 3$)

Excess drug added to analyte (%)	Theoretical content	% Recovery	% RSD
80	90	98.75	0.33
100	100	99.60	0.21
120	110	100.45	0.28

Table 9.10 Robustness of the method for artemether and arteether ($n = 3$)

Excess drug added to analyte (%)	Drugs	Theoretical content	Recovery (%)	% RSD
80	Artemether	144	98.89	0.211
	Arteether	135	98.93	0.250
100	Artemether	160	99.34	0.203
	Arteether	150	99.48	0.230
120	Artemether	176	100.24	0.314
	Arteether	165	100.11	0.150

Analysis of Artemisinin and Artsunate in Marketed Formulations

To determine the content of artemisinin in capsules, 20 capsules were weighed and content finely powdered. Powder equivalent to 250 mg of artemisinin was weighed accurately and extracted with 40 ml of ethanol. The solution was sonicated for 30 min and volume made up to 100 ml with ethanol. The resulting solution was centrifuged at 3,000 rpm for 5 min and filtered twice. Final concentration of 2,500 ng/μl was obtained and 1 μl of the solution was spotted on the plate and developed, sprayed, and scanned optomechanically.

To determine the content of artesunate in injection, the contents of the vial were pooled and mixed with methanol. The solution was sonicated for 30 min and volume was made up to 100 ml with methanol. The resulting solution was filtered twice. Final concentration of 600 ng/μl was obtained and 1 μl of this solution was spotted on plate, developed, and scanned. Both experiments were done six times.

The developed HPTLC method was applied to the analysis of artemisinin in capsule dosage form (Table 9.11). The mean % recovery value of 99.34 was obtained with % RSD value of 0.19. The drug content and % RSD was found to be 98.88%, 0.55 for artesunate tablets and 98.83%, 0.60 for artesunate injection, respectively (Table 9.12).The values of SD. or % RSD and coefficient of correlation are within the prescribed limit of 2% showing the high precision of the method. Hence, this indicates the suitability of the method for the routine analysis of artemisinin in bulk and pharmaceutical dosage forms.

Analysis of Artemether and Arteether in Marketed Formulations

To determine the content of artemether in injection (labeled claim: 80 mg/ml), the contents of the vials were pooled and mixed with methanol. The solution was sonicated for 30 min and volume was made up to 100 ml with methanol. The resulting solution was filtered twice. Final concentration of 800 μg/ml was obtained

Table 9.11 Analysis of marketed formulation of artemisinin ($n = 3$)

S. No.	Theoretical content (mg/capsule)	Amount of drug recovered (mg)	% Recovery
1	250	248.93	99.57
2	250	248.30	99.32
3	250	247.80	99.12

Table 9.12 Analysis of marketed formulations of artesunate ($n = 6$)

S. No.	Theoretical content	Amount of drug recovered (mg)	% Recovery
1	50 mg tab	49.44	98.83
2	60 mg/ml vial	59.30	98.83

Table 9.13 Analysis of marketed formulations of artemether and arteether ($n = 6$)

Drugs	Theoretical content (mg/ml vial)	Amount of drug recovered (mg/ml vial)	% Recovery
Artemether	80	74.27	98.81
Arteether	75	79.05	99.03

and 1 μl of this solution was plotted on plate, developed, and scanned. To determine the content of arteether in injection (labeled claim: 150 mg/2 ml), the contents of the vial was pooled and mixed with methanol. Same procedure was followed as above to obtain final concentration of 750 ng/μl, and 1 μl of this solution was spotted on plate, developed, and scanned. Both experiments were done six times. The % drug content was found to be 98.81% with %RSD of 0.64 in case of artemether and 99.03% with %RSD of 0.32 for arteether (Table 9.13)

Notes

This HPTLC method is precise, specific, and accurate. Statistical analysis proved that the method is repeatable and selective for the analysis of artemisinin and its derivatives (artesunate, artemether and arteether) in bulk drug and pharmaceutical formulations. The method can be used to determine the purity of the drug obtained from different sources by detecting related impurities. It may be extended to the determination of the degradation kinetics of artemisinin and its derivatives (artesunate, artemether, and arteether) in biological fluids.

References

Agarwal SP, Ali A and Ahuja S. Estimation of artemether and arteether by HPTLC. Asian J Chem 2007; 19, 4407–4414.

Agarwal SP, Ali A and Ahuja S. HPTLC determination of artesunate as bulk and in pharmaceutical formulation. Indian J Pharm Sci 2008; 69, 841–843.

Agarwal SP, Ali A, Dua Y and Ahuja S. Determination of artemisinin in bulk and pharmaceutical dosage forms by HPTLC. Indian J Pharm Sci 2009; 71, 100–103.

Bartsch H, Eiper A and Frank HK. Stability indicating assays for the determination of piroxicam – comparison of methods. J Pharm Biomed Anal 1999; 20, 531–541.

Batty KT, Ileit KF, Davis T and Davis ME. Chemical stability of artesunate injection and proposal for its administration by intravenous infusion. J Pharm Pharmacol 1996; 48, 22–26.

Brossia A, Venugopalan B, Gerpe LD. Yeh HJC, Flippenanderson JE, Buchs P, Luo XD, Milhous W and Peters W. Arteether, a new antimalarial drug: synthesis and antimalarial properties. J Med Chem 1988; 31, 645–650.

Davidson Jr DE. Role of arteether in the treatment of malaria and plans for further development. Trans R Soc Trop Med Hyg 1994; 88, 51–52.

Dhingra V, Rao KV and Narasu ML. Current status of artemisinin and its derivatives as antimalarial drugs. Life Sci 2000; 66, 279–300.

Ferreira JFS and Janick J. Immunoquantitative analysis of artemisinin from *Artemisia annua* using polyclonal antibodies. Phytochemistry 1996; 41, 97–104.

Ferreira JFS, Charles DJ, Wood K, Simon JE and Janick J. A comparison of GC and HPLC for artemisinin analysis. Phytochem Anal 1994; 5, 116–120.

Green MD, Mount DL, Todd GD and Capomacchia AC. Chemiluminescent detection of artemisinin: novel endoperoxide analysis using luminal without hydrogen peroxide. J Chromatogr A 1995; 695, 237–242

Haynes RK, Vonwiller SC. Extraction of artemisinin and artemisinic acid: preparation of artemether and new analogues. Trans R Soc Trop Med Hyg 1994; 88, S23–S26.

Jastrebova J, Nyholm L, Marides KE and Bergqvist Y. Online deoxygenation for reductive electrochemical detection of artemisinin and dihydroartemisinin in liquid chromatography, Analyst 1998; 123, 313–317.

Jaziri M, Diallo B, Vanhaelen M, Homes J, Yoshimatsu K and Shimosura K. Detection of artemisinin by ELISA method. Phytochemistry 1993; 33, 821–826.

Kotiyan PN and Vavia PR. Stability indicating HPTLC method for the estimation of estradiol. J Pharm Biomed Anal 2000; 22, 667–671.

Li Y, Yu PL, Chea YX, Li LQ, Cai YZ, Wang DS and Zheng YP. Studies on analogs of artemisinine. I. The synthesis of ethers, carboxylic esters and carbonates of dihydroartemisinine. Acta Pharmaceutica Sinica 1981; 16: 429–439.

Puthli SP and Vavia PR. Stability indicating HPTLC determination of piroxicam. J Pharm Biomed Anal 2000; 22, 673–677.

Salako LA, Walker O, Sowunmi A, Omokhodion SJ, Adio R and Oduola AMJ. Artemether in moderately severe and cerebral malaria in Nigerian children. Tans R Soc Trop Med Hyg 1994; 88, 13–15.

Thaitong S and Beale GH. Susceptibility of Thai isolates of *P. falciparum* to artemisinin and artemether. Bull World Health Org 1985; 63, 617–619.

Webster HK and Lehnert EK. Chemistry of artemisinin – an overview. Trans R Soc Trop Med Hyg 1994; 88, 27–29.

Xinyi N, Liyi HO, Zhihong R and Zhenyu S. Metabolic fate of qinghaosu in rats: a new TLC densitometric method for its determination in biological material. Eur J Drug Metab Pharmacokinet 1985; 10, 55–59.

Zhao KC, Song ZY. The pharmacokinetics of dihydroqinghasu given orally to rabbits and dogs. Yao Xue Xue Bao 1990; 25, 147–149.

Chapter 10
TLC/HPTLC in Biomedical Applications

A. Mohammad and A. Moheman

Abstract The main objective of this chapter is to encapsulate the applications of thin-layer chromatography (TLC) and high-performance thin-layer chromatography (HPTLC) as used in the analysis of compounds of pharmaceutical importance. The chapter discusses the advantages of using TLC or HPTLC for biomedical applications and summarizes important information on stationary and mobile phases, adopted methodology, sample application, zone detection, and identification and quantification of amino acids and proteins, carbohydrates, lipids, bile acids, drugs, vitamins, and porphyrins in biological matrices such as blood, urine, feces, saliva, cerebrospinal fluid, body tissues, etc. Among the stationary phases, silica gel has been the most preferred layer material in combination of mixed aqueous– organic or multicomponent organic solvent systems as mobile phase. For quantitative determination of analyte in various matrices, densitometry has been more commonly used. According to the literature survey, the interest of chromatographers in using the TLC/HPTLC has been in the following order: drugs > amino acids and proteins > lipids > bile acids > carbohydrates/vitamins > porphyrins.

Of the chromatographic techniques, thin-layer chromatography (TLC) is the most simple, robust, economical, and rapid technique. TLC is characterized by high selectivity, and it enables separation of the analyte from interfering substances. The special advantages of TLC are its versatility due to the availability of a large number of different sorbents in commercial form, the possibility of plate spraying with specific and general visualizing reagents, and coupling TLC with different specific detectors, as well as the ability to use a broad range (as regards polarity and selectivity of solvent) of mobile phases. Another attractive feature of TLC is that many samples can be simultaneously analyzed on a single plate. In modern TLC,

A. Mohammad (✉) and A. Moheman
Department of Applied Chemistry, Faculty of Engineering and Technology, Aligarh Muslim University, Aligarh 202002, Uttar Pradesh, India
e-mail: alimohammed08@gmail.com

MM. Srivastava (ed.), *High-Performance Thin-Layer Chromatography (HPTLC)*, DOI 10.1007/978-3-642-14025-9_10, © Springer-Verlag Berlin Heidelberg 2011 sample application, development, and recording of the chromatogram are realized by fast automated procedures. There are several books (Fried and Sherma 1996, 1999; Bladek and Zdrojewski 2000), reviews, and research papers (Anderson and Van Lente 1995; Bladek and Neffe 2003; Sherma 1996) that comprehensively summarize the applications of TLC in biomedical and clinical chemistry. According to the survey of literature on biomedical TLC of the last 25 years (Fig. 10.1), more emphasis on TLC analysis of drugs and amino acids has been given as compared to other biomedical substances. The objective of this chapter is to discuss the utility of TLC in biomedical analysis.

Chromatographic Condition in Biomedical Application

Thin-layer chromatography (TLC) is a physicochemical separation method, in which the components to be separated are distributed between two phases. One of these phases constitutes a stationary phase of large surface area and the other is a mobile phase that traverses along the stationary phase. The components to be separated are differentially attracted to the stationary phase because of variation of their physicochemical properties. This distinctive influence is manifested by different rates and distances migrated. Substances that are less strongly held on the stationary phase would tend to move faster in the mobile phase and vice versa. In other words, the speed of migration of a particular component would depend on its sportive affinity.

Operationally, TLC is very simple. In this technique, the basic steps followed for analysis of biomedical samples are shown in Fig. 10.2.

Qualitative analysis of separated components in TLC is based on a comparison of rates of migration. The retention factor, R_F value, is used to characterize and compare components for various samples.

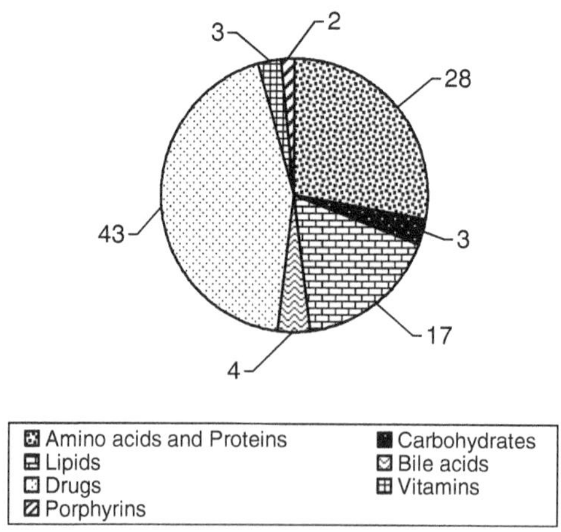

Fig. 10.1 Percentage of TLC/HPTLC publications appeared during last 25 years on biomedical substances

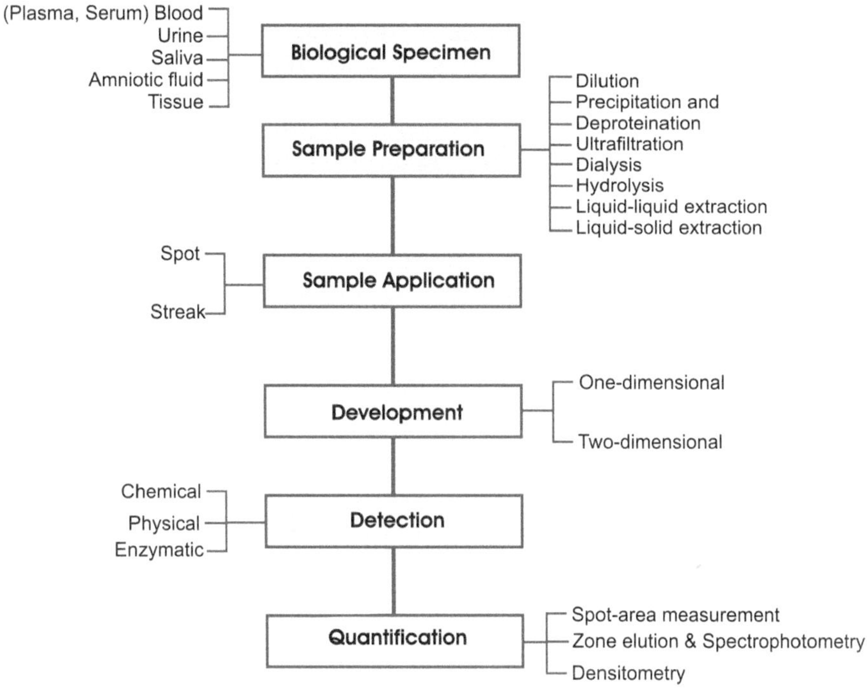

Fig. 10.2 Basic steps in TLC for separation of biomedical compounds retention parameter in TLC

Retention Parameter in TLC

The R_F value is defined as follows:

$$R_F = \frac{\text{Distance: start line – centre of the spot}}{\text{Distance: start line – solvent front}}$$

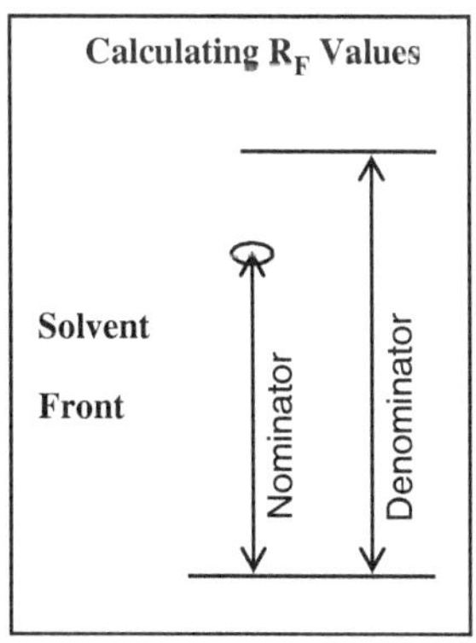

The value of the retention factor is between 0 (solute remains on the start) and 1 (solute move with the front of the mobile phase). In order to get reproducible R_F values, the atmosphere in the developing chamber must remain constant.

Chromatographic Systems

The probability of separation is greatly enhanced by the proper selection of a chromatographic system for particular analytes. This term refers to both stationary and mobile phases. If the stationary phase is of higher polarity than the mobile phase, the chromatographic system is called the normal phase (NP) system; if the mobile phase is of higher polarity, the system is named the reversed phase (RP). Representative separations of biomedical interest in both NP and RP systems are tabulated in Table 10.1.

Stationary Phases

Stationary phases are generally chemically well-defined inorganic (sometimes organic) materials. The adsorbents used as stationary phases in TLC are silica gel, alumina, cellulose, and Kieselguhr, alone or in their combination. Most of the workers have used laboratory-made TLC plates (layer thickness > 2.0 mm) obtained by applying the slurry of silica, alumina, Kieselguhr, and cellulose or their combinations in different proportions on glass plates (20×20, 10×20 or 5×10 cm) with the aid of layer spreading equipment. Precoated HPTLC plates have also been used. The particle size of most commercially available sorbents is about 20–50 μm for TLC and 5–15 μm for HPTLC. An inorganic binder such as gypsum ($CaSO_4$) is usually added to the sorbents to adhere the stationary phase to the backing. Fluorescent additives may be useful if detection of separated spots by examination in UV light is intended.

During the last decade, bonded stationary phases have led to a renewed interest in reversed-phase (RP) TLC. These include C_{18}, C_{12}, C_8, C_2, aminopropyl, diphenyl, and cyanopropyl bonded stationary phases. Furthermore, complex-forming stationary phases or chiral phases have also been introduced for separation of components in biological or environmental samples. Generally, RP-TLC has proved useful for the chromatography of polar compounds. Advantages of RP-TLC over normal phase (NP)TLC are higher recoveries of materials from the plate, ease of optimization of solvent system, and minimum decomposition of sensitive compounds. Although there are many similarities between RP-TLC and RP-HPLC, it is unlikely that reverse phase will achieve the dominant position in TLC that it occupies in HPLC. However, RP-TLC represents a useful extension of technique of TLC, especially for polar compounds.

Mobile Phases

With a particular stationary phase, the selection of an appropriate mobile phase system in TLC exerts a decisive role on the separation of biomedical compounds.

Table 10.1 Representative examples of separations in NP and RP systems

Type of analyte	Matrix	Chromatographic system	Fundamental goal of analysis	Ref.
Biologically active amines	Urine	NP	Determination of metabolism of catecholamines and serotonin	Dworzak and Hauk (1971)
Carbohydrates	Glycosides	NP	Use of two-dimensional thin-layer chromatography for the determination of different sugars	Wang and Ma (1989)
Lipids	Serum	RP	Evaluation of chromagenic and fluorogenic reagents for detection and densitometric quantification of lipids and phospholipids	Sherma and Bennett (1963)
	Lung tissue	NP	Separations of lung phospholipids from matrix and their determinations	Krahn (1987)
	Serum	RP	Determination of cholesterol sulfate and dehydroepiandro-sterone sulfate	Serizawa et al. (1987)
	Brain	NP	Micropreparative isolation and purification of ganglosides	Muthing and Heltmann (1993)
Bile acids	Feces	NP	Determinations of major unconjugated bile acids (cholic, chenodeoxycholic, deoxycholic, urodeoxycholic, and lithocholic acids) in human stool specimens	Kindel et al. (1989)
	Bile	RP	Separation of unconjugated bile acids and their glycine-and taurine-amidated, 3-sulfated, 3-glucosylated and 3-glucuronidated conjugates	Momose et al. (1998)
Drugs	Urine and liver	NP and RP	TLC of acidic and neural drugs on RP, and of basic, amphoteric, and quaternary drugs NP and RP	Ojanpera et al. (1999)
	Serum	NP	Determination of carbamazepine, phenytoin, and phenobarbitone	Patil and Bodhankar (2005)

(*continued*)

Table 10.1 (continued)

Type of analyte	Matrix	Chromatographic system	Fundamental goal of analysis	Ref.
Vitamins	Plasma	NP	Use of HPTLC and densitometry for the detection and quantitative estimation of retinol, α-tocophenol, and tocophenyl acetate in human plasma	Chavan and Khatri (1992)
	Fluids	NP and RP	Screening, separation, and preliminary identification of hydrophilic vitamins, and lipophilic vitamins in body fluids	Cimpoiu and Hosu (2007)
Porphyrins	Urine	NP	Description of a new detection method for the determination of porphyrins present in human urine	Huie and William (1989)

Organic or inorganic solvent and even solutes of a strong acid or base, ion-pairing agents, ion-exchanger agents, etc., can be used in TLC as a mobile phase. Solvents with low boiling point, low viscosity, and low toxicity are preferable for TLC application. It is also worth stressing that the mobile phase is evaporated before detection, and it does not interfere with the determination of the position of solute spots. A number of mixtures with various solvent percentages can be used as mobile phase in biomedical applications, with their compositions depending on the nature of separated analytes. For example, a ternary mixture butanol–glacial acetic acid–water (60.14:18.77:21.09) has proved experimentally the optimum mobile phase for the separation and identification of seven amino acids on microcrystalline cellulose (Rezic et al. 2007). Lipids can be excellently separated by a mobile phase of strong elution strength (first elution) and by a less-polar mixture in the second run. The four (Romano et al. 1994) solvent systems (1) ethyl acetate–methanol–30% ammonia (85:10:15, v/v), (2) cyclohexane–toluene–diethylamine (65:25:10, v/v), (3) ethyl acetate–chloroform (1:1, v/v), and (4) acetone have been proposed for separation of drugs and their metabolites on silica gel. The fat-soluble vitamins A and E and water-soluble vitamins B-1, B-2, B-6, and B-12 were separated by silica HPTLC using fractional elution; benzene was the first mobile phase and a 0.02 M aqueous micellar solution of sodium dodecyl sulfate the second (Kartsova and Koroleva 2007). Mixing the additive in the eluent used as a mobile phase can also modify the chromatographic system (Dynamic modification), but the use of modified adsorbents has led to an improvement of resolution. Example works include that by Armstrong and Zhou (1994), who used a macrocyclic antibiotic on the chiral selector for enantiomeric separations of acids, racemic drugs, and dansyl amino acid on biphenyl-bonded silica.

Amino Acids and Proteins

Amino acids are organic compounds containing the amino group (NH_2) and the carboxyl group (COOH), occurring naturally in plant and animal tissues. These compounds are particularly important in biomedical chemistry, where this term refers to α-amino acids (the amino and carboxylate groups are attached to the same carbon atom). The general formula is $H_2NCHRCOOH$, where R is organic substituent. Amino acids are the main building material in the synthesis of specific tissue proteins, enzymes, peptide hormones, and other physiologically active compounds. The amounts required also depend on the age and health of the individual, so it is hard to make general statements about the dietary requirement for some amino acids. The separations and analysis of amino acids in blood, urine, and tissues are crucial because an early estimation of free amino acids in biological fluids and tissues may prevent neurological damage and mental retardation in young infants with inborn errors of amino acids metabolism (Wu 1991). Proteins are compounds of high molecular weight and contain carbon, hydrogen, nitrogen, and sulfur. They consist of α-amino acids joined by peptide linkages. Different amino acids are commonly found in protein, each protein having unique, genetically defined amino acid sequence, which determines its specific shape and function. They serve as enzymes, structural elements, hormones, immunoglobulins, etc. and are involved in oxygen transport, muscle contraction, electron transport, and other activities.

TLC has proved useful for screening and quantitative analysis of amino acids, peptides, and proteins in biological fluids and tissues. Applications of the ninhydrin reaction for analysis of amino acids, peptides, and proteins in biomedical science were reviewed (Friedman 2004). Almost all separations of untreated amino acids are performed in NP systems. The stationary phases commonly used for separations are cellulose, silica gel, impregnated adsorbents, and ion exchangers. In the case of silica gel, two- or three- component mixtures of solvents usually carryout separation. Acetone, methanol, water, acetic acid, and formic acids, sometimes with the addition of ammonia or pyridine, are most commonly used as the components of mobile phases. Three new solvent systems, pyridine–benzene (2.5:20), methanol–carbon tetrachloride (1:20), and acetone–dichloromethane (0.3:8), provided improved resolution and identification of amino acids compared to previously reported systems (Bhushan et al. 1994a). Separations of 18 amino acids were compared on HPTLC silica gel, cellulose, and C-18 bonded silica gel layers, the identification of amino acids in the hemolymph and digestive gland–gonad complex of *Biomphalaria glabrata* snails were studied, and alanine and aspartic acid were quantified in hymolymph by scanning densitometry (Norfolk et al. 1994).

A simple and rapid method was developed for predicting the separation of D and L optical isomers of amino acids and hydroxy acids on a chiral stationary phase (Pyka 1993). TLC on chiral plates with acetonitrile–methanol–water (4:1:1) mobile phase and ninhydrin detection reagent was used in the quality control of L-tryptophan (Jork and Gang 1994), and thyroxine enantiomers were separated by chiral TLC to determine optical purity in bulk materials (Aboul-Enein and Serignese

1994). Amino acid racemates were separated on silica plates impregnated with a complex of copper and L-proline (Bhushan et al. 1994b) and on borate-gelled guran-impregnated silica gel (Mathur et al. 1994). RP-TLC separations of dansyl amino acid enantiomers were carried out by the use of β-CD mobile-phase additive (LeFevre 1993). While 2-*O*-(*R*)-2-hydroxypropyl)-β-cyclodextrin mobile phase additive gave improved chiral recognition compared to β-CD for enantiomers of six amino acids (Hao et al. 1995), Ohtake et al. (1995) have made an attempt to investigate homocysteine (HCys) in dried blood using TLC. It easily conjugates with plasma proteins, which makes detection crucial; as a result, the diagnosis of homocystinuria in the newborn period is impeded. A new compound, namely, *S*-(2-hydroxyethylthio) homocysteine was isolated in the study using TLC technique. Total HCys was extracted from dried blood spots and the extract was applied to a cellulose plate. After the development of the chromatogram (butanol–acetic acid–water, 4:1:1, v/v), the unknown fraction was harvested by scraping and analyzed using LC/MS. It was observed that *S*-(2-hydroxyethylthio) homocysteine was formed equally at room temperature and at 37°C. Examples of TLC applications for separations and identification of amino acids in biological specimen are presented in Table 10.2.

Carbohydrates

Carbohydrates are the main energy source for the human body. Chemically, carbohydrates are one of the most abundant biomolecules in which carbon, hydrogen, and oxygen bond together in the ratio: $C_x(H_2O)_y$, where x and y are whole numbers. All carbohydrates are made up of units of sugar (also called saccharide units). Mono-, di-, tri-, oligo-, and polysaccharides, ketoses, tri-, tetra-, pento-, and hexoses as well as reducing and nonreducing have great importance in the life sciences. Several diseases are accompanied by the increased elimination of sugars of various groups in the urine and feces.

In biomedical research and clinical laboratories, all chromatographic techniques are applied for the separation of carbohydrates in various biological samples. TLC has been extensively used in the analysis of carbohydrates. Various stationary phases are used in the TLC of carbohydrates. Silica gel, alumina, Kieselguhr, and cellulose are the most popular stationary phases. Separation of sugars (sucrose and fructose) in biological samples has been recommended on silica gel with ethyl acetate–isopropanol–water (60:30:10, v/v) and isopropanol–*n*-butanol–0.5% aqueous boric acid (50:30:30, v/v) followed by *n*-butanol–acetone–0.5% boric acid (40:50:10, v/v), respectively. The retention properties of α-, β-, and γ- cyclodextrins were examined using water-wet table RP-C-18 layers and mobile phases composed of acetonitrile, THF, methanol, ethanol, or propanol with water or methanol in acetonitrile (Zarzycki et al. 1995). The retention of aromatic *N*-glycosides was studied using overpressured development on silica gel layers impregnated with tricaprylmethylammonium chloride and methanol–water mobile phases (Szilagyi et al. 1993).

Table 10.2 Examples of TLC application for amino acids and proteins analyses

Analyte	Matrix	Chromatographic system		Visualization and quantification	Ref.
		Stationary phase	Mobile phase (v/v)		
Amino acids	Protein hydrolysates	Cellulose	Butanol-2-butanone–NH_3–water (5:3:1:1) and isopropanol–formic acid (20:1.5)	Ninhydrin	Zhu (1984)
Chain and aromatic amino acids	Serum	Polyamide	Two-dimensional: (i) Benzene–acetic acid (9:17), (ii) Formic acid: water (1.5:100)	UV_{254} detection fluorometric determination after elution	Li et al. (1985)
6 amino acids	Cerebrum	Polyamide	(a) Benzene–acetic acid (9:1), (b) 1.3% aqueous formic acid, (c) ethylacetate–methanol–acetic acid (20:1:1)	Detection by dansylation quantification by fluorometry	Ren et al. (1987)
3-Phenyl-2-thiohydantoine derivatives of 16 amino acids	–	Silica gel	Three phases: (i) Pyridine–benzene (5:40), (ii) Methanol–carbon tetrachloride (1:20), (iii) Acetone–dichloromethane (3:80)	Iodine vapors	Bhushan et al. (1994b)
Homocysteine	Blood	Cellulose	2-Mercaptoethanol	Ninhydrin reagent	Ohtake et al. (1995)
Proline and hydroxyproline	Different biological samples	Silicic impregnated glass fiber sheets	Isopropanol–water (7:3)	Detection by spraying with 0.25% solution of ninhydrin in 95% ethanol and heating at 105°C for 5 min, autoradiography	Demeglio and Svanberg (1996)
L-Arginine, its metabolites	Urine	Silica gel 60 GF254 and octadecylsilane phase RP-18 F254s	Methanol–50% acetic acid (3:1) and 5% acetic acid–methanol–acetonitrile (50:35:15)	1% ethanolic solution of ninhydrin	Baranowska et al. (2009)
Free amino acids	DGG and haemolymph	Silica gel HPTLC	*n*-Butanol–acetic acid–water (3:1:1)	Ninhydrin reagent quantification at 570 nm by colorimetry	Preet and Gupta (2009)

The glycosyl sequence of glycosides was determined by enzymatic hydrolysis followed by separation of the hydrolyzate by TLC and direct analysis of the spots by FABMS (Yao et al. 1994). Determination of sugars in food or clinical samples was performed by AMD with a polarity gradient based on acetonitrile–acetone–water on buffered amino layers and scanning densitometry for quantification (Lodi et al. 1994). A densitometric method using HPTLC silica gel 60 plates impregnated with sodium bisulfite and pH 4.8 citrate buffer, triple development with acetonitrile–water (85:15), and zone detection with α naphthol reagent was described for the determination of sugars in the hemolymph and digestive gland–gonad complex of infected *B. glabrata* (Perez et al. 1994) and *Helisoma trivolvis* (Conaway et al. 1995) snails. Other examples of TLC applications in carbohydrate analysis are presented in Table 10.3.

Lipids

Lipids are naturally occurring molecules, which include fats, waxes, sterols, fat-soluble vitamins (such as vitamins A, D, E, and K), monoglycerides, diglycerides, phospholipids, etc. The main biological functions of lipids include (a) energy storage, (b) structural components of cell membranes, and (c) important signaling molecules. All lipids are hydrophobic or amphiphilic in nature. The amphiphilic nature of some lipids allows them to form structures such as vesicles, liposomes, or membranes in an aqueous environment. Biological lipids originate from two distinct types of building blocks: Ketoacyl and isoprene groups. On this basis, lipids may be divided into eight categories, fatty acyls, glycerolipids, glycerophospholipids, sphingolipids, saccharolipids, polyketides, sterol lipids, and phenol lipids. Lipids are among the most widely analyzed compounds by TLC because of several important inherent advantages. In TLC analysis, many good lipid detection reagents are available, and layers can be impregnated with various reagents to improve selectivity, such as argentation TLC to fractionate lipid mixtures on the basis of the number, configuration, and position of double bonds in their components. Argentation TLC was used for the following separations: positional isomers of monosaturated fatty acids as their phenacyl derivatives (Nikolova-Damyanova et al. 1994); polyunsaturated fatty acids with detection by selective self-staining after development with toluene–acetone mixtures (Martinez-Lorenzo et al. 1994); positionally isomeric triacylglycerols on silica gel impregnated with low concentrations of silver ion and development with chloroform–methanol (Nikolova-Damyanova et al. 1993); and molecular species of triacylglycerols from highly unsaturated plant oils by successive argentation and RP-TLC (Tarandjiiska et al. 1995). TLC analysis of glycosphingolipid (GSL) was reviewed (Schnaar and Needham 1994), and the following GSL TLC studies were published: detection at the 5 pmol level using 5-hydroxy-1-tetralone as fluorescent-labeling reagent (Watanabe and Mizuta 1995); TLC and TLC-immunostaining analysis of neutral and sulfated glucuronyl GSLs purified from human motor and sensory nerves and myelins (Ogawa-Goto et al. 1993); and purification of GSLs

Table 10.3 Examples of TLC application for carbohydrates studies

Analyte	Matrix	Chromatographic system		Visualization and quantification	Ref.
		Stationary phase	Mobile phase (v/v)		
Oligosaccharides	Gangliosides	Polyamide, NH_2-silica	Ethyl acetate, acetic acid, methanol, and water mixtures in different percentages	Enzyme immunostaining	Higashi et al. (1987)
Glucose	Human urine Serum	NH_2-bonded HPTLC silica	Twofold development: (i) Acetonitrile–water (70:30), (ii) Ethyl acetate–pyridine–water–acetic acid–propionicacid (50:50:10:5:5) (i) Acetonitrile–water (80:20), (ii) Ethylacetate–pyridine–water–glacial acetic acid–propionic acid (50:50:10:5:5)	UV detection or heating at 150°C fluorescence determination	Klaus (1991)
Carbohydrates	Biological materials	NH_2-bonded HPTLC silica	Different solvent systems	Heating at 150°C. Quantification by densitometry in fluorescence mode at 366 nm	Klaus et al. (1991)
Reducing sugars	Protein hydrolyzates	Silica gel	Two dimensional: (i) Butanol–pyridine–water (16.5:4) over 3 cm. (ii) Ethylacetate–methanol–acetic acid–water (4:1:1:1) over 7 cm	*O*-Toludine reagent densitometry at 480–360 nm	Morcol and Velander (1991)
Neutral sugars	Cell walls	Silica gel impregnated with phosphate buffer	Acetonitrile–amyl alcohol–water (6:2:2)	*N*-(1-Naphthyl)-ethylene diamine reagent and heating at 100°C	Batisse et al. (1992)
Sugars and maltodextrins	Human milk, biological fluid	Cellulose	Multiple development: (i) Butanol–ethanol–water, (ii) Pyridine–ethyl acetate–acetic acid–water (5:5:1:3)	Silver nitrate, sodium thiosulfate, or Elson Morgan reagent	Boscher-Reig et al. (1992)
Oligosaccharides	Human milk	Anion exchange	Ethyl acetate–acetic acid–water (5:5:4)	FAB-MS	Rudloff et al. (1996)
Carbohydrates	Biomphalaria glabrata snails (tissue and blood)	Silica gel	Ethyl acetate–acetic acid–methanol–deionized water	1-Naphthol-sulfuric acid reagent. Densitometry at 515 nm	Eidam et al. (2001)

and phospholipids by the TLC blotting technique (Taki et al. 1994). Improved HPTLC separation of gangliosides was obtained by utilizing AMD (Muething 1994), and ganglioside detection was simplified and made more sensitive by the use of chemiluminescence (Arnsmeier and Paller 1995). Extraction, purification, separation, and quantitative analysis of membrane phospholipids by TLC were reviewed (Cartwright 1993). Densitometric quantification was applied to the lipid determinations, carried out by single development, one-dimensional (1D) TLC or HPTLC on unmodified silica gel layers. Few examples to cite include: human stratum corneum lipids by sequential 1D TLC with detection by charring (Schuerer et al. 1995); four phospholipids for identification of fetal lung maturity with chloroform methanol–ethanol–acetic acid–water (24:4:2:6:0.5) mobile phase, detection by exposure to iodine vapor, and scanning at 460 nm (Lin et al. 1994a); gangliosides by use of 2D HPTLC (Wiesner and Sweeley 1995). Endogenous, human-plasma-derived lipids that inhibit the platelet-stimulating activity of the platelet-activating factor (PAF) have been identified by Woodard and coworkers (Woodard et al. 1995). This is an important problem from the medical point of view, because the PAF is unique class of acetylated phospholipid autacoids with the most potent mediators of allergy and inflammation. PAF-acether has been identified in various biological fluids as a preformed lipoprotein-bound compound (lipo-PAF). Aliquot components obtained after the purification of blood (drawn from healthy volunteers) were separated on silica gel with chloroform–methanol–water (65:35:6, v/v). The areas of the bands were located by iodine staining. A method was developed by Raith et al. (2000) for the profiling of human stratum corneum ceramides. The method enables the investigation of the role of ceramides in maintaining the barrier function of stratum corneum. TLC using automated multiple development was modified for semipreparative purposes. The fractionation of complex lipid extracts using this method ensured specific, sensitive, reliable analysis by electrospray tandem mass spectrometry. Further representative examples of TLC applications in the separations of various lipid classes are listed in Table 10.4.

Bile Acids

Bile acids are 24-carbon steroid derivatives. They are formed by the conversion of cholesterol to cholic and chenodeoxy cholic acids (primary bile acids). They have different structural characteristics such as a 24-carbons, carboxyl group, β-linked A and B rings of the steroid nucleus, a C_3 hydroxyl group, further hydroxyl groups at C_6, C_7, or C_{12} only, and conjugation of C_{24} carboxyl moiety with amino acid glycine or taurine only. They are formed by the conversion of cholesterol to cholic and chenodeoxy cholic acids (primary bile acids). These are then conjugated with glycine or taurine via amide linkage in the liver. After conjugation, they are eliminated together with the bile. Most of the bile acids entering the gut are reabsorbed in the terminal ileuum, while some of them undergo bacterial deconjugation. In this process, secondary bile acids (deoxycholic, lithocholic, and ursodeoxycholic acids)

Table 10.4 Examples of TLC application for lipids

Analyte	Matrix	Chromatographic system		Visualization and quantification	Ref.
		Stationary phase	Mobile phase (v/v)		
Acidic lipids	Human liver	Silica	Chloroform–methanol–0.25% KCl (60:35:8)	Bromothymol blue	Hara and Taketomi (1982)
Lipids	Blood	Silica	Ammonium sulfate with petrol ether–ether–acetic acid (82:18:1)	Heating at 180°C for 10 min quantification by Densitometry at 580 nm	Fei et al. (1988)
Neutral lipids	Brain extracts	Silica	Chloroform–methanol–water (65:25:4)	Exposure to iodine vapor or spraying with molybdenum blue reagent	Ansari and Shoeman (1988)
Cholesterol palmitate, oleate, linolate, and linoleate	Human serum	HPTLC silica	Carbon tetrachloride	Phosphomolybdic acid in ethanol and heating 2 min at 100°C. Densitometry by absorbance at 546 nm	Kovacs et al. (1989)
Lipids	Human skin and blood plasma	HPTLC silica	Hexane–ether (6:4) and after intermediate drying, with chloroform–acetone–methanol (19:1:5–20:25 mm), then again after drying, with chloroform–acetone–methanol–formic acid (85:10:5:0.6–90 mm)	Benzidine reagent, quantification by transmittance scanning densitometry	Colarow (1990)
Phospholipids and phospholipids hydroperoxides	Meat extracts	HPTLC silica	Chloroform–ethanol–methanol–triethylamine–water (30:25:10:35:8)	Dipping in a freshly prepared solution at 1.0 g *N-N*-dimethyl-*p*-phenylenediamine and 1 ml of acetic acid in 50% aqueous methanol. Quantification by densitometry at 654 resp. 547 nm	Bruun-Jensen et al. (1995)
Sebaceous neutral ^{14}C-lipids	Human sebaceous gland	Silica	Hexane–ether–acetic acid (80:20:1)	Direct quantification in the range of 20–2,000 dpm by radio scanning	Christelle et al. (1997)

and a number of keto bile acids occurring in feces are formed. The main function of bile acids is to facilitate the formation of micelles, which promotes processing of dietary fat. The determination of bile acids and their metabolites in various biological samples e.g., bile, serum, urine, liver biopsies, gastrointestinal fluid, feces, and hepatic is becoming increasingly important for the diagnosis of several diseases and disorders. In biomedicine, concentration of bile acids in human feces is used for diagnosis of bile acid malabsorption and chologenic diarrhea. In liver disorders, serum levels of bile acids are elevated and their measurement is a sensitive index of liver disease. Bile acids are not found in urine owing to efficient uptake by the liver and excretion into the intestine. In hepatocellular disease and obstructive jaundice, however, their urinary excretion increases.

Thin-layer chromatography (TLC) has been used extensively for the separation of bile acids. Prior to TLC analysis of bile acid, purification of the biological specimens (e.g., bile, serum, duodenal contents, and crude fecal extracts) is essential. Sample cleanup steps would, however, depend upon the type of biological specimens being analyzed. Isolation of individual free bile acids in serum involves protein separation, alkaline hydrolysis, solvolysis of conjugated bile acids, and purification of lipids. Kindel et al. (1989) separated five predominant bile acids (cholic, chenodeoxycholic acid, deoxycholic acid, lithocholic acid, and ursodeoxycholic acid) in human stool specimens on a silica gel plate using two different solvent systems. The plate was first developed in isooctance–2-propanol–acetic acid (30:10:1, v/v) for 40 min, dried, and developed again with isooctane–ethyl acetate–acetic acid (10:10:2, v/v) for 65 min. For quantitative analysis, the plate was dipped in 0.2% 2,7-dichlorofluorescein ethanolic solution, and bile acid fluorescence was measured by TLC with direct scanning fluorometry (TLC–DSF). In clinical and biomedical practice, this TLC procedure constitutes a reliable, simple, and time-saving alternative to gas chromatography. The influence of several important experimental conditions on the separation of ten standard bile acids on silica gel and reversed phases (RP-8 and RP-18) were carried out by Rivas-Nass and Mullner (1994). The influence on migration distance, band shape and intensity, and resolving power exerted by conditions such as temperature, pH, support, and the addition of modifiers has been evaluated. In general, fluorimetry is the most sensitive detection technique, for quantitative analysis of bile acids. Bile acids are converted into fluorescent derivatives by reaction with 5% perchloric acid in methanol. However, spraying with 10% sulfuric acid and subsequent heating at 90°C for 10 min yields fluorescent, light blue spots on a dark-violet background, which has proved very successful in detection. Example applications of TLC for such analysis are presented in Table 10.5.

Drugs

Drug analysis in body fluids and tissues has always been of interest to biomedical chemists. It is carried out for two purposes: the identification and quantification of drugs taken for nontherapeutic purposes, usually analgesics, sedatives, tranquilizers,

Table 10.5 Examples of TLC application for bile acids

Analyte	Matrix	Chromatographic system		Visualization and quantification	Ref.
		Stationary phase	Mobile phase (v/v)		
Bile acids (CA, CDCA, DCA, LCA)	Human stomach	Silica	Isooctance–isopropylether–acetic acid–butanol–water (10:5:5:3:1)	Densitometry by absorbance at 365 nm	Yishen and Aihua (1985)
Taurocholic acid, taurodeoxycholic acid, taurochenodeoxycholic acid, glycocholic acid, glycodeoxycholic acid, and glycochenodeoxycholic acid	Gallstone	Silica	Chloroform–isopropanol–isobutanol–acetic acid–water (30:20:10:1:0.7)	Spraying with 10% phosphomolybdic acid Quantification by densitometry at 590 nm	Deng et al. (1986)
Bile acids	Bile	Silica	Chloroform–methanol–acetic acid–water (65:24:15:9)	Quantification by densitometry	Sambaiah et al. (1986)
Free bile acids	Bile	Silica	Isooctance–isopropylether–acetic acid–butanol–water (10:5:5:3:1)	Spraying with a solution of phosphomolybdic acid quantification by densitometry	Wang et al. (1989)
Cholic acid, deoxycholic acid, chenodeoxycholic acid	Bezoar	Silica	Isooctane–butyl acetate–acetic acid (4:2:1)	Quantification by densitometry at 380 nm	Zhang et al. (1990)
Bile acids	Gastric juice	Silica	Isoocatane–ethylacetate–acetic acid–butanol (30:10:3:3)	Spraying with 10% phosphomolybdic acid in ethanol and heating at 110°C for 10 min. Quantification by densitometry at 400 nm	Li et al. (1993)
Taurocholic acid, taurochenodeoxycholic acid, tauroursodeoxycholic acid	Bear bile	Silica	Chloroform–isopropanol–acetic acid–water (25:30:4:1)	Spraying with 10% phosphomalybdic acid in ethanol. Quantification by densitometry at 590 nm	Song et al. (1993)
Cholic acid, deoxycholic acid	Bile salts	Silica	Isooctane:ethyl butyrate–acetic acid–water (18:5:3)	Spraying with 30% sulfuric acid in ethanol and heating at 110°C. Quantification by densitometry at 385 nm	Lin et al. (1994b)
Bile acids	Biological matrices	Silica	1) Water saturated chloroform–methanol–acetic acid (170:40:17) 2) Isooctane–ethylacetate–butanol–acetic acid (30:10:3:3) 3) Diethyl ether–acetic acid–water 100:1 4) Chloroform–isopropanol–isobutanol–acetic acid–water (30:20:10:2:1)	Spraying with a solution of 0.4 g $MnCl_2 4H_2O$ and 4 ml H_2SO4 conc. and heating at 120°C for 2 min	Dax and Mullner (1998)

cannabinoids, and stimulants, and therapeutic monitoring of drug levels to ensure drug dosage. The former is usually confined to the emergency investigation of the unconscious or semiconscious patient and drug screening is increasingly being sought by drug dependence treatment centers. Detection of drugs of abuse in biological fluids provides concurrent and independent validation of drug abuse. These tests supplement and confirm clinical and biomedical history of drug intake, monitor treatment efficacy with regard to illicit drug use, contribute to the clarification of the medical–legal problems and promote pharmacokinetic and metabolic research. Therapeutic drug monitoring has also become a routine part of many biomedical laboratories. Plasma drug concentrations at any given point in time are used to adjust and optimize dosage on an individual basis.

Several kinds of techniques are employed for drug analysis. The optimal choice of technique depends on many factors, which include cost, workload, and program needs. Analysis of drugs by TLC or HPTLC saves time and reduces cost, when compared to HPLC analysis, particularly during handling of a large number of samples. Multiple numbers of samples can be analyzed in a single run allows one to save time and thus cost of analysis. A number of body fluids and tissues can be used for drug abuse analysis. Urine is a simpler matrix than other biological specimens because of the usual absence of protein and cellular constituents, which simplifies its sample preparation and analysis. As for the analysis of other compounds, sample pretreatment is necessary prior to TLC. The sample pretreatment step requires pH-dependent extraction followed by purification and concentration of the drug from the biological sample. In some laboratories, commercially available prepacked columns are used for the purification step in drug analysis. HPTLC and TLC plates coated with silica gel are used for detection of drug. For toxicological drug screening, silica gel G with a fluorescent indicator is commonly used for separating abused drugs. A large number of mobile phases have been tested for development of TLC plates. The most commonly described solvent system is ethyl acetate–methanol–ammonium hydroxide (85:10:5, v/v). Chloroform–acetone (9:1, v/v), in another common mixture, is used especially for separation of barbiturates and other acidic drugs. A new TLC method for detecting μg amounts of cocaethylene and cocaine involved SPE of urine buffered to pH 9.3, silica gel TLC with hexane–toluene–diethylamine (13:4:1) mobile phase, and iodoplatinate spray reagent (Bailey 1994). Silica gel TLC with acetone–ethanol–ammonia mobile phase and fluorescamine detection reagent were used to detect hygromycin B in bovine plasma and swine urine (Medina et al. 1993). Asprindine and nadoxolol were detected in human plasma, respectively, by C-18 SPE and ascending or horizontal sandwich chamber silica gel TLC with diethyl ether–methanol (7:3) and ethyl acetate-concentrated ammonia (8:2) mobile phases and detection with kiefer reagent (Misztal and Gumieniczek 1994). Benzodiazepines and zopiclone were detected in serum at microgram level by HPTLC (Otsubo et al. 1995). Ivermectin was determined in cattle serum at 0.1 ng by TLC after derivatization with trifluoroacetic anhydride-1-methylimidazole and visual examination of fluorescence under long-wavelength UV light (Taylor et al. 1994). Quantification of oxyphenbutazone and ibuprofen in the presence of paracetamol and dextropropoxyphene in pharamaceuticals involved separation by TLC, scraping, and extraction of the compounds with methanol, and UV spectrophotometry (Parimoo et al. 1994).

The following compounds were quantified by TLC or HPTLC by the use of absorption or fluorescence scanning densitometry: Cannabinoids in plasma by C-18 SPE derivatization with dansyl chloride, and scanning at 340 nm (Alemany et al. 1993); halofantrine in biological samples by C-18 SPE, hexane–ethanol–triethylamine (92.5:2.5:5) mobile phase, and scanning at 256 nm (Cenni and Betschart 1994); escin in horse chestnut extracts (HPLC and densitometric TLC gave accurate and reproducible results in excellent agreement) (Kockar et al. 1994); iovastain lactane and free hydroxyl acid from fermentation broths with dicholoromethane–acetic acid (85:15) mobile phase and scanning at 233 nm (Konfino et al. 1993); prednisolone in preparations and aqueous humor from rabbit eyes by stepwise-gradient TLC. Cyclophosphamide and its metabolites were determined in plasma and urine of pediatric patients (Tasso et al. 1992), drugs in equine urine samples (TLC was compared favorably to HPLC) (Taylor and Westwood 1993). A breakdown of some interesting examples of TLC applications for drug analysis is presented in Table 10.6.

Vitamins

Vitamins are natural substances found in plants and animals and known as essential nutrients for human beings. The name vitamin is obtained from "vital amines" as it was originally thought that these substances were all amines. Human body uses these substances to stay healthy and support its many functions. There are two types of vitamins: water-soluble and fat-soluble. In human, there are 13 vitamins: 4 fat-soluble (A, D, E, and K) and 9 water-soluble (8 B vitamins, and Vitamin C). Each vitamin has multiple functions. Vitamins do not provide energy (calories) directly, but they do help regulate energy producing processes. With the exception of vitamin D and K, vitamins cannot be synthesized by the human body and must be obtained from the diet. Vitamins have to come from food because they are not manufactured or formed by the body. Deficiencies of vitamins are classified as either primary or secondary. A primary deficiency occurs when an organism does not get enough of the vitamin in its food. A secondary deficiency may be due to an underlying disorder that prevents or limits the absorption or use of the vitamin, due to "life style factor" such as smoking, excessive alcohol consumption, or the use of medications that interfere with the absorption or use of the vitamin. Well-known human vitamin deficiencies involve thiamine (beriberi), niacin (pellagra), vitamin C (Scurvy), and vitamin D (rickets). In much of the developed world, such deficiencies are rare, this is due to fortification. TLC has great advantages (simplicity, flexibility, speed, and relative inexpensiveness) for the separation and analysis of hydrophilic vitamins. The quantification of the separated vitamins can be performed by using densitometry. Because amounts of most hydrophilic vitamins are low in food and natural products, bioautography or derivatization is employed

Table 10.6 Examples of TLC application in drugs analyses

Analyte	Matrix	Chromatographic system		Visualization and quantification	Ref.
		Stationary phase	Mobile phase (v/v)		
19 Drugs	Urine	Silica	Toluene–isopropanol–NH_3 (3:6:1)	Trichlorobenzoquinoneimine, Dragendorff reagent, diphenylcarbazonel mercury II, UV 366 nm	Clerc (1986)
Flunitrazepam (Rohypnol R)	Urine	Silica	Chloroform–acetone–NH_3 (60:30:5)	Bratton–Marshall reagent	(Edler and Schluter (1994)
Antipyrin and paracetamol	Plasma	Silica	Chloroform–methanol (9:1)	Quantification by densitometry	Chatterjee and Singh (1996)
82 Basic drugs, and metabolities	Autopsy urine	HPTLC silica	Trichloroethylene-methyl ethyl ketone-n-butanol-acetic acid-water (17:8:25:6:4), and butyl acetate-ethanol-tripropylamine-water (340:37:20:3)	Utilizing automated identification by hRF/UV. Quantification by densitometry	Ojanpera et al. (1999)
Puerarin	Yuquan pills	Silica gel	Ethyl acetate–acetone–water–toluene (65:45:12:7)	Detection under UV 365 nm. Quantification by densitometry at 270 nm	Zhang and Lin (2000)
Tanshinone IIA	Jinbuhuan pills	Silica gel	Benzene–ethyl acetate (19:1)	Detection under daylight Quantification by fluoro densitometry at 465 nm	Che (2001)
Ecdultin	Huanchun buli pills	Silica gel	Benzene–ethylacetate (30:1)	Detection under UV 365 nm. Quantification by densitometry at 325 nm	Feng and Li (2001)
Taurine	Human blood	Silica gel	Butanol–acetic acid–ethanol–water (4:2:1:1)	Detection by spraying with 0.5% ninhydrin in ethanol and heating at 110°C for 8 min. Quantification by densitometry at 513 nm	Guo (2001)

Ephedrine	Xichuan pills	Silica gel	Chloroform–methanol–NH_3 (100:8:1)	Detection by spraying with 5% ninhydrin in ethanol and heating at 105°C. Quantification by densitometry at 520 nm	He and Wang (2001)
Carbamazepine, phenytoin, and phenobarbitone	Serum	Silica gel HPTLC	Toluene–acetone (5:2)	Quantification by absorbance measurement at 217 nm	Patil and Bodhankar (2005)

before densitometry. Each spot of various vitamins and related compounds separated by TLC is removed from the plates, reextracted, and assayed or further purified by HPLC. Various high-qualities precoated silica gel, cellulose, or various RP-plates are available for TLC. Various dyes were evaluated as detection reagents for vitamins D, A, and E after separation by TLC (Wardas and Pyka 1995). Separation of tocophenol isomers and enantiomers and the relationship between R_F values and topological indexes were studied (Sliwiok et al. 1993). TLC, HPTLC, and HPLC methods were developed for identification of retinoic acids, retinol, and retinyl acetate in topical facial creams and solutions (Bargagna et al. 1991). Vitamin B complex and folic acid were separated on silica gel layers impregnated with transition metal ions and detected by exposure of the layer to iodine vapor (Bhushan and Parshad 1994). Vitamin K was assayed in bovine liver by use of RP-TLC separation with dichloromethane, methanol (7:3) mobile phase for separation, densitometric quantification, and MS confirmation (Madden and Stahr 1993). Nanogram levels of thiamine, riboflavin, and niacin were quantified by fiber-optic fluorodensitometry after silica gel HPTLC with methanol–water (7:3) mobile phase (Diaz et al. 1993). Some of the applications of TLC for vitamin analysis are presented in Table 10.7.

Table 10.7 Examples of TLC application for vitamin analyses

Chromatographic system		Fundamental goal of the experiment	Ref.
Stationary phase	Mobile phase (v/v)		
Silica gel HPTLC	Chloroform–cyclohexane (55:45)	Determination of retinol and α-tocopherol in human plasma	Chavan and Khatri (1992)
Silica mixed with GDX	Hexadecyltrimethylammonium bromide	Study of the use of surfactants in the separation of acid, alkaline, and water soluble food dyestuffs and water soluble vitamins B_2, B_6, B_{12}	Yin et al. (1994)
Silica gel and chemical bonded silica gel	14 mobile phases, three of which giving best separation: (1) Butanol–chloroform–acetic acid–ammonia–water (7:4:5:1:1) (2) Benzene–methanol–acetone–acetic acid (14:4:1:1) (3) Chloroform–ethanol–acetone–ammonia (2:2:1:1)	Analysis of hydrophilic vitamins e.g., thiamine (B_1), riboflavin (B_2), niacin (B_3), pyridoxine (B_6), carbalamine (B_{12}) ascorbic acid (c) and folic acid in standards and Helisoma trivolvis snails	Ponder et al. (2004)
Silica gel	Mixtures of methanol and benzene	Identification and separation of eight hydrophilic vitamins (B_1, B_2, B_3, B_5, B_6, B_9, B_{12}, and C)	Cimpoiu et al. (2005)
Silica gel	(Cyclohexane–ether 1:1, 17:3, and ethyl acetate)	Use of TLC to investigate possible chemical interactions of vitamins A and D with drugs	Ramic and Marica (2006)

Porphyrins

The analysis of porphyrins and porphyrin-related compounds is of great biomedical significance. Analyses of porphyrins have therefore a great diagnostic value (Table 10.8). Porphyrias are diseases characterized by excessive production and excretion of porphyrins because of defects associated with the enzymes of the heme biosynthetic pathways. These disorders may be inherited, such as those occurring naturally within the red blood cell and the liver, or they may be induced due to exposure to certain toxicants, such as lead or polychlorinated biphenyls. Porphyrins and metalloporphyrins are being actively investigated by chemists and biomedical researchers world wide for the purposes of imagining and photodestruction of concerns tumors.

Table 10.8 Separation of porphyrins and porphyrin-related compounds

Chromatographic system	Visualization and quantification	Objective of the investigation	Ref.
Octadecyl-bonded silica Buffered mobile phase containing *N*-acetyl-*N*,*N*,*N*,-1 trimethylammonium bromide	Detection under UV 366 nm. Quantification by fluorodensitometry at 404 nm	Ion-pair chromatography and quantitative analysis of urinary porphyrins	Junker-buchhiet and Jork (1988)
Silica gel HPTLC Toluene–ethylacetate–methanol (75:16:2, v/v/v)	Detection by laser fluorometry	Separation and detection of porphyrin methyl esters present in biological materials	Huie and Williams (1989)
Diol modified silica gel Dichloromethane–ethylacetate (97:3, v/v)	Quantitative by densitometry at 420 nm	Identification and quantitative estimation of porphyrins	Habdas and Matysik (2003)
Silica gel *n*-Hexane–chloroform (7:3, v/v) Rp-18 Methanol or methanol–water (9:1, v/v)	Visual detection	Separation of porphyrin, selected porphyrins, and thin metalloderivatives	Stefaniak et al. (2004)
Silica gel, polyamide, and cellulose methanol, ethanol, propanol, butanol, acetonitrile, and tetrahydrofuran	Iodine vapor; UV light at 254 nm	Influence of different adsorbents on the retention behavior of Uroporphyrin I, Uroporphyrin III, Coporphyrin I, coproporphyrin III, and protoporphyrin IX (as methyl esters)	Nowakowska (2006)
Rp-18 Binary mixtures (Carbon tetrachloride–methanol, chloroform–methanol, dichloromethane–methanol, and chlorobenzene–methanol)	Visual detection	Investigation of the effect of two component mobile phases on the separation of porphyrins	Podgorna (2007)

In the 1970s, various chromatographic methods have been employed extensively for the separation of various porphyrins. TLC is the most widely used technique for the routine analysis of porphyrins. Individual prophyrin esters have been successfully separated by TLC (Podgórna et al. 2002; Petryka and Watson 1979). Porphyrins were extracted from urine or feces and separated on silica gel with methanol, chloroform, methylene chloride, carbon tetrachloride, ethyl acetate, benzene, and toluene mixtures. The separated porphyrins were observed by viewing the plate under long-wave fluorescent light. Luo and Lim (1995) have studied porphyrin metabolisms in human porphyna cutanea tarda (PCT) and in rats treated with hexachlorobenzene (HCB) by using HPTLC, HPLC, and LSIMS. The analyses of porphyrin metabolites in the urine, feces, and liver biopsies of patients with PCT (porphyria cutanea tarda) have shown that apart from uroporphyrin I and III and their expected decarboxylation intermediates and products, a complex mixture of many other porphyrins are present. The new porphyrins (meso-hydroxyuroporphyrin III, β-hydroxypropionic acid uroporphyrin III, hydroxyacetic acid uroporphyrin III, peroxyacetic acid uroporphyrin III, β-hydroxypropionic acid heptacarboxylic porphyrin III, hydroxyacetic acid hepatocarbomylic porphyrin III, and peroxyacetic acid pentacarboxylic porphyrin III were identified.

Notes

The future trends in biomedical analysis by TLC/HPTLC will be about the use of precoated TLC plates rather than the use of handmade plates. There is a need of searching out ecofriendly green solvents for their use as mobile phase by replacing volatile organic solvent systems currently in use. For accurate quantification, the emphasis will be on the use of densitometry and coupled techniques. Currently, the need for development of highly sensitive new detection reagents is strongly felt.

References

Aboul-Enein JY and Serignese V. Optical purity determination of thyroxins enantiomers in bulk materials by chiral thin layer chromatography. Biomed Chromatogr, 1994, 8, 317–318.

Alemany G, Gamundi A, Nicolau MC and Saro D. A simple method for plasma canabinoid separation and quantification. Biomed Chromatogr, 1993, 7, 273–274.

Anderson DJ and Van Lente F. A review on clinical chemistry. Anal Chem, 1995, 67, 377R–524R.

Ansari KA and Shoeman D. Rapid separation of free fatty acids from neutral lipids in brain extracts. J Chromatogr, 1988, 439, 453–458.

Armstrong DW and Zhou Y. Use of a macrocyclic antibody as the chiral selector for enantiomeric separations by TLC. J Liq Chromatogr, 1994, 17, 1695–1707.

Arnsmeier SL and Paller AS. Chemiluminescence detection of gangliosides by thin-layer chromatography. J Lipid Res, 1995, 36, 911–915.
Bailey DN. Thin-layer chromatographic detection of cocaethylene in human urine. Am J Clin Pathol, 1994, 101, 342–345.
Baranowska I, Markowski P, Wilczek A, Szostek M and Stadniczuk M. Normal and reversed-phase thin-layer chromatography in the analysis of L-arginine, its metabolites, and selected drugs. J Planar Chromatogr, 2009, 22, 89–96.
Bargagna A, Mariani E and Dorato S. TLC, HPTLC and HPLC determination of cis-and trans-retinoic acids, retinol and retinyl acetate in topically applied products. Acta Technol Legis Med, 1991, 2, 75–86.
Batisse C, Dauraade MH and Buonias M. TLC separation and spectrodensitometric quantitation of cell wall neutral sugars. J Planar Chromatogr, 1992, 5, 131–133.
Bhushan R and Parshad V. Separation of vitamin B complex and folic acid using TLC plates impregnated with some transition metal ions. Biomed Chromatogr, 1994, 8, 196–198.
Bhushan R, Mahesh VK and Verma A. Improved thin layer chromatographic resolution of PTH amino acids with some new solvent systems. Biomed Chromatogr, 1994, 8, 69–72.
Bhushan R, Reddy GP and Joshi S. TLC resolution of DL amino acids on impregnated silica gel plates. J Planar Chromatogr Mod TLC, 1994b, 7, 126–128.
Bladek J and Neffe S. Application of thin-layer chromatography in clinical chemistry. Sep Purif Rev, 2003, 32, 61–120.
Bladek J and Zdrojewski A (2000). Clinical Chemistry: Thin-Layer (Planar) Chromatography. In: Wilson, I.D., Cooke C, Poole C (eds). Encyclopedia of Separation Science, published by Academic Press, London.
Boscher-Reig F, Marcote MJ and Minana MD. Separation and identification of sugars and maltodextrins by thin-layer chromatography. Application of biological fluids and human milk. Talanta, 1992, 39, 1493–1498.
Bruun-Jensen L, Colarow L and Skibsted LH. Detection and quantification of phospholipids hydroperoxides in turkey meat extracts by planar chromatography. J Planar Chromatogr, 1995, 8, 475–479.
Cartwright IJ. Separation and analysis of phospholipids by thin layer chromatography. Methods Mol Biol (Totowa NJ), 1993, 19, 153–167.
Cenni B and Betschart B. The determination of halofantrine in biological samples by high performance thin-layer chromatography. J Planar Chromatogr-Mod TLC 1994; 7, 294-296.
Chatterjee S and Singh BP. A high performance thin-layer chromatographic (HPTLC) estimation of antipyrine and paracetamol from plasma. Indian Drugs, 1996, 33, 355–357.
Chavan JD and Khatri JM. Simultaneous determination of retinol and λ-tocopherol in human plasma by HPTLC. J Planar Chromatogr, 1992, 5, 280–282.
Che X. Determination of tanshinone IIA Jinbuhuan pills by thin-layer chromatography. J Chinese Trad Herb Drugs (Zhongcaoyao), 2001, 32, 802–803.
Christelle C, Vingler P, Boycra N, Galey I and Bernard BA. Direct quantitative digital autoradiography TLC coupling for the analysis of neutral ^{14}C-lipids neosynthesized by the human sebaceous gland. J Planar Chromatogr, 1997, 10, 243–250.
Cimpoiu C and Hosu A. Thin-layer chromatography for the analysis of vitamins and their derivatives. J Liq Chromatogr Rel Technol, 2007, 30, 701–728.
Cimpoiu, C, Casoni D, Hosu A, Miclaus, V, Hodisan T and Damian G. Separation and identification of eight hydrophilic vitamins using a new TLC method and Raman Spectroscopy. J Liq Relat Technol, 2005, 28, 2551–2559.
Clerc JT. Simple qualitative determination of pharmaceutical active ingredients in urine. Swiss Pharma, 1986, 8, 6–17.
Colarow L. Quantitation of ceramides with essential fatty acid moieties in the human skin surface and blood plasma lipids. J Planar Chromatogr, 1990, 3, 126–132.
Conaway CA, Fried B and Sharma J. High performance thin-layer chromatographic analysis of sugars in *Helisoma trivolvis* (Pennsylvania strain) infected with larval *Echinostoma trivolvis*

and in uninfected *H. trivolvis* (Pennsylvania and Colorado strains). J Planar Chromatogr-Mod TLC, 1995, 8, 184–187.

Dax CI and Mullner S. Convenient and optimized method for sample pretreatment for the analysis of bile acids in biological matrices. Chromatographia, 1998, 48, 681–689.

Demeglio DC and Svanberg GK. A thin layer chromatographic procedure for the separation of proline and hydroxyproline from biological samples. J Liq Chrom Rel Technol, 1996, 19, 969–975.

Deng Zh, Wang F, Zheng L and Luo M. Determination of conjugated bile acids in gallstone by thin-layer chromatography. Chin J Anal Chem, 1986, 14, 382–384.

Diaz AN, Paniagua AG and Sanchez FG. Thin-layer chromatography and fibre-optic fluorimetric quantitation of thiamine, riboflavin and niacin. J Chromatogr, 1993, 655, 39–43.

Dworzak E and Hauk H. Quantitative determination of metabolism of catecholamines and serotonin through fluorescence quenching on thin-layer chromatography. J Chromatogr, 1971, 61, 162–169.

Edler M and Schluter R. Detection and identification of flunitrazepam (Rohypnol R) in urine by TLC. T + K, 1994, 61, 74–79.

Eidam PM, Schariter JJ, Fried B, Sherma J. HPTLC analysis of tissue and blood of Biomphalaria glabrata snails to assess the effects of a diet of romaine lettuce leaf versus midrib on the concentrations of lipids, pigments, and carbohydrates. J Liq Chrom Rel Technol, 2001, 24, 1467–1478.

Fei Z, Zhan Q and Mei M. Determination of lipids in blood by thin-layer chromatography. J Adv Biochem Biophys, 1988, 15, 305–307.

Feng Sh and Li H. Determination of ecdultin in Huanchun Buli pills by thin-layer chromatography. J Chin Trad Herb Drugs (Zhongcaoyao), 2001, 32, 803–804.

Fried B and Sherma J (1996) Practical Thin-Layer Chromatography: A Multidisciplinary Approach, published by CRC Press Inc., Boca Raton, FL.

Fried B and Sherma J (1999). Thin-Layer Chromatography: Techniques and Applications. 4th edition, published by Marcel Dekker Inc., New York.

Friedman MJ. Applications of the ninhydrin reaction for analysis of amino acids, peptides, and proteins to agricultural and biomedical sciences. Agric Food Chem, 2004, 52, 385–406.

Guo L. Determination of taurine in human blood by thin-layer chromatography. J Chin Trad Herb Drugs, (Zhongcaoyao), 2001, 32, 217–218.

Habdas J and Matysik G. Determination of the yields of meso-tetraphenylporphyrins by densitometry. J Planar Chromatogr, 2003, 16, 289–292.

Hao AY, Tong LH, Zhang FS, Gao XM and Inoue Y. Direct thin-layer chromatographic separation of enantiomers of six selected amino acids using 2-O-[(R)-2-hydroxypropyl]-β-CD as d mobile phase additive. Anal Lett, 1995, 28, 2041–2048.

Hara A and Taketomi T. Isolation and determination of cholesterol glucuronide in human liver. Lipids, 1982, 17, 515–518.

He Y and Wang Y. Determination of ephedrine in Xichuan pills by thin-layer chromatography. J Chin Trad Herb Drugs (Zhongcaoya), 2001, 32, 324–325.

Higashi H, Hirabayashi Y, Ito M, Yamagata T, Matsumoto M, Uedas and Kato S. Immunostaining on thin-layer chromatograms of oligosaccharides released from gangliosides by endoglycoceramidose. J Biochem (Tokyo), 1987, 102, 191–196.

Huie CW and William WR. Laser fluorometric detection of porphyrin methyl esters for high-performance thin-layer chromatography. Anal Chem, 1989, 61, 2288–2292.

Jork H and Gang J. Opportunities and limitations of modern TLC/HPTLC in the quality control of L-tryptophan. L-Tryptophan. Curr Prospects Med Drug Saf, 1994, 338–350.

Junker-buchhiet and Jork H. Urinary prophyrins: Ion-pair chromatography and fluorimetric determination. J Planar Chromatogr, 1988, 1, 214–219.

Kartsova LA and Koroleva OA. Simultaneous determination of water-and fat-soluble vitamins by high-performance thin-layer chromatography using an aqueous micellar mobile phase. J Anal Chem, 2007, 62, 255–259.

Kindel M, Ludwig-Koehn H and Lembeke M. New and versatile method for determination of fecal bile acids by thin-layer chromatography and *in-situ* spectrofluorimetry. Anal Biochem, 1989, 134, 135–138.

Klaus R. Identification, qualitative and quantitative analysis of uric acid, creatine and creatinine together with carbohydrates in biological material: HPTLC-NH_2. Planar chroamtogr, 1991, 157–172.

Klaus R, Fischer W and Hauck HE. Qualitative and quantitative analysis of uric acid, creatine and creatinine together with carbohydrates in biological material by HPTLC. Chromatographia, 1991, 32, 307–316.

Kockar OM, Kara M, Kara S, Bozan B and Baser KHC. Quantitative determination of escin. A comparative study of HPLC and TLC-densitometry. Fitoterapia, 1994, 65, 439–443.

Konfino M, Deltcheva S and Mindjova K. The isolation of iovastatin and its determination by densitometric TLC and by HPLC. J Planar Chromatogr -Mod TLC, 1993, 6, 404–406.

Kovacs L, Martos E, Pick F and Pucsok J. Cholesterol ester mapping of human serum by HPTLC. J Planar Chromatogr, 1989, 2, 155–157.

Krahn J. Variables affecting resolution of lung phospholipids in one dimensional thin-layer chromatography. Clin Chem, 1987, 33, 135–137.

LeFevre JW. Reversed-phase thin-layer chromatographic separations of enantiomers of dansyl-amino acids using p-cyclodextrin as a mobile phase additive. J Chromatogr, 1993, 653, 293–302.

Li L, Shen G, Jiang J, Cheng X, Zhan X, Zhang CH and Zhao G. Thin-layer chromatographic and fluorometric determination of the ratio of serum free branched chain amino acids to aromatic amino acids. Chin J Med, 1985, 68, 298–301.

Li J, Li X, Su Zh and Shi J. Determination of bile acids in gastric juice by thin-layer chromatography. Chin Anal Chem, 1993, 21, 244.

Lin L, Zhang J, Huang S, He C and Tang J. Identification of fetal lung maturity by one-dimensional thin layer chromatographic determination of several phospholipids in amniotic fluid. J Planar Chromatogr -Mod TLC, 1994a, 7, 25–28.

Lin J, Yan D and Lu J. Determination of bile acids in commercial bile salts by thin-layer chromatography. Chin J Pharma Anal, 1994b, 14, 27–39.

Lodi G, Betti A, Brandolini V, Menziani E and Tosi B. Automated multiple development HPTLC Analysis of sugars on hydrophilic layers: I. Amino Layers. J Planar Chromatogr -Mod TLC, 1994, 7, 29–33.

Luo J and Lim CK. Isolation and characterization of new porphyrin metabolites in human porphyria cutanea tarda and in rats treated with hexachlorobenzene by HPTLC and liquid secondary ion mass spectrometry. Biomed Chromatogr, 1995, 9, 113–122.

Madden UA and Stahr HM. Reverse phase thin-layer chromatography assay of vitamin K in bovine liver. J Liq Chromatogr, 1993, 16, 2825–2834.

Martinez Lorenzo MJ, Marzo I, Naval J and Piniero A. Self-staining of polyunsaturated fatty acids in argentation thin-layer chromatography. Anal Biochem, 1994, 220, 210–212.

Mathur V, Kanoongo N, Mathur R, Narang CK and Mathur NK. Resolution of amino acid racemates on borate-gelled guaran-impregnated silica gel thin layer chromatographic plates. J Chromatogr A, 1994, 685, 360–364.

Medina MB, Unruh JJ, Jones E and Bueso CE. Residues Vet Drugs Food, Proc. Euro Residue Conf., 2nd 1993, 2, 490–494.

Misztal G and Gumieniczek A. Chromatographic analysis (TLC) aprindine and nadoxolol in human plasma. Acta Pol Pharm, 1994, 51, 215–218.

Momose T, Mure M, Lida T, Goto J and Nambara T. Method for the separation of unconjugates and conjugates of chenodeoxycholic acids and deoxycholic acids by two-dimensional reversed-phase thin-layer chromatography with methyl β-cyclodextrin. J Chromatogr, 1998, 811, 171–180.

Morcol T and Velander WH. An O-toluidine method for detection of carbohydrates in protein hydrolyzates. Anal Biochem, 1991, 195, 153–159.

Muething J. Improved thin-layer chromatographic separation of gangliosides by automated multiple developments. J Chromatogr B: Biomed Appl, 1994, 657, 75–81.

Muthing J and Heltmann D. Nondestructive detection of ganglioxides with lipophilic fluorochromes and their employment for preparative high performance thin-layer chromatography. Anal Biochem, 1993, 208, 121–124.

Nikolova-Damyanova B, Chobanov D and Dimov S. Separation of isomeric triacylglycerols by silver ion thin-layer chromatography. J Liq Chromatogr, 1993, 16, 3997–4008.

Nikolova-Damyanova B, Christie WW and Herslof B. Improved separation of some positional isomers of monounsaturated fatty acids, as their phenacyl derivatives, by silver-ion thin layer chromatography. J Planar Chromatogr-Mod TLC, 1994, 7, 382–385.

Norfolk E, Khan SH, Fried B and Sharma J. Comparison of amino acid separations on high performance silica gel, cellulose, and C-18 reversed phase layers and application of HPTLC to the determination of amino acids in *Biomphalaria glabrata* snails. J Liq Chromatogr, 1994, 17, 1317–1326.

Nowakowska J. The retention behavior of selected porphyrins on silica gel, polyamide, and cellulose TLC plates. J Planar Chromatogr, 2006, 19, 393–397.

Ogawa-Goto K, Ohta Y, Kubota K, Funamoto N, Abe T, Taki T and Nagashima K. Glycosphingolipids of human peripheral nervous system myelins isolated from cauda equina. J Neurochem, 1993, 61, 1398–1403.

Ohtake H, Hase Y, Sakemoto K, Oura T, Waday and Kodama H. A new method to detect homocysteine in dry blood spots using thin layer chromatography. Screening, 1995, 4, 17–26.

Ojanpera I, Ojansivu RL, Nokua J and Vuori E. Comprehensive TLC in 661089.8 Toxicology: Comparison of findings in urine and liver. J Planar Chromatogr, 1999, 12, 38–41.

Ojanpera K, Goebel E and Vuori E. Toxicological drug screening by overpressured layer chromatography. J Liq Chrom Rel Technol, 1999, 22, 161–171.

Otsubo K, Seto H, Futagami K and Oishi R. Rapid and sensitive detection of benzodiazepines and zopiclone in serum using high-performance thin-layer chromatography. J Chromatogr B: Biomed Appl, 1995, 669, 408–412.

Parimoo P, Bharathi A and Shajahan M. Indian Drugs, 1994, 31, 139–143.

Patil KM and Bodhankar SL. High performance thin-layer chromatography method for monitoring of anti-epileptic drugs in serum. Indian Drugs, 2005, 42, 665–670.

Perez M K, Fried B and Sharma J. High performance thin-layer chromatographic analysis of sugars in *Biomphalaria glabrata* (gastropoda) infected with *Echinostoma caproni* (trematoda). J Parasitol, 1994, 80, 336–338.

Petryka ZJ and Watson CJ. Separation and spectrodensitometric quantitation of porphyrin esters on thin-layer chromatograms. J Chromatogr, 1979, 179, 143–149.

Podgorna M. Effects of the composition of two-component mobile phases on the separation of selected prophyrins by partition thin-layer chromatography. J Planar Chromatogr, 2007, 20, 259–260.

Podgórna M, Świerk A and Śliwiok J. The influence of steric effects on the TLC separation of porphyrins. J Planar Chromatogr, 2002, 15, 466–469.

Ponder EI, Fried B and Sherma J. Thin-layer chromatographic analysis of hydrophilic vitamins in standards and *Helisoma trivolvis* snails. Acta chromatogaphica, 2004, 14, 70–80.

Preet S and Gupta S. Thin layer chromatographic analysis of free amino acids in the haemolymph and DGG of *Indoplanorbis exustus* (Mollusca: Gastropoda) naturally infected with digenetic trematodes. R J Parasitol, 2009, 4, 87–93.

Pyka A. A new optical topological index (Iopt) for predicting the separation of D and L optical isomers by TLC Part III. J Planar Chromatogr Mod TLC, 1993, 6, 282–288.

Raith K, Zellmer S, Lasch J and Neubert RHH. Profiling of human stratum corneum ceramides by liquid chromatography electrospray mass spectrometry. Anal Chem Acta, 2000, 418, 167–173.

Ramic A, Marica MS, Turina and Jasprica I. TLC detection of chemical interactions of vitamins A and D with drugs. J Planar Chromatogr, 2006, 19, 27–31.

Ren J, Zhao T and Nan G. Determination of some amino acids in cerebrum of rat by thin-layer chromatography. Chin J Adv Biochem Biophys, 1987, 2, 37–40. (Chinese).

Rezic I, Rezic, T and Bokic L. Optimization of the TLC separation of seven amino acids. J Planar Chromatogr-Mod TLC, 2007, 20, 165–171.

Rivas-Nass A and Mullner S. The influence of critical parameters on the TLC separation of bile acids. J Planar Chromatogr -Mod TLC, 1994, 7, 278–285.

Romano G, Ceruso G, Musumarra G, Pavone D and Cruciani G. Qualitative organic analysis. Part 3. Identification of drugs and their metabolites by PCA of standardized TLC data. J Planar Chromatogr Mod TLC, 1994, 7, 233–239.

Rudloff S, Pohlentz G, Diekmann L, Egge H and Kunzc. Urinary excretion of lactose and oligosaccharides in pre-term infants fed human milk or infant formula. Acta Pediatr, 1996, 85, 598–603.

Sambaiah K, Ganesh B and Chandroaekhara N. Elimination of phospholipid interference in biliary bile acid determination by thin layer chromatography-densitometry. J Chromatogr, 1986, 380, 235–237.

Schnaar RL and Needham LK. Thin-layer chromatography of glycosphingolipids. Methods Enzymol, 1994, 230, 371–389.

Schuerer NY, Schliep V and Barlag K. Quantitative screening of human stratum corneum lipids by sequential one-dimensional thin layer chromatography. Exp Dermatol, 1995, 4, 46–51.

Serizawa S, Nagai T and Sato Y. Simplified method of determination of serum cholesterol sulfate by reverse phase thin-layer chromatography. J Invest Dermatol, 1987, 89, 580–587.

Sherma J. A review on planar chromatography. Anal Chem, 1996, 68, 1R–19R.

Sherma J and Bennett S. Comparison of reagents for lipid and phospholipid detection and densitometric quantification on silica gel and C18-reversed phase thin-layers. J Liq Chromatogr, 1963, 6, 1193–1211.

Sliwiok J, Kocjan B, Labe B, Kozera A and Zalejska J. Chromatographic studies of tocopherols. J Planar Chromatogr -Mod TLC, 1993, 6, 492–494.

Song Zh, Deng Zh and Han J. Determination of bile acids in bear bile by thin-layer chromatography. Chin J Pharma Anal, 1993, 13, 194–196.

Stefaniak M, Kus P and Sliwiok J. TLC and RPTLC Separation of porphine, selected prophyrins, and their metaloderivatives. J Planar Chromatogr, 2004, 17, 72–74.

Szilagyi J, Kovacs-Hadady K and Kovacs A. OPLC study of retention of aromatic N glycosides on layers impregnated with tricaprylmethylammonium chloride. J Planar Chromatogr -Mod TLC, 1993, 6, 212–215.

Taki T, Handa S and Ishikawa D. A simple and quantitative purification of glycosphingolipids and phospholipids by thin-layer chromatography blotting. Anal Biochem, 1994, 223, 232–238.

Tarandjiiska R, Marekov I, Nikolova-Damyanova B and Amidzhin B. Determination of molecular species of triacylglycerols from highly unsaturated plant oils by successive application of silver ion and reversed phase TLC. J Liq Chromatogr, 1995, 18, 859–872.

Tasso MJ, Boddy AV, Price L, Wyllie RA, Pearson AD and Idle JR. Pharmacokinetics and metabolism of cyclophosphamide in paediatric patients. Cancer Chemother Pharmacol, 1992, 30, 207–211.

Taylor MR and Westwood SA. An investigation of drug detection in equine urine samples by modern TLC. J Planar Chromatogr -Mod TLC, 1993, 6, 415–418.

Taylor WG, Danielson TJ and Orcutt RL. Thin-layer chromatographic detection of ivermectin in cattle serum. J Chromatogr B: Biomed Appl, 1994, 661, 327–333.

Wang SH and Ma L. Determination of combined sugar in glycosides by 2-dimensional thin-layer chromatography. J Chin Herb Med, 1989, 20, 155–157.

Wang J, Gao Zh, Li Zh, Wang Z and Liu Sh. Determination of free bile acids in bile by thin-layer chromatography. Shaanxi Provin J Med, 1989, 18, 28–30.

Wardas W and Pyka A. New visualizing agents for fatty vitamins in TLC. Chem Anal (Warsaw), 1995, 40, 67–72.

Watanabe K and Mizuta M. Fluorometric detection of glycosphingolipids on thin-layer chromatographic plates. J Lipid Res, 1995, 36, 1848–1855.

Wiesner DA and Sweeley CC. Characterization of gangliosides by two-dimensional high-performance thin-layer chromatography. Anal Chim Acta, 1995, 311, 57–62.

Woodard DS, Ostrom KK and McManus LM. Lipid inhibitors of platelet-activating factor (PAF) in normal human plasma. J Lipid Mediat Cell Signal, 1995, 12, 11–28.

Wu JT. Screening for inborn errors of amino acid metabolism. Ann Clin Lab Sci, 1991, 21, 123–126.

Yao Z, Liu G, Wen H, He W, Zhao S and Hu W. Determination of the glycosyl sequence of glycosides by enzymatic hydrolysis followed by TLC-FABMS. J Planar Chromatogr -Mod TLC, 1994, 7, 410–412.

Yin P, Li H and Yan CH. Use of surfactants in thin-layer chromatography with plates of silica mixed with GDX. Chin J Chromatogr (Sepu), 1994, 12, 35–36.

Yishen Y and Aihua Z. Determination of refluxed bile acids in human stomach by thin-layer chromatography. Chin J Anal Chem, 1985, 13, 40–42.

Zarzycki PK, Nowakowska J, Chmielewaka A and Lamparczyk H. Retention properties of cyclodextrins in reversed phase HPTLC. J Planar Chromatogr -Mod TLC, 1995, 8, 227–231.

Zhang Z and Lin Q. Determination of puerarin in Yuquan pills by thin-layer chromatography. J Chin Trad Herb Drugs (Zhongcaoyao), 2000, 31, 519–520.

Zhang Q, Li Sh, Cheng J, Yan K and Tian S. Determination of free bile acids in bezoar by thin-layer chromatography. J Chin Trad Med, 1990, 15, 360–362.

Zhu W. Separation and identification of amino acids in protein hydrolysates by thin layer chromatography. J Nanjing Univ (Nat Sci), 1984, 3, 524–526.

Chapter 11
Analytical Aspects of High Performance Thin Layer Chromatography

Gunawan Indrayanto

Abstract High performance thin layer chromatographic method is applied for qualitative and quantitative analysis. Qualitative analysis was used for identification of drugs/chemicals, or certain toxic residuals compounds in food or other preparations, and finger printing of herbal drugs/biological matrixes. Quantitative analysis was used for the determination of drugs/metabolites in pharmaceutical preparations or biological matrixes, impurities, residual toxic compounds, drug stability testing, and quality control purposes. For obtaining reliable analysis results, the (high performance) thin layer chromatographic method should be validated first, before using it as a routine quality control tools. The validation parameters, which should be evaluated, are stability of the analyte, specificity/selectivity, linearity, accuracy, precision, range, detection limit/decision limit, quantification limit/detection capability, and robustness/ruggedness.

Chromatography means a method of analysis in which a mobile phase passes over a stationary phase and the mixture of compounds can be separated into its components. Chromatographic methods have become methods of choice for various applications including pharmaceuticals, biotechnology, natural products, plant analysis, bioanalysis, toxicology, and many more. Some methods of chromatograph are high performance liquid chromatography, gas chromatography, thin layer chromatography (TLC), counter current chromatography, super critical fluid chromatography, ion chromatography, etc. TLC is an old method of analysis that has been used for about 40 years in practice and was first introduced by Egon Stahl in 1956. According to the name "thin layer," the chromatography processes occur on a stationary phase that consists of a thin layer(layer thickness 250 μm; particle size 10–15 μm), which was applied to a solid substrate (e.g., glass, aluminum foil, plastic) as a support material, therefore TLC can be referred to as "planar

G. Indrayanto
Plant Biotechnology Research Group and Assessment Service Unit, Faculty of Pharmacy, Airlangga University, Surabaya 60286, Indonesia

MM. Srivastava (ed.), *High-Performance Thin-Layer Chromatography (HPTLC)*,
DOI 10.1007/978-3-642-14025-9_11, © Springer-Verlag Berlin Heidelberg 2011

chromatography." TLC can be applied for preparative (isolation and purification), qualitative and quantitative analysis. It can be used for all class of compounds, i.e., ionic, strongly polar, medium polarity, and nonpolar. The many advantages of TLC are that the method is relatively fast, cheap, and very simple. Other advantages are that the substances remain on the layer and can be recovered again, and simultaneous detections are also possible. If the particle size of the stationary phase is reduced to about 3–5 μm (layer thickness 100–200 μm), the separation capability of the method will be increased significantly compared to conventional TLC, and it is named as "high performance thin layer chromatography" (HPTLC). By combining HPTLC with computer controlled scanning densitometry, it becomes a modern powerful analytical technique with separation power performance and reproducibility superior to conventional TLC. Before using a qualitative/quantitative method as routine tools, the method should be validated first. Without any validation studies, the reliability of the method cannot be proved.

Qualitative Analysis

The objective of a qualitative method is to confirm the presence or absence of a certain analyte(s) in the samples. If a certain analyte(s) is absent in samples, it means that the concentration of the analyte in the sample is below its detection limit. The presence of analyte(s) in samples indicates that its concentration is above the DL (see section "Validation Method of Analysis").

The simple parameters, which can be used for qualitative analysis by HPTLC/TLC method, are the retardation factor (R_f) value and the color of the spot after reaction with detection reagent. R_f can be defined as the ratio of distance of spot from start to the distance of mobile phase migration from start (Lepri and Cincinelli 2005). R_f can be expressed also with hR_f ($100 \times R_f$). If an internal standard (IS) is used, the parameter of R_f will be replaced with RR_f (ratio of R_f of the analyte's spot with the R_f of IS). The analyte spot can be also visualized using UV 254/366 nm (if the compounds have UV absorbing functional group). It should be always kept in mind that if the R_f values and/or the colors of two spots are different, it means that the two spots are consisted of two different compounds. If the R_f values and colors of two spots are identical, it does not mean that the spots are consisted of same compounds. One single spot does not mean that the spot contains only one compound. By using a scanning densitometry, the HPTLC/TLC chromatogram can be converted into a densitogram, in which all spots will be observed as peaks.

Identity of the spot/peak can be evaluated by measuring its in situ UV/VIS spectrum and compared to the authentic standard reference compound (Figs. 11.1 and 11.2). Identical compounds should have correlation factor (r) ≥ 0.999. By using CATS software (Camag, Muttenz), the UV/VIS spectrum of a sample spot/peak will be compared to UV/VIS spectra of two standard peaks, and the test value $\acute{z}$ will calculated. The sample peak is identical with standard, if the value of $\acute{z}$ was ≥ 2.327 ($p = 0.01$). Better result can be achieved by coupling HPTLC with a

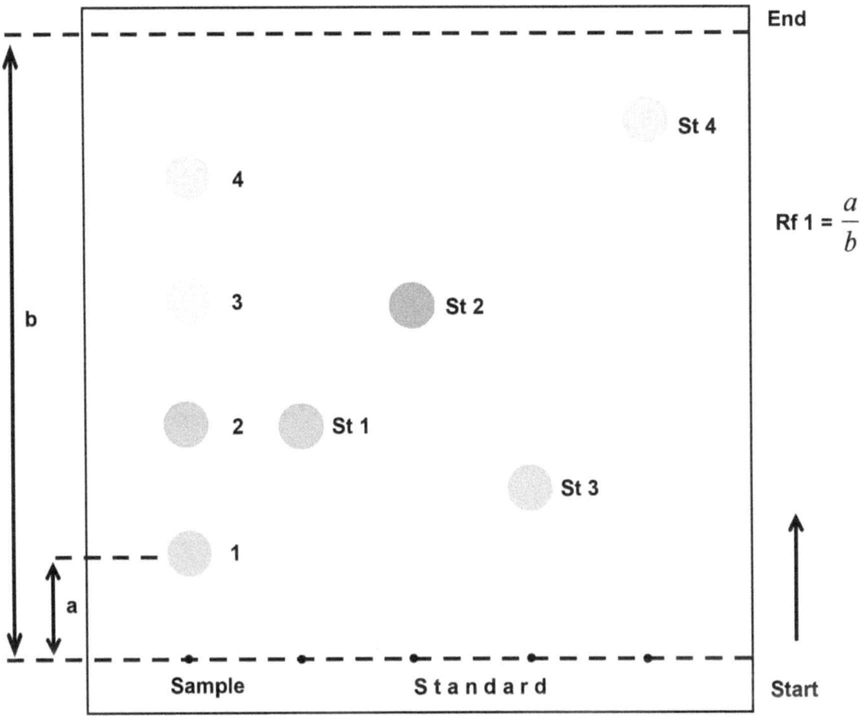

Fig. 11.1 Calculation of R_f-value. Spot 2 has an identical R_f and color with standard 1 (St 1), but it does not mean that spot 1 was always identical with St 1. Spot 3 has an identical R_f – value with St 2, but their colors were not identical, so Spot 3 and St 2 are different compounds. Although Spot 1–4 showed as one single spot, it does not mean the spots contained only one compound

MS detector. A direct MS measurement on the TLC plate can be performed using TLC-MALDI-MS, and a recent review of this topic was published (Crecelius et al. 2004; Fuchs et al. 2009). Recently, Camag produced a semiautomatical equipment to extract the spot directly and connect to a LC-MS system. By using MS detector, the identity of the compound can be determined almost absolutely (Camag Flash 2009). A brief discussion on TLC-MS has been described (Rozylo 2005).

Purity of a spot can be confirmed using two-dimensional TLC using the system mobile phase in the first direction and mobile phase of different selectivity in the second direction (Ferenczi-Fodor et al. 2001). A densitometer can be used easily for proving whether the spot contains one compound or more by measuring its UV/VIS spectrum at the up slope, apex, and down slope of the peak. If the value of the correlation coefficient of up slope to apex (r_{sm}) and apex to down slope (r_{me}) are more than 0.999, it can be concluded that the spot/peak is pure. From the two values of r, the test value z' can be calculated(CATS software, Camag). If the value of $z' \geq 2.576$, the peak is pure ($p = 0.01$).

Confirmation of the identity of an analyte's spot can be also determined by two-dimensional TLC with full scan UV/VIS and cochromatography. The analyte's spot

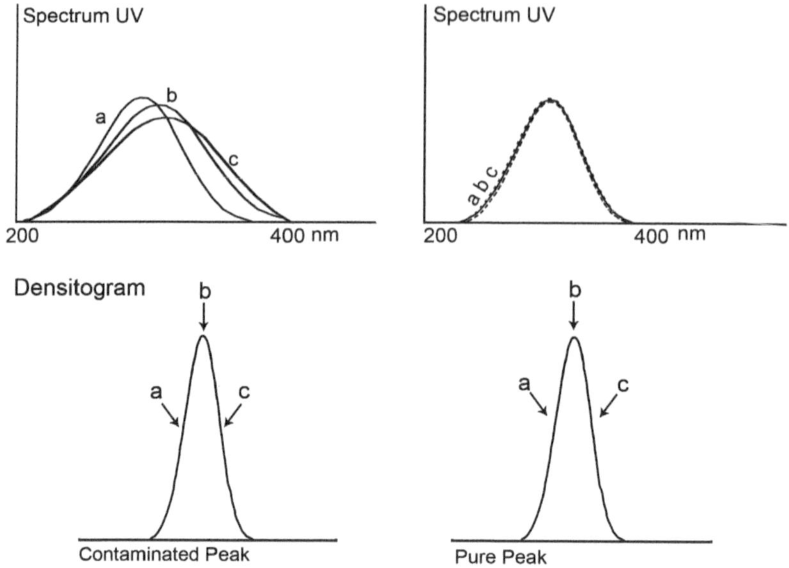

Fig. 11.2 Peak purity determination by using a scanning densitometer. The UV spectra of up-slope, apex, and down-slope of the peak of the densitogram were determined. Pure peak showed identical spectra

should be intensified by doing a cochromatography, and new spot should not appear and the visual appearance shall not change; R_f value of the analyte shall agree with the reference standard within ±5%. This method can be used only for group B substances residual analysis (European Commission 2002). HPTLC method can be used also for qualitative analysis of herbal drugs or medicinal plants. In this case, a Botanical Reference Material (BRM) is required, instead of a reference standard compound. The HPTLC finger-print of the sample will be compared to its BRM finger-print. The sequence (number of spots, color, intensity, and position) in a densitogram finger-printing will be evaluated (Reich et al. 2008). Evaluation can be performed visually or by using a scanning densitometer. Common mobile phases for qualitative screening purposes for botanical drugs by using silica gel TLC plates were reported by Schibli and Reich (2005). Widmer et al. (2008) reported the differentiation of *Hoodia gordonii* with its close species, e.g., *Hoodia currorii*, *Hoodia parvifolora*, and with common adulterant *Opuntia ficus-indica* by using HPTLC. The same technique was used by Nowak (2007) to identify some species of *Rosa*. Other examples of the application of planar chromatography for differentiating herbal drugs and the detection of the adulterant were presented in Figs. 11.3–11.5. These examples showed that by using HPTLC finger printing, identity or purity of an herbal drug can be confirmed. A large collection of HPTLC methods for analyzing herbal drugs can be found at http://www.camag.com. By combining the chemical densitogram finger-printing of herbal drugs/medicinal plants and multivariate analysis, i.e., Principle Component Analysis (PCA), Soft Independent Modeling of Class Analogy (SIMCA), and Partial Least

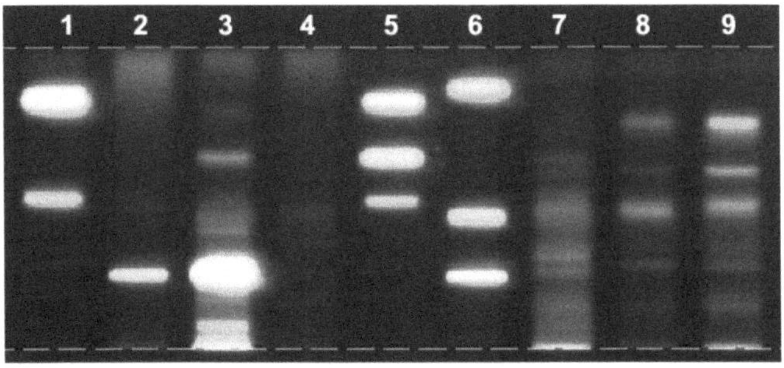

Fig. 11.3 The application of HPTLC for differentiating some *Echinaceae sp.*, and its adulterant. Stationary phase: HPTLC plate 60F254 (Merck), mobile phase: ethyl acetate, ethyl methyl ketone, formic acid, water (15–9–3–3), detection UV 366 nm after reacted with NP reagent. *Echinaceae purpurea* root (1), *E. pallida* root (2), *E. angustifolia* root (3), *E. atrorubens* root (4), Caftaric acid, cynarine, chicoric acid (5; with increasing R_f), echinacoside, chlorogenic acid, acid (6, with increasing R_f), *Liatris punctata* root (7; adulterant), *Parthenium integrifolium* (8, 9; adulterant) Reproduced with permission from Application Note F-24a (Camag)

Comparison with different Ginseng species:

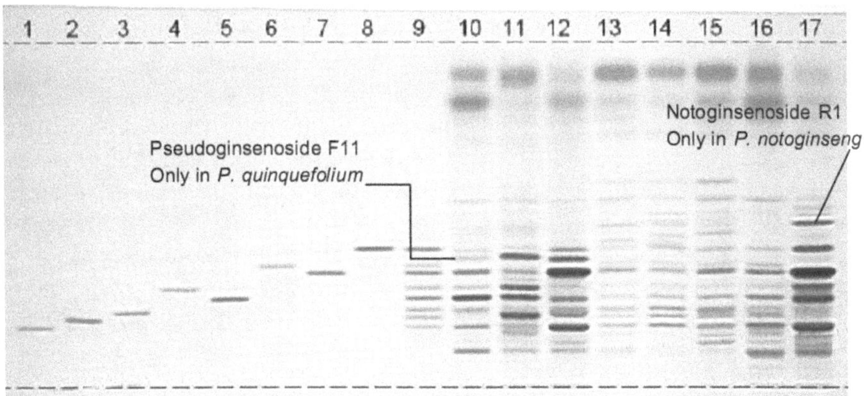

Fig. 11.4 The application of HPTLC for differentiation of Ginseng. Stationary phase: HPTLC plate 60F254 (Merck); mobile phase:chloroform, ethyl acetate, methanol, water (15–40–22–9); detection: Sulfuric acid reagent (100°, 5 min). Ginsenoside Rb1 (1), Ginsenoside Rb2 (2), Ginsenoside Rc (3), Ginsenoside Rd (4), Ginsenoside Re (5), Ginsenoside R_f (6), Ginsenoside Rg1 (7), Ginsenoside Rg 2(8), Ginsenoside mixtures (9), Cultivated *Panax quinquefolium* (10), Leaves of *P. quinquefolium* (11), Wild *P. quinquefolium* (12), Kirin ginseng/*Panax ginseng* (13), Shih cu ginseng/*P. ginseng* (14), White Ginseng/*P. ginseng* (15), Wood grown Ginseng/*P. ginseng* (16), *Panax notoginseng* (17). Reproduced with permission from Application Note F-33 (Camag)

Square Discriminant Analysis (PLS-DA), the quality of medicinal plants or herbal drugs can be evaluated more accurately. By this method, species, age of plants, and place of cultivation of a medicinal plant can be detected easily. The adulterant of medicinal plants can be also determined easily (Verpoorte et al. 2008; Choi et al. 2007).

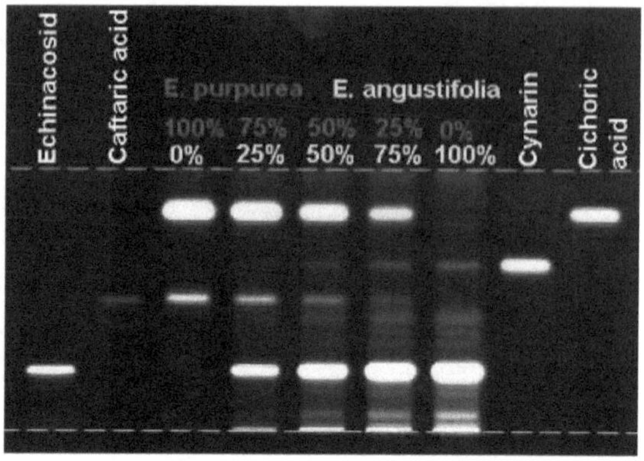

Fig. 11.5 Application of HPTLC for purity-check of the herbal drugs (root of *Echinaceae*). Reproduced with permission from presentation of Modern TLC (Camag)

Quantitative Analysis

A quantitative method generates information on the concentration or amount of the analyte(s) in the samples. Generally, two common methods are available for quantitative analysis using HPTLC, i.e., densitometry and video densitometry. For calibration purposes, both external- and internal-standardization methods can be applied. Before the method(s) can be used as a routinely tools, the method(s) should be first validated (see below). The identity and purity check of all analyte peaks in each sample should be tested before a quantitative assessment can be performed. All values of correlation coefficients should be in the acceptance range. For quantitative purposes, R_f values of the analyte(s) should be in the range of 0.3–0.8 (Frey and Zieloff 1993). It is strongly recommended to use "parallel" measurement on every lane of each sample and standard instead of using "trans-verse direction" scan (measures only the analyte spots) (Hahn-Dienstrop 2000).

External and Internal Standardization Methods

This method is the most general method for determining the concentration of an analyte in samples. It is called external standardization because the standards and the unknown samples were prepared and analyzed separately. It involves the construction of a calibration plot (area or height versus analyte concentrations) by using some different concentrations (minimum 3–4 concentrations) of reference standard compound; reference standards and samples should be spotted on the same plate. The concentration of the analyte can be determined by "interpolation" using the calibration curve on the same plate. The use of peak area is preferred compared to the use of peak height for making calibration curve (Wätzig 1995). Response factor (RF) of the analyte can be calculated from the ratio of standard peak area (A_i) to standard concentration (c_1).

$$\mathrm{RF} = \frac{A_\mathrm{i}}{c_\mathrm{i}}.$$

For bioanalytical studies and residual analysis in food, the calibration plot should be performed in the same biological matrices. It is recommended to use a working range WR (ratio of highest concentration to lowest concentration of the calibration plot) not more than 10 (Kuss 2006). If WR is more than 10, a weighted regression (or other calibration functions) is recommended.

The principle of the internal standardization involves the addition of a known quantity of the internal standard (IS) to the analyzed sample(s) and to the reference standard(s). This is to compensate errors mostly by variation of extraction, spotted amounts of standard and samples, and derivatization. The calibration plot is constructed by using of the ratio of peak area or height of standard(s) and internal standard(s) against concentrations of standard(s) or the ratio of concentrations of standard(s) and internal standard(s) (minimum 3–4 points). The concentration of the analyte(s) can be determined by "interpolation" of the calibration curve(s), which has been constructed on the same plate. In this case, RF can be calculated as follows:

$$\mathrm{RF} = \frac{A_\mathrm{i}}{c_\mathrm{i}} x \frac{c_\mathrm{IS}}{A_\mathrm{IS}}.$$

Standard Addition Method

If the concentration of the analyte is below the (linear) range of the calibration curve, the external- and internal-standardization methods cannot be applied. In this case, one should add certain amount of the analyte in the sample (spiking). It is recommended to add 3–4 concentrations of the analyte into the sample (ca. 30–70% of the original concentration of the analyte in the sample). A linear regression curve is constructed between the peak areas (Y-axis) against the added concentrations of the analyte in the sample (X-axis). The concentration of the analyte can be calculated from the intercept of the regression curve with the X-axis. Standard addition method can also be used to prove that there is no proportional error of the method. If the slopes of the regression curves (with and without addition) are equal, it means no proportional error was observed. This standard addition method can also be used if the analyte in the matrix which is to be quantified affects the chromatographic behavior (Vial and Jardi 2005).

Reporting the Results of Analyses

For reporting routine analytical results, the value of confidence interval (Funk et al. 1992) or uncertainty (Kromidas 1999) or standard certainty (Kaiser 2005) should be included. Two simples method are presented here.

Method of Bromides (Kromidas 1999):

$$\text{Result} = \text{Mean} \pm \frac{t \cdot \text{SD}}{\sqrt{N}},$$

where N is the number of replicate, SD is standard deviation, and t is value from t table; if N is more than 10, the equation can be simplified as

$$\text{Result} = \text{Mean} \pm 2x\text{SD}\,(\text{for } p = 0.05),$$
$$\text{Result} = \text{Mean} \pm 3x\text{SD}\,(\text{for } p = 0.01).$$

Kaiser (2005) suggested four replicates for analyzing samples; in this case, standard certainty can be estimated as three times standard deviation (for $n = 4$).

$$\text{Result} = \text{Mean} \pm 3x\text{SD}.$$

Interested readers can refer the work of Gonzales and Herrador (2007) for a detailed discussion on the measurement of uncertainty in analytical method.

Validation Method of Analysis

According to The United States Pharmacopoeia (2007) (USP), validation of an analytical method is the process by which it is established, by laboratories studies, so that the performance characteristics of the method meet the requirements for the intended analytical applications. Therefore, validation is an important step in determining the reliability and reproducibility of the method because it could confirm that the method is suitable to be conducted on a particular system. The performance parameters that should be determined in validation studies include specificity/selectivity, linearity, accuracy/trueness, precision, detection limit (DL), quantitation limit (QL), range, and ruggedness/robustness. The parameters that require to be validated depend on the type of the analyses; thus, different test methods require different validation schemes. The types of analyses are: identification test, quantitative determination of active ingredients or major component (assay), and determination of impurities (limit test and quantitation) (Ferenczi-Fodor et al. 2001; Hahn-Dienstrop 2000; The United States Pharmacopoeia 2007). European Commission decision 2002/657/EC (European Commission 2002) classified the analytical method as a qualitative and quantitative method (Tables 11.1 and 11.2).

Besides validation method of analysis, a fundamental requirement to perform a good laboratory practice is that the analytical instruments (TLC-scanner, sample application system, video system, automated multiple development), the software, and the glasswares that will be used should be validated and calibrated. The validation of the instrument can be divided into design qualification (DQ), installation qualification (IQ), operation qualification (OQ), and performance qualification (PQ). The user

Table 11.1 Data elements required for analytical method validation, modified from Hahn-Dienstrop (2000), Yuwono and Indrayanto (2005)

Analytical procedure	Assay		Impurity testing				Performance characteristic		Identification	
			Quantitative		Limit test					
Parameter	USP	ICH	USP	ICH	USP	ICH	USP	ICH	USP	ICH
Specificity	+	+	+	+	+	+	*	na	+	+
Linearity	+	+	+	+	–	–	*	na	–	–
Accuracy	+	+	+	+	*	–	*	na	–	–
Precision	+	na	+	na	–	na	+	na	–	na
Repeatability	na	+	na	+	na	–	na	na	na	–
Intermediate precision	na	+	na	+	na	–	na	na	na	–
Detection limit	–	–	–	–	+	–	*	na	–	–
Quantitation limit	–	–	+	+	–	–	*	na	–	–
Range	+	+	+	+	*	–	*	na	–	–

+: determination mandatory; *: maybe required, depending on the nature of the specific test; –: no need to determined; na: not available

Table 11.2 Classification of analytical methods by the performance characteristics that have been determined (European Commission 2002)

Parameters		Detection capability (CCβ)	Decision limit (CCα)	Trueness/ recovery	Precision	Selectivity/ specificity	Applicability/ ruggedness/ stability
Qualitative method	S	+	–	–	–	+	+
	C	+	+	–	–	+	+
Quantitative method	S	+	–	–	+	+	+
	C	+	+	+	+	+	+

Where; S = screening method; C = confirmatory methods; +: determination mandatory; –: no need to determined

of the instrument is obliged to maintain all documents; these include documentations of DQ, IQ, OQ, and PQ, SOP for instrument calibration, SOP for cleaning and maintenance, user book, log book, operating instruction, and software/hardware validation certificates (Hahn-Dienstrop 2000). The availability of authentic standard reference compounds or BRM is very essential.

Before validation methods can be started, the stability of analyte in the selected solvent, on the plate and during development, and stability of the visualized chromatogram after reaction with detection reagent should be tested (Hahn-Dienstrop 2000). The analyte (as standard and in sample) should be stable during sample preparations for (at least 30 min), on the sorbent surface before development (at least 30 min), and during development (at least 1 h) (Ferenczi-Fodor et al. 2001). The age/reuse mobile phase should also give the same value of R_f, T (tailing factor) and N (theoretical plate). The acceptance criterion for the variation of the peak-area of the analyte is within ±2% compared to peak-area of the fresh prepared-solution. Reconstituted samples must be remained stable in the solvents at the working temperature until spotted. For finger-print of herbal drugs, Reich et al. (2008)

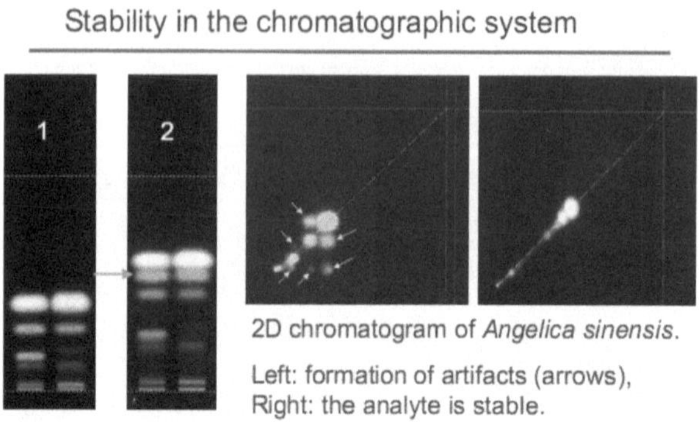

Fig. 11.6 The degradation of sample in solution can be observed easily by using two-dimensional HPTLC. Reproduced with permission from presentation Modern Planar Chromatography (Camag)

recommended to test the stability of analytes (BRM) in solution and on the plate over at least 3 h.

Stability during chromatography should be tested using two-dimensional development. If the samples are stable, all components can be detected on the diagonal line connecting the start-position and the intersection of the two solvent fronts. The presence of spots off this line indicated the possibility of degradation (Fig. 11.6). If visualization requires a derivatization step, the stability must be evaluated; they suggested taking one image immediately after 5 min, 10 min, 20 min, 30 min, and 1 h of derivatization, and then the images are compared visually and by using video-densitometry. If the prevalidation testing (stability-testing) does not show an acceptable result in the desired times, the researcher should try to find the more appropriate solvents before continuing to perform the validation method of analysis.

Specificity

The terms specificity and selectivity are often interchangeably used. A method is said to be "specific" if it provides a response for only a single analyte. The term "selective" means that the method provides responses for a number of chemical entities that may be distinguished from each other. Selectivity also means the ability to separate the analyte from the degradation products, metabolites, and coadministrative drugs (Yuwono and Indrayanto 2005). USP and ICH used the term specificity, while IUPAC and AOAC used term selectivity. The analytical method should have the capability to differentiate the desired analyte with closely related substances as impurities, isomers, metabolites, degradation products, endogenous compounds, and matrix constituents. A consequence of this requirement is that the resolution (Rs) of the desired analyte with other possible interfering components should be more than 1.0. Other parameters should ideally be in the following range: R_f ($0.1\text{–}0.3 \le R_f \le 0.8\text{–}0.9$) tailing factor ($0.9 \le T \le 1.1$) (Renger et al. 1995).

It is also well known that spots/peaks that have identical R_f value and color may or may not consist of same identical compounds. One single spot does not mean that the spot contains only one compound. That is why the measurement of the UV/VIS spectra of all analyte spots is essential to prove the identity and selectivity of the analytical method. For identity check, the UV/VIS spectrum of the analyte peak should be compared to the spectrum of reference standard, which developed on the same plate (r should be ≥ 0.999; z'value ≥ 2.327 for $p = 0.01$). For determination, the purity of the analyte peak, the UV/VIS spectra of up-slope, the apex, and the down-slope should be measured. The peak is pure if all r (r_{sm} and r_{me}) values are ≥ 0.999 and z'value ≥ 2.576 ($p = 0.01$) (Fig. 11.2) (Ferenczi-Fodor et al. 2001; Indrayanto et al. 2010). In a Quality Control Laboratory of Pharmaceutical Industry, the selectivity can be proved easily by spotting the standard, blank sample, possible impurities or degradation products, in the proposed system, the analyte spot must not be interfered by other spots, which can be proved by calculating the resolution (Rs >1). If references for impurities and degradation products are not available, the samples should be exposed to stress condition such as heat (50–80°C), oxidant (3% H_2O_2), UV-light (2,000–3,000 lux), acid (0.1–1 N HCl), and base (0.1–1 N NaOH). After incubating samples in various stress conditions at certain time, the samples were extracted than spotted; the analyte peak should have a good Rs values (>1) with possible degradation products, and also pass the identity ($r > 0.999$) and purity check (r up-slope to apex and apex to down-slope >0.999) using a densitometer.

For dissolution studies, the dissolution media or excipients should not give a peak/spot that has an identical R_f with the desire analyte (Chan et al. 2004). The identity and purity check of the analyte should be checked in order to prove that there is no degradation of the analyte during the experimental time.

In the bioanalytical method, analyses of blank samples (plasma, urine, or other biological matrix) should be obtained from at least six sources. Each blank sample should be tested for the possible interference of the endogenous substances, metabolites, degradation products. The response of the interfering spots/peaks at the R_f of the analyte should be less than 20% of the response of a lower limit of quantitation standard (QL), and should be less than 5% of the response of the internal standard that was used (Garofalo 2004). For residual analysis of banned substances, the Commission decision (European Commission 2002) recommended to use 20 representatives of blank samples; additionally, the representative of blank samples should be fortified at a relevant concentration with substances that are likely to interfere with the identification of the target analyte.

For proving the selectivity for analysis of botanical drugs, the availability of BRM is essential. It will be very nice if at least three authenticated BRM from different origin of each species can be used. This can be used to illustrate natural variability. A method is specific if, during validation, a sample representing target species showed identical finger print to that of BRM and samples representing other species give different fingerprints. Evaluation can be performed visually or by using a densitometer (Reich et al. 2008).

Linearity of the Response

According to ICH definition, the linearity of a method is its ability to provide measurement results (response) that are directly proportional to the concentration (amount) of the analyte (Lee 2004). The linearity is usually documented as the ordinary least squares (OLS) curve or simply as linear regression curve of the measured responses (peak area or height) as a function of increasing analyte concentrations. Peak area shall be preferred compared to peak heights for making the calibration curve. The linearity of the detector (system precision) can be obtained by diluting the analyte stock solution, while the linearity of the analytical method can be determined by making a series of concentrations of the analyte from independent preparations (weighing, spiking) The primary objective of linearity study of the method is to validate or verify the calibration model for quantitative purposes (Ermer 2005). It is also essential that the linear curve should be performed by using independent samples and not by samples that have been prepared by dilution and spotted on one HPTLC plate. The linearity range to be tested depends on the purpose of the test method (Table 11.3). If the target concentration is known, it is recommended to use a calibration range ±20% of the target concentration (n = 4–5 levels).

In the bioanalytical studies, calibration study should be prepared in the same biological matrix as in the sample in the intended study, by spiking the matrix with known concentration of the analyte. In this case, a blank sample, a zero sample (blank + internal standard), and 6–8 nonzero samples covering the expected range, including QL, should be performed in the linearity calibration study (Guidance for Industry and Bioanalytical method validation 2001). For analysis of banned compounds in food (Verdon et al. 2006; Van Locco et al. 2007), the calibration range must be consisted of a blank sample 0.25, 0.5, 1.0, 1.5, and 2.0 MPRL (minimum required performance limit) or its permitted limit (PL); the standards to be spiked in the respective matrix.

The linear calibration line can be evaluated by using several parameters, e.g., the relative process standard deviation value (V_{xo}), the Mandel's test, and Xp value.

Table 11.3 Linear range and acceptance criteria [modified from Dong (2006)]

Test	Linearity	
	Level and ranges	Acceptance criteria
Assay and content Uniformity	5 levels, 50–150% of label claim	$r > 0.999$ % Y intercept not more than 2%
Dissolution	5–8 levels, 10–150% label claim	$r > 0.99$, % Y intercept not more than 5%
Related substances	5 levels, QL to acceptance criteria	$r > 0.99$
Cleaning surface validation	QL to 20 QL	$r > 0.99$
Bioanalytical	6–8 levels covering the dynamic range	$r > 0.99$

Value of V_{xo} (in %) and Xp of the linear regression line $Y = a + bX$ can be calculated with the following equations (Funk et al. 1992):

$$V_{xo} = \frac{Sxo}{X} \cdot 100\% \quad Sxo = \frac{Sy}{b}; Sy = \sqrt{\frac{\sum (Yi - \hat{Y}i)^2}{N-2}},$$

where

$$(\hat{Y}i = a + bXi),$$

$$Xp = 2Sxo \cdot t_{\text{table}} \cdot \sqrt{\frac{1}{N} + 1 + \frac{(Yp - \hat{Y})^2}{b^2 \cdot Qxx}} \quad Yp = a + Sy \cdot t_{\text{table}} \cdot \sqrt{\frac{1}{N} + 1 + \frac{X^2}{Qxx}},$$

$$\text{Qxx} = \sum Xi^2 - \frac{1}{N}\left(\sum Xi\right)^2,$$

$$(t_{\text{table}} = \text{student} - t - \text{factor}; f = N - 2; p = 0.05),$$

The value V_{xo} should be not more than 5%. In addition, the lowest concentration that has been used for the linear calibration curve should not be less than the value of Xp (see Fig. 11.7). The permitted lowest concentration of the calibration curve is the quantitation limit (QL). Additionally, it also should be proved that the intercept

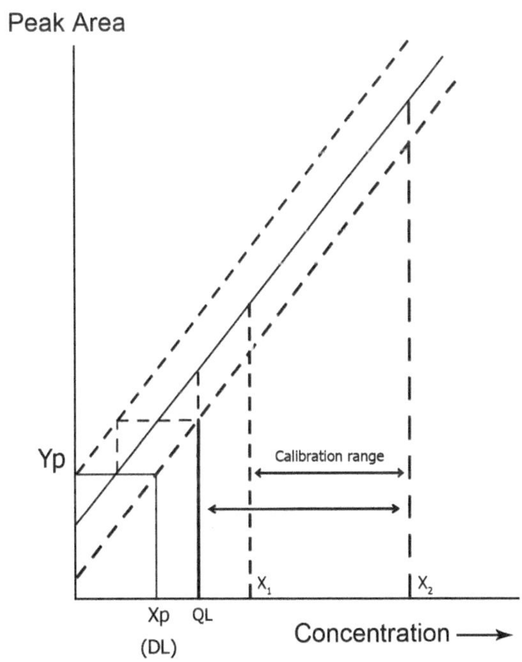

Fig. 11.7 Securing the lower limit of a calibration curve. Value of Xp must be $<X_1$ (lower limit of calibration range), while X_2 is the upper limit of calibration curve. The permitted lowest concentration of calibration curve is QL. DL value = Xp. Modified from Funk et al. (1992) and Ermer and Burgess (2005). The mathematical equation for calculating of QL value was described in Ermer and Burgess (2005)

of the linear calibration curve is not significantly different from zero ($p > 0.05$), or at least it must be about 2–5% of the target concentration (Kromidas 1999; Dong 2006). The curve of RF against concentration should have showed a zero slope in the linear calibration range.

It is recommended that the correlation coefficient (r) alone is not used anymore. Correlation coefficient described the relation between two random parameters; it showed no relevance for the analytical calibrations. The correlation coefficient does not indicate the linearity or lack thereof, unless $r \geq 0.99(9)$ (Analytical Method Committee 1988). Camag (Muttents) described the other parameter, i.e., "sdv" (residual standard deviation of the standard point), for expressing the fit of a calibration curve for HPTLC by using its CATS software. The lower the value means, the closer the points to the curve. This sdv value is a measure for the quality of all types of calibration curves. The acceptance criterion of sdv (according to our experiences) is not more than 5.0. The sdv can be calculated from the deviation between each measured standard value and the calibration curve. The equation of sdv is as follows:

$$\text{sdv} = 100\frac{\sqrt{\text{var}Yi}}{\overline{Yi}}.$$

The variance homogeneity (homoscedasticity) over the whole range of the linear calibration line should also be proved using the F-test (Funk et al. 1992). When heteroscedasticity is confirmed, and the calibration range cannot be reduced, weighting factor should be used, e.g., equal to $1/C$ or $1/C^2$, where C = concentration or other weighting factors. If the concentration range of the calibration curve is very broad, some deviations of the linear model might occur. In this case, other models such as polynomial regression (for low concentration with wide range), Michaelis–Menten (for high concentration with wide range), or weighted regression should be selected. For example, if the calculated value of V_{xo} of the linear model $< V_{xo}$ of second degree model, or the test value PW in Mandel-test > F table, the second degree model calibration should be used in the whole experiments. If someone works using CATS software (Camag), the smallest value of sdv should be selected to determine the most suitable regression model that will be used for the calibration (Yuwono and Indrayanto 2005).

Accuracy

Accuracy or trueness of an analytical method is given by the extent by which the value obtained deviates from the true value. Firstly, accuracy can be determined by analyzing a certified reference material (CRM). If the CRM is not available, then accuracy can be determined by comparing the test results obtained from the new method to the results from the existing validated method that is known to be accurate. The other approaches are based on the percent recovery of known analyte spiked into blank matrices or products. The last technique is known as the standard addition method. For spiked samples into blank matrices, it is recommended to prepare the sample in five different concentrations at the level of 80–120% or 75–125% of the target concentration. These preparations used for accuracy studies usually called "synthetic mixtures" or "laboratory-made preparations". On the other hand, if the blank matrix is not available, the standard addition method can be used, the spiking concentrations are 50–150% of the label-claimed value in this case. The accuracy of the method should be performed using at least nine determinations of three levels of concentrations. The concentration range for accuracy testing is presented in Table 11.4. The acceptance criteria are dependent on the analyte concentrations in samples (Table 11.5).

For bioanalytical method, accuracy should be performed for at least three quality control (QC) samples (low, medium, and high) in triplicate, and the accuracy was expressed as % recovery. QC sample was prepared by spiking standards into the biological matrix. The concentrations of QC samples are dependent on its expected concentrations in the biological fluids (Shah et al. 2000). For analyzing banned compounds in food, accuracy was performed by fortifying 6 matrixes each at 1, 1.5, and 2 MRPL or 0.5, 1, and 1.5 PL, so there will be 18 samples (European Commission 2002). The term "absolute recovery" in bioanalysis means the ratio of concentration of samples fortified before extraction and samples fortified after extraction, usually also expressed as percentage (Cronly et al. 2009).

Table 11.4 Concentration range for accuracy testing for pharmaceuticals (modified from Brutto and Patel)

Type of analytical procedure	Minimum range
Assay of API	80–120% of label claim
Content uniformity	70–130% of Label claim
Dissolution	±30% of specified range For controlled release product: 20% (1 h) to 90% (24 h); or 10%–130%
Degradation products and impurities	Reporting level to 130% specification limit

Table 11.5 The acceptance criteria of accuracy and precision studies at different analyte concentrations [modified from Gonzales and Herrador (2007) and Yuwono and Indrayanto (2005)]

Analyte concentration (%)	Unit	Mean recovery (%)	Precision (RSD) %	Horwitz's precision (%RSD)
100	100%	98–102	1.3	2
≥10	10%	98–102	2.7	2.8
≥1	1%	97–103	2.8	4
≥0.1	0.1%	95–105	3.7	5.7
0.01	100 ppm	90–107	5.3	8
0.001	10 ppm	80–110	7.3	11.3
0.0001	1 ppm	80–110	11	16
0.00001	100 ppb	80–110	15	22.6
0.000001	10 ppb	60–115	21	32
0.0000001	1 ppb	40–120	30	45.3

To prove whether systematic errors are not occurred, a linear regression of recovery curve of X_f (concentration of the analyte measured by the propose method) against X_c (nominal concentration of the analyte) should be constructed. The equation of the recovery curve is:

$$X_f = a_f + b_f \cdot X_c.$$

The confidence range of the intercept (Cr a_f) and slope (Cr b_f) from the recovery curves are calculated for $p = 0.05$. These confidence ranges of intercept and slope should not be significantly different with 0 (zero) and 1 (one), respectively, and can be calculated with following equations: (Funk et al. 1992)

$$(t_{\text{table}} = \text{student} - t - \text{factor};\ f = N - 2; p = 0.05),$$

$$Crb_f = b_f \pm \frac{t_{table} \cdot Syf}{\sqrt{Qxx}},$$

$$Cra_f = a_f \pm t_{\text{table}} \cdot S_{\text{yf}} \cdot \sqrt{\frac{1}{N} + \frac{\hat{X}c^2}{Qxx}} \quad S_{\text{yf}} = \sqrt{\frac{\sum [X_{\text{if}} - (a_f + b_f - X_i)]^2}{N - 2}}.$$

Precision

Precision of the analytical method can be divided into three categories, i.e., repeatability, intermediate precision, and reproducibility. *Repeatability*, or intra-assay within-day precision, is determined when the analysis is done in one laboratory by one analyst with same conditions (equipment, TLC plate, reagents) and performed within 1 day work. *Intermediate precision* is obtained when the analysis was performed within a laboratory by different analysts, equipments, reagents, and plates over a number of days or weeks. *Reproducibility* represents the precision obtained from some laboratories with the aim to verify whether the method can yield the same results in different laboratories.

For the determination of the repeatability for pharmaceuticals, at least six independent analyses of three levels of concentration should be performed (80, 100, 120% of target concentration); so for performing an intermediate precision study, the minimum samples to be analyzed should be at least 6 × 3 different concentrations × 3 different times (total = 54 samples). All the determination should be performed using real samples or laboratory-made preparation (Yuwono and Indrayanto 2005).

For bioanalytical method, it is recommended to test the precision using a minimum of five determinations per concentration. A minimum of three levels of concentrations in the expected range is recommended; RSD is not permitted more than 15%, and at the maximum limit, lower concentration RSD was 20% (Garofalo

2004). For residual analysis of banned compounds, the concentrations are same with accuracy studies (European Commission 2002).

The realistic standard deviation in HPTLC analyses is ca. 0.2% on multiple scanning of one spot, 0.8–1.5% on multiple spotting of the same sample solution, and 1.5–2% on multiple analysis of the same sample (Renger et al. 1995). As a general rule, the standard deviation of a method should be lower than 1/6 of the specification range, or the relative standard deviation (RSD) value should not be more than 2% (Edwardson et.al. 1990; Carr and Wahlich 1990). The acceptance criteria are dependent on the concentrations of the analyte (Table 11.5). For interlaboratory studies (reproducibility evaluation), Horwitz devised an equation for calculating RSD of the method (Gonzales and Herrador 2007).

$$\mathrm{RSD} = 2^{(1-0.5\log C)}.$$

Other parameters that should be tested in the precision study are David-, Dixon- or Grubbs-, and Neumann-Test. David-Test is performed for testing whether the precision data are normal distributed. Outlier-test of the data is performed by Dixon-Test (if $n < 6$–8) or Grubbs-Test (if $n > 6$–8), while trend test of the data was performed by Neumann-test. A detailed method has been described in the book written by Kromidas (1999). Reich et al. (2008) reported a detailed method to evaluate precision for qualitative HPTLC finger print of herbal drugs. They recommended to use at least three portions of the BRM and spotted onto three different plates. The fingerprint must be identical; the variability of R_f values of three markers should not exceed 0.01 across each plate, 0.02 for repeatability, 0.05 for intermediate precision (1–3 days), and 0.07 for reproducibility.

Detection Limit, Quantification Limit

According to ICH, detection limit of an individual analytical procedure is the lowest concentration of an analyte in a sample, which can be detected but not necessarily quantitated as an exact value. The quantitation limit (QL) of an individual analytical procedure is the lowest in a sample that can be quantitatively determined with suitable accuracy and precision. Generally, QL can be estimated as 2–3 times of DL. DL and QL for instrumental (chromatographic) analytical methods can be defined in terms of the signal-to-noise ratio (2:1–3:1 for DL and 10:1 for QL) or in terms of the ratio of the standard deviation of the blank response, the residual standard deviation of the calibration line, or the standard deviation of intercept (s) and slope (S) can be used (Yuwono and Indrayanto 2005; Lee 2004)

$$\mathrm{DL} = 3.3x\frac{s}{S},$$

$$\mathrm{QL} = 10x\frac{s}{S}.$$

According to Carr and Wahlich (1990), DL and QL can be calculated with the following equation:

$$C = kx\frac{S_b}{S},$$

where S is the slope of response versus concentrations, k is constant, $k = 3$ for DL, $k = 10$ for QL, and the value of S_b can be calculated using $S_b = N/5$, where N is largest peak-to-peak fluctuation corresponding the blank chromatogram in the region of 20 times with of the analyte peak. In this case, the value of QL can be estimated as 3 DL value.

By constructing a linear regression of relatively low concentrations of analyte, and calculating the Xp value, DL can be determined (DL = Xp). The authors recommend using 5–10 relatively low concentrations of analyte for determining of Xp by diluting until no responses was detected. In this case, the requirements of the linearity parameters (V_{xo}, r, Xp-value, etc.) of the regression line should be fulfilled before DL can be estimated using the value of Xp. See Fig. 11.7. (Funk et al. 1992), QL can be estimated 3 DL or can be calculated as described in Fig. 11.6. Detailed discussion can be referred in the book of Ermer and Burgess (2005).

Other similar parameter was defined by the EU Commission decision, i.e., decision limit (CCα) and detection capability (CCβ). CCα defined as the limit at and above which it can be concluded with an error probability of α that a sample is noncompliant. CCβ means the smallest content of the substance that maybe detected, identified, and/or quantified in a sample with en error probability of β. In case of substance for which no permitted limit has been established, the detection capability is the lowest concentration at which a method is able to detect truly contaminated samples with a statistic certainty of $1 - \beta$. A detailed discussion can be referred from (European Commission 2002; Verdon et al. 2006; Van Locco et al. 2007). Recently, Janiga et al. (2008) reported that the value of CCα is almost identical with DL, while CCβ is identical with QL.

For determining DL or CCα, the linear regression of the analyte should be performed in the blank sample or matrix. For example, for determining the value of DL or CCα in a certain fish, a series of very dilute desired analytes should be spiked to the same matrix and processed as real samples. DL- or CCα -value should be below its maximum permitted level of the banned compounds or its minimum required performance limit.

Range

According to the USP 30 (USP), range can be defined as the range of an analytical method in the interval between upper and lower levels (in the Pharmaceutical Industry usually a range from 80 to 120% of the target concentrations was tested) of the analyte that has been demonstrated to be determined with a acceptable level of precision, accuracy, and linearity using the method to be validated. Routine analyses should be conducted in this permitted range. A wide range should be tested for pharmacokinetic measurements; the maximum value should exceed the highest expected body fluid concentration, while the minimum value is the QL.

Robustness/Ruggedness

Robustness can be defined as a measure of the capability of the method to remain unaffected by small, but deliberate, variations in method parameters. It provides an indication of its reliability during normal usage. Ruggedness of a method is the degree of reproducibility of test results obtained by the analysis of same samples under a variety of conditions such as different laboratories (interlaboratory study), analysts, and instruments; different lot of reagents; different days; etc. Some important parameters for testing of the robustness evaluation for HPTLC/TLC methods are presented in Table 11.6 (Van der Hyden et al. 2001).

In our laboratory, the design and analysis of effect of the robustness data were performed and calculated by using a multivariate statistically software. As an example, we studied the effect of small variation in the composition of mobile phase on R_f, TF, and recovery of the active ingredient as described in Cholifah et al. (2007) and Melianita et al. (2009). When no significant difference (e.g., R_f, T and % recovery) was observed due to small variation of the analytical conditions, the robustness of a proposed method can be proved.

For robustness evaluation of botanical drugs by planar chromatography, the acceptance criteria are that the R_f values of all marker should lie within the acceptance criteria of the intermediate precision ($R_f \leq 0.05$) and the system suitability test (SST) be passed. Typically specific features of the finger print of the sample (critical separation of two zones) are used to define the SST (Reich et al. 2008). Detailed discussion on robustness/ruggedness evaluation can be referred from the publication of Van der Hyden et al. (2001) and Ermer (2001).

Table 11.6 Potential factors to be examined in the robustness evaluation [modified from Van der Hyden et al. (2001)]

Variables which should be tested
Mobile phase composition
pH of the mobile phase
Temperature
Development distance
Spot shape
Spot size
Batch of the plates
Volume of samples
Drying condition (temperature, time)
Condition of spot visualization (spraying reagents, dipping reagents, UV detection)

Table 11.7 Scheme for the determination of analyte in solution (European Commission 2002)

	−20°C	+4°C	+20°C
Dark	10 aliquots	10 aliquots	10 aliquots
Light	–	–	10 aliquots

Stability for Bioanalytical Method

The stability of analyte(s) in biological matrix should be evaluated. It is important to prove that there is no degradation of the analyte(s) between the time of collection of the samples and their analysis. Analyte instability can arise from various factors such as interaction with container surface, reaction with air, thermal decomposition, evaporation of volatiles, and photolytic decomposition. To minimize the degradation, there are some treatments such as lowering the temperature (4°C) or freezing the samples, protecting from light or adding some stabilizing agent, adjusting the pH, making chemical derivatization, and adding enzyme inhibitors (Yuwono and Indrayanto 2005). For determining the stability of banned compounds in a matrix, 40 aliquots of samples should be kept for about 20 weeks (Table 11.7) and analyses performed on the 1st, 2nd, 4th, and 20th week and calculated as remaining % compared to the fresh solution (European Commission 2002).

Application of HPTLC

HPTLC can be used for qualitative/quantitative analysis of chemicals/natural products/drugs and its impurities/degradation-products and residual toxic compounds in food/drug preparations. Sherma (2003, 2005) summarized the application of TLC to analyze pesticides and other agro chemical compounds. Determinations of amino acids, drugs, carbohydrates, lipids, toxins, vitamins, indoles, antibiotics, peptides, pigments, phenols, bile, acids, and coumarins in sample matrices such as blood, urine, feces, saliva, cerebrospinal, fluid, body tissues, and other biologics by TLC has been reviewed (Sherma and Fried 2005). Morlock and Schwack (2007) reported application of HPTLC in food analysis. Pharmacokinetic study of ursolic acid in *Alstonia scholaris* by using HPTLC was also reported. (Shetty et al. 2007). Qualitative identification of mycotoxin by TLC was also described in detail (Grzywnowicz and Nowicka 2007). Gorog (2004) reviewed analysis of steroid compounds using planar chromatography. The separation of antibiotics on thin layers of titanic silicate inorganic ion-exchanger with organic, aqueous, and mixed aqueous–organic mobile phases was reported (Husain et al. 2004).

In our laboratory, HPTLC method was used to determine qualitatively the presence of some drugs or chemical in traditional medicine and some banned compounds such as chloramphenicol (detection limit was 0.7 ppm) and nitrofurans derivatives (detection limit was 0.09 ppm) in food preparations. Our experiences showed that the unit-cost for determination of API in pharmaceuticals using

HPTLC is lower compared to HPLC-DAD or GC-FID/MS. The advantage of using HPTLC combined with a densitometer compared to HPLC-fixed UV and GC-FID/TCD is its capability to determine the identity and purity of the analyte's spot/peak. The stability of pharmaceuticals after being exposed to stress conditions such as acid, base, heat, and oxidant can be also observed easily by using a densitometer. An excellent complete database of HPTLC methods for quantitative/qualitative analysis was provided by Cumulative Camag Bibliography Service (CBS); this data base can be downloaded at http://www.camag.com.

Notes

With the development of thin layer stationary phase combined with computer controlled scanning densitometry, HPTLC method of analysis can be applied qualitatively and quantitatively for almost various applications and samples. The many advantages of this planar chromatography method are that it is fast, simple, and relatively cheap compared to other chromatography methods such as HPLC and GC. Validation method must be performed to ensure that under routine use, the proposed method will be specific/selective, accurate, precise, robust, and reproducible for the specified range for which the analyte(s) will be determined. It is essential that the method be validated first before it can be used as routine analytical tools. Without doing the validation method, the reliability of the analysis results cannot be proved.

Acknowledgments The author would like to thank Mr. Yahya Kurniawan, General Manager, PT. Abadinusa Usahasemesta (authorized dealer of Camag, Muttenz, Switzerland), Jakarta, Indonesia for the permission for reproducing some Figures of the Camag's CD, and information on detailed mathematical equations of the CATS software. The preparation of some figures by Mr. Fajar Zulkarnain (Assessment Service Unit, Faculty Pharmacy, Airlangga University) is highly appreciated.

References

Analytical Method Committee, Use (proper and improper) of correlation coefficient. Analyst 1988; 113, 1469–1471.

Camag Flash March 2009. http://www.camag.com.

Carr GP and Wahlich JC. A practical approach to method validation in pharmaceutical analysis. J Pharm Biomed Anal 1990; 8 (8), 613–618.

Chan CC, Pearson N, Rabelo-Cameiro A and Lee YC (2004). Dissolution Method Validation. In: Chan CC, Lam H, Lee YC, Zhang, X-U. (eds.) Analytical Method Validation and Instrument Performance Verification, published by Wiley-Interscience, Hoboken, NJ.

Choi YH, Kim HK and Verpoorte, R (2007) Metabolomic. In: Kayser O, Quax WJ, (eds). Medicinal Plant Biotechnology. From Basic Research to Industrial Application, published by Wiley-VCH Verlag GmbH & Co. KGaA, Weinheim.

Cholifah S, Noviansari A, Kartinasari WF and Indrayanto G. Densitometric determination of Fenbendazole in veterinarian suspension: validation of the method. J Liq Chromatogr R & T 2007; 30 (4), 489–498.

Crecelius A, Clench MR and Richards DS. TLC-MALDI in pharmaceutical analysis. Curr Trends Mass Spectr 2004; 19 (5S), 28–34.

Cronly M, Behan P, Foley B, Malone E and Regan L. Development and validation of a rapid method for the determination and confirmation of 10 niroimidazoles in animal plasma using liquid chromatography tandem mass spectrometry. J Chomatogr B 2009; 877, 1494–1500.

Dong MW. (2006) Modern HPLC for Practicing Scientist, published by Wiley Interscience, Hoboken, NJ.

Edwardson PAD, Bhaskar G and Fairbrother JE. Method validation in pharmaceutical analysis. J Pharm Biomed Anal 1990; 8 (8), 929–933.

Ermer J. Validation in pharmaceutical analysis: Part 1. An integrated approach. J Pharm Biomed Anal 2001; 24, 755–767.

Ermer J (2005). Linearity. In: Ermer J, Miller McB. (eds.) Method validation in Pharmaceutical Analysis, published by Wiley-VCH, Weinheim.

Ermer J and Burgess C (2005) Detection and Quantitation limit In: In: Ermer J, Miller McB. (eds.) Method validation in Pharmaceutical Analysis, published by Wiley-VCH, Weinheim.

European Commission (2002). Commission Decision EC No 657/2002.

Ferenczi-Fodor K, Vegh Z, Nagi-Turak A and Renger B. Validation and quality assurance of planar chromatographic procedures in pharmaceutical analysis. J AOAC Int 2001; 84 (5), 1265–1276.

Frey HP and Zieloff K (1993). Qualitative und Quantitative Dünnschict-chromatographie, published by VCH, Weinheim, NewYork, Basel, Cambridge.

Fuchs B, Sûβ R, Nimptsch A and Schiller J. MALDI-TOF-MS directly combined with TLC: a review of the current state. Chromatogr Suppl 2009; 69, S95–S105.

Funk W, Dammann W and Donnevert G (1992). Qualitätssicherung in der Analytischen Chemie, published by VCH Verlagsgesellschaft mbH, Weinheim, New York, Basel, Cambridge.

Garofalo F (2004). BioAnalytical Method Validation. In: Chan CC, Lam H, Lee YC, Zhang, X-U. (eds.). Analytical Method Validation and Instrument Performance Verification, published by Wiley-Interscience, Hoboken, NJ.

Gonzales AG and Herrador MA. A Practical guide to analytical method validation, including measurement uncertainty and accuracy profiles. Trends Anal Chem 2007; 26 (3), 227–238.

Gorog S. Recent advances in the analysis of steroid hormones and related drugs. Anal Sci 2004; 20 (5), 767–783.

Grzywnowicz K and Nowicka M. TLC Identification of occupationally relevant mycotoxins. J Planar Chromatogr Mod TLC 2007; 20 (1), 69–71.

Guidance for Industry, Bioanalytical method validation (2001), published by US Department of Health Services, Food and Drug Administration, Center for Drug Evaluation and Research, Center for Veterinary Medicine.

Hahn-Dienstrop E. (2000). Applied Thin Layer Chromatography, published by Wiley-VCH. Weinheim, New York, Chichester, Brisbane, Singapore, Toronto.

Husain SW, Ghoulipour V and Sepahrian H. Chromatographic behavior of antibiotic on thin layers of in-organic ion exchangers. Acta Chromatographica 2004; 14, 101–110.

Indrayanto G, Yuwono M and Suciati (2010). TLC: Validation of Analyses. In: Cazes. J. (ed.) Encyclopedia of Chromatography, 3rd edition, published by Taylor & Francis Group, London.

Janiga I, Mocak J and Garay I. Comparison of minimum detectable concentration with IUPAC detection limit. Meas Sci Rev 2008; 8 (5), 108–110.

Kaiser RE. Methods for detecting and reducing Systematic errors in quantitative planar chromatography. Part 1. Fundamentals of systematic quantitative errors. J Planar Chromatogr Mod TLC 2005; 18 (1), 51–56.

Kromidas S. (1999). Validierung in der Analytik, published by VCH Verlagsgesellschaft mbH, Weinheim, New York, Chichester, Brisbane, Singapore,Toronto.

Kuss HJ (2006). Optimization of the evaluation in Chromatography. In: Kromidas S. (ed.) HPLC Made to Measure, published by Wiley-VCH, Weinheim.

Lee YC (2004). Method validation for HPLC analysis of related substances in Pharmaceutical Products. In: Chan CC, Lam H, Lee YC, Zhang, X-U. (eds.) Analytical Method Validation and Instrument Performance Verification, published by Wiley-Interscience, Hoboken, NJ.

Lepri L and Cincinelli A (2005). R_f. In: Cazes J, Ed. Encyclopedia of Chromatography, 2nd edition, published by Taylor & Francis Group, Boca Raton, Florida.

Lo Brutto R and Patel T. Method Validation. In: Kazakevich Y, Lobrutto R. (eds.) (2007) HPLC for Pharmaceutical Scientist, published by John Wiley & Sons Inc. Hoboken, NJ.

Melianita F, Witha J, Arifin JS, Kartinasari WF and Indrayanto G. Simultaneous determination of 6-gingerol, 8-gingerol, 10-gingerol and 6-shogaol in some commercial gingers. J Liq Chromatogr R & T 2009; 32, 567–577.

Morlock GE and Schwack W. The contribution of planar chromatography in food analysis. J. Planar Chromatogr Mod TLC 2007; 20 (6), 399–406.

Nowak R. TLC Finger printing for analysis European dog Rose. J Planar Chromatogr Mod TLC 2007; 20 (1), 43–48.

Reich E, Schibli A and De Batt A. Validation of high performance thin layer chromatographic methods in a cGMP environment. J AOAC Int 2008; 91 (1), 13–20.

Renger B, Jehle H, Fischer M and Funk W. Validation of analytical procedure in pharmaceutical analytical chemistry; HPTLC assay of theophylline in an effervescent tablet. J Planar Chromatogr Mod TLC 1995; 8 (4), 269–278.

Rozylo Ik (2005). TLC/MS, In: Cazes.J. Ed. Encyclopedia of Chromatography, 2nd edition, published by Taylor & Francis Group, Boca Raton, Florida.

Schibli A and Reich E. Modern TLC: a Key technique for identification and quality control of botanicals and dietary supplements. J Planar Chromatogr Mod TLC 2005; 18 (1), 34–38.

Shah VP, Midha KK, Findlay JWA, Hill AM, Hulse JD, McGilveray IJ, McKay G et al. Bioanalytical method validation – a revisit with a decade progress. Pharm Res 2000; 17 (12), 1551–1558.

Sherma J. Recent advances in the thin layer chromatography of pesticides: a review. J AOAC Int 2003; 86 (3), 602–612.

Sherma J. Thin layer chromatography of pesticides, a review of application for 2002–2004. Acta Chromatographica 2005; 15, 7–28.

Sherma J and Fried B. thin layer of biological samples: a review J Liq Chromatogr R & T 2005; 28 (15), 2297–2314.

Shetty PG, Mangaonka KV, Sane RT, Jaripatke KK and Singh S. Pharmacokinetic analysis of ursolic acid in *Alstonia scholaris* R. Br. by high-performance thin-layer chromatography. J Planar Chromatogr Mod TLC 2007; 20 (2), 117–120.

The United States Pharmacopoeia 30 – The National Formulary 25 (2007), published by The United States Pharmacopeial Convention, Inc., Rockville, MD.

Van der Hyden Y, Nijhuis A, Smayers-Verbeke J, Vandeginste BMG and Massart DL. Guidance for robustness/ruggedness test in method validation. J Pharm Biomed Anal 2001; 24, 723–753.

Van Locco J, Janosi A, Impens S, Fraselle S, Cornet V and Degroodt JM. Calculation of the decision limit and detection capability for banned substances: the imperfect marriage between the quantitative and the qualitative criteria. Anal Chem Acta 2007; 586, 8–12.

Verdon E, Hurtaud-Pessel D and Sandersa P. Evaluation of the limit performance of an analytical method based on a statistical calculation of its critical concentration according to ISO standard 11843: application to rotinecontrol of banned veterinary drug residues in food according to European Decision 657/2002/FC. Accred Qual Assur 2006; 11, 58–62.

Verpoorte R, Choi YH, Mustafa MR and Kim HK. Metabolomic: back to basis. Phytochem Rev 2008; 7, 525–537.

Vial J and Jardi A (2005). Quantitation by Standard Addition. In: Cazes. J. (ed.) Encyclopedia of Chromatography, 2nd edition, published by Taylor & Francis Group., Boca Raton, Florida.

Wätzig H. Appropriate calibration functions for capillary electrophoresis I: precision and sensitivity using peak areas and heights. J Chromatogr A 1995; 700, 1–7.

Widmer V, Reich E and De Batt A. Validated HPTLC method for identification of *Hoodia gordinii*. J Planar Chromatogr Mod TLC 2008; 21 (1), 21–26.

Yuwono M and Indrayanto G (2005). Validation of Chromatographic Methods of Analysis. In: Brittain G. (ed.) Profiles of Drugs Substances, Excipients and Related Methodology vol. 32, published by Academic Press Elsevier Inc, San Diego.

Chapter 12
Quantitative Analysis and Validation of Method Using HPTLC

Pinakin C. Dhandhukia and Janki N. Thakker

Abstract High performance thin layer chromatography is an emerging alternative analytical technique in comparison with conventional column chromatography because of its simplicity, rapidity, accuracy, robustness, and cost effectiveness. Choice of vast array of supporting matrices and solvent systems resulted in separation of almost all types of analytes except volatiles. First step of a robust method development for routine quantification is to check the stability of analyte during various steps of chromatographic development followed by preparation of calibration curves. Thereafter, various validation aspects of analysis namely peak purity, linearity and range, precision, limit of detection, limit of quantification, robustness, and accuracy have to be measured.

Preference of HPTLC over Column Chromatography

Thin Layer Chromatography (TLC) is the most simple and inexpensive technique used for a long time for separation of the variety of compounds. Recently developed High Performance Thin Layer Chromatography (HPTLC) provides an efficient, fast, and reliable alternative for quantitative determination of natural products (Hans-Deinstrop 2000; Khoobdel et al. 2007; Pawar et al. 2008). TLC and HPTLC complement gas chromatography (GC) and high-performance column liquid chromatography (HPLC) for separation, detection, identification, and quantification

P.C. Dhandhukia (✉)
Ashok and Rita Patel Institute of Integrated Study & Research in Biotechnology and Allied Sciences, New Vallabh Vidyanagar 388121, Gujarat, India
e-mail: pinakin.dhandhukia @gmail.com

J.N. Thakker
Department of Biotechnology, PD Patel Institute of Applied Science, Charutar University of Science & Technology, Education Campus, Changa 388421, Gujarat, India

MM. Srivastava (ed.), *High-Performance Thin-Layer Chromatography (HPTLC)*,
DOI 10.1007/978-3-642-14025-9_12,

of analyte because of their following unique advantages over column chromatography: (1) Single use of the layer simplifies sample preparation procedures. This ensures no carry forward of the samples as unidentified peaks as well as intermittent baseline correction and column stabilization are not required (Sherma 2005). (2) Simplicity of development by dipping the plate into a mobile phase in a chamber is particularly advantageous in dealing with different type of supporting matrices and mobile phases without any change over period. (3) High sample throughput with low operating cost because multiple samples can be run simultaneously with standards on a single plate using a very low volume of solvent. Using GC-MS and HPLC, only 10 and 20 samples per day could be analyzed, respectively, whereas using HPTLC, more than 100 samples per day can be analyzed (Anderson 1985; Raverkar 2001; Mueller and Brodschelm 1994). (4) High resolution through multiple development or two-dimensional (2D) development on a plate with a single adsorbent or dual adsorbents (Sherma 2005). (5) Selective and sensitive postchromatographic detection and identification with a very wide variety of chromogenic, fluorogenic, and biological reagents and coupled spectrometric techniques (Sethi 1996; Ueda and Miyamoto 1994). (6) High resolution and accurate and precise quantification achieved on HPTLC plates, especially with automated sample application, development, and densitometric scanning methods. (7) Samples do not require to dissolve in the solvent (mobile phase), which is used for development, as after loading the sample, the solvent gets dry leaving only the sample behind, therefore, solvent peak will be absent. Even samples extracted with different solvents can be compared on the same chromatographic plate. (8) In GC, most of the samples have to be methylated prior to separation to lower its boiling temperature; however, in case of certain compounds, which are also found in the methylated form in nature, distinguishing determination between the analyte and its methyl form both in a same sample is not possible (Mueller and Brodschelm 1994). (9) Other compounds/impurities with the same retention time may interfere; also retention time varies with the concentration of analyte, e.g., methyl jasmonate in packed columns such as SE-30 and OV-1 (Ueda and Miyamoto 1994); this has been corrected in the HPTLC as standard has always been loaded along with the sample; hence, any change in R_f value occurs due to change in environmental parameter or change in absorbent layer could be identical to standard. (10) Detection and determination of the poor UV absorbing compounds is difficult in HPLC. Due to poor UV absorption, derivatization with fluorescent hydrazides such as densyl hydrazide and dimethylaminobenzohydrazide is used to develop a sensitive method for quantification of poor UV absorbing compounds (Anderson 1986). Although this approach gives sensitive results, the dimethylaminobenzohydrazide is very toxic, requires great care in handling, and moreover, it takes in excess of 12 h for derivatization (Ueda and Miyamoto 1994). However, poor UV absorbing compounds can be detected and determined as black spot on HPTLC plates against its fluorescent background. (11) Also, simultaneous chromatography of samples and standards under identical conditions in the same foil leads to results with excellent accuracy and precision, and solvent usage per sample basis is also very low (Wagner and Sherma 2001). (12) Some incomparable advantages of HPTLC over other methods are that once the separation is achieved, sample can be reextracted from the solid support, or each band can be scanned with absorbance. Same plate can also be derivatized and rescanned. Even after development of the plate, it can be stored for many days before scanning without significant decrease in concentration (depends on the compound of interest).

Quantification

Earlier, a typical approach for quantification of separated compounds was scrapping the separated zone from support material and then extracting with suitable solvent. Eluted compound could be analyzed by any convenient analytical method viz., spectrophotometry, fluorometry, or by suitable color development method (Hahn-Deinstrop 1991). To compensate interference from the support matrix, usually a blank area of layer was eluted simultaneously and used as blank in the analysis. Prewashing of the TLC plate using methanol, methanol: chloroform (1:1 v/v), or methyl chloride successfully lowered the Blank values (Ravi et al. 2006). However, in spite of wide utilization of this method because of its simplicity and inexpensiveness, it suffered from some limitations such as compound may irreversibly bind to TLC support; elution step may cause some chemical transformation; or loss of compound during extraction procedure resulted in erroneous quantification. Nowadays, layers containing gypsum as binder are available, which are considerably softer and especially used for the preparative chromatography (Sethi 1996; Hahn-Deinstrop 1991). In situ quantitative TLC is a semiquantitative procedure, based on visual comparison of size and/or intensity of colored, fluorescent, or UV (254 nm) absorbing compounds. This method is simple, quick, and does not require much instrumentation and laboratory facilities; however, it suffered from low accuracy (±10%) and hence used for quick assessment of quality of analyte. To overcome these limitations, accurate in situ quantification (without elution) of separated compounds is possible with the advent of the technology based on densitometry. Densitometry is the in situ instrumental measurement of visible, UV absorbance, fluorescence, and fluorescent quenching directly on the layer without resorting to scrapping and elution. For densitometric evaluation of a TLC/HPTLC chromatogram, its tracks are scanned with a light beam in the form of a slit selectable in length and width. Diffusely reflected light is measured by the photo sensor. As a rule, sample and standard should be chromatographed on the same plate under similar conditions. For quantification, the difference between the optical signal from the sample free background and that from a sample zone (fraction) is correlated with the amount of the respective fractions of calibration standards chromatographed on the same plate (Dhandhukia and Thakkar 2008). Densitometric measurements can be made by absorbance or by fluorescence. Alternatively to classical densitometry, planar chromatograms can be evaluated by video technology. Requirement for various steps for quantification using TLC/HPTLC are very stringent. Accurate and precise application of sample is the most critical step. Further, the chromatographic development should clearly and completely separate all the compounds of interest with no loss by decomposition, evaporation, or irreversible biding during application or development. Accurate application of the sample can be performed using automated TLC applicator under flow of N_2, e.g., Linomate applicator – the detail of the applicator function has been discussed in the instrumentation section. In this chapter, we emphasize more on the quantitative utilization of the HPTLC instrument and validation of the developed method.

Stability of the Analyte

Analyte should not decompose during development of the chromatogram and should be stable in solution and on the sorbent for at least 30 and 15 min, respectively. The intensity of the spot on chromatogram should be constant for at least 60 min. While optimizing the extraction/purification procedure, one must keep in mind the chemical properties and purity of the extraction solvent. Chemical reactions of the solvent and their impurities can cause artifacts, which may produce extra band or spot/peak, thus leading to false assay value. Other important factor is pH of the aqueous phase used for the extraction/purification, which may lead to hydrolysis, oxidation, and isomerization; evaporation of organic solvent to dryness may also lead to formation of artifacts, so complete removal of organic solvent should be avoided.

Stability of the Sample in Solution

Many solutes may decompose in sample solution prepared prior to TLC/HPTLC investigation. Solutions of same concentration are prepared from stock sample solution and stored at RT for several hours and then applied on the same chromatoplate. After development and scanning, the plates are observed for change in peak areas and appearance of additional peaks. The chromatogram should not contain any spot absent in the initial chromatogram, and any spot present in the initial chromatogram should not be absent from final chromatogram. Ravi et al. (2006) observed that the plates were stable up to 3 h, after which there was a significant change found in peak areas and R_f values.

Stability on the Absorbent Layer Prior to Development

Further possible source of error is the decomposition of the analyte due to active site of the stationary phase. Sample application, spotting technique, size and shape of the spot, temperature of the air stream used for drying the spot, chemical effect of light, air, and the time the sample is left to stand on the sorbent prior to chromatographic development are some of the significant factors, which can influence the stability of the separated spots and are required to be investigated for validation.

Usually, it is recommended that the spot should be stable for at least 15 min on the sorbent layer. To find out any decomposition occurring during spotting and development, 2D chromatography using the same solvent system should be carried out (Sethi 1996).

Stability during Development

Stability to environmental factors: Many compounds are very sensitive to the conditions, most importantly oxidative and hydrolytic decomposition primarily due to dissolve oxygen, light, and temperature used for chromatographic development during method development or mobile phase optimization. (a) To evaluate any detrimental effect of temperature and exposure to light, a separate experiment can be designed at room temperature, cool place, as well as both in reflected light and protected light. (b) Moreover, oxidative decomposition can be prevented by using protective gasses such as nitrogen and argon both in chamber as well as by removing dissolved oxygen from eluent by flushing or using stabilizers (antioxidants), e.g., BTH in TLC of vitamin D_2 and danazole used to protect the sample from oxidative decomposition.

Identifying decomposition during spotting and chromatographic development: Evidence can be confirmed by resorting it to 2D chromatography using same solvent system in both directions. (a) If decomposition occurs during spotting and spot drying only, the chromatogram obtained in the second direction of the eluent flow will be free from any impurity peak, i.e., single peak chromatogram shall be obtained. (b) If decomposition occurs during development, peak(s) of decomposition product(s) shall be obtained for the analyte both in the first and second direction of the run. The shape of the peak of decomposition product usually shows significant tailing compared to both the directions being similar to each other. (c) Alternatively, the decomposition may occur both during spotting/spot drying and chromatographic development. The situation is similar to one encountered above, and the peak(s) of decomposition products in the first direction exhibit less tailing than that obtained with second direction.

Stability after Chromatographic Development

When sensitive substances are subjected to HPTLC, intensity of the spot sometime may change on the developed chromatogram as the time passes. This difficulty is often encountered when eluent cannot be easily evaporated from the sorbent layer. Constant spot signal, usually up to 30 min, during densitometric evaluation can indicate the stability of the spot(s) after chromatographic development and adequate for quantitative densitometric evaluation. The stability of drugs on developed plates was checked up to 24 h and was found to be stable up to 9 h. Hence, the plates should be scanned within 9 h after development (Ravi et al. 2006).

Purity of Reagent and Solvents

The substances are in dissolved form during the most stages of chromatographic procedures, making it essential to know the quality of the solvents employed and types of additives present to stabilize the solvents. These stabilizers may have a positive or negative effect on separation. Usually, large quantities of solvents are used for extraction and subsequently concentrated to a small volume. Therefore, the solvent must be free from residual water, and nonvolatile or not readily volatile impurities as these impurities, if present, become enriched on concentration and can cause serious contamination problems. Most of the commonly used organic solvents contain stabilizers, which can influence the polarity of the solvents.

Densitometry

Optimization of Wavelength

Classical densitometry with the CAMAG TLC Scanner 3 accesses the complete spectral range between 190 and 800 nm with high spectral selectivity. Absorption spectra used for identification of substances can be acquired within the complete spectral range. UV absorbing compounds are determined by in situ densitometric scanning after chromatographic development (Fig. 12.1). The light intensity remitted by chromatographic zone is usually lower that the absorbent layer around it. Therefore absorption spectra of a compound can be directly determined on HPTLC plate itself compared to substance-free portion of sorbant layer. A similarity has usually been observed between spectra of substance as recorded on the plate and

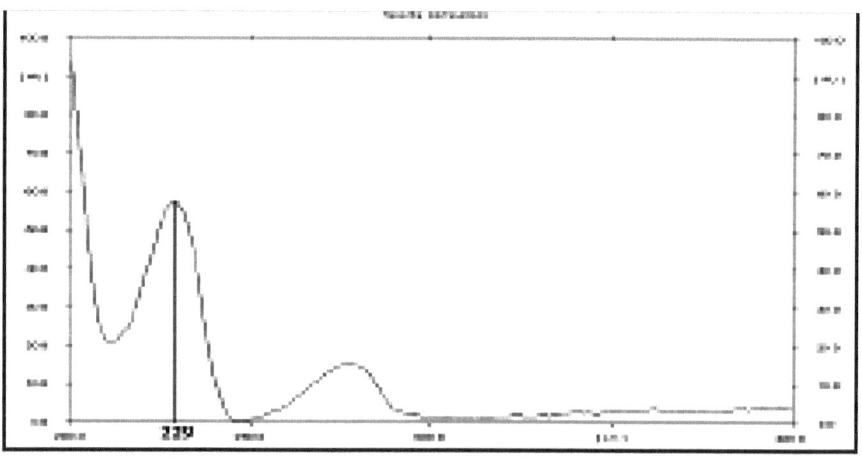

Fig. 12.1 The in situ UV spectrum of bisoprolol fumarate (Kakde et al. 2008)

one that was recorded when compound is in solution, although in case of modified sorbents, it may change due to hypochromic shifts. The main reason for scanning at wavelength of maximum absorbance is that difference between absorption by the chromatographic zone and blank region is at its largest. For more than one compound in the sample to be analyzed, wavelength must be very near to their maximum; although it gives satisfactory results, the variation of the LOD and LOQ is needed to be corrected.

Purity of the Separated Band

It is essential to know that the separated band from biological matrix is pure, i.e., to ensure the purity of the band under analysis. Separated band is scanned for purity, in which the band is scanned from three spots viz., lower, middle, and upper edges for the entire spectrum. The band is considered to be pure only if all three spectral patterns match.

Calibration Curve

To carry out HPTLC densitometric analysis, three or four standards and purified samples are applied on the same plate. After development and detection, the chromatogram is scanned. A calibration curve consisting of scan area of standard versus amount of analyte in the sample represented by scan area is interpolated from the standard curve (Fig. 12.2). Depending upon the analytical problem, one can resort the single level or multilevel calibration.

Single Level Calibrations

Single level calibration is most useful when known target values are to be verified such as stability testing and dissolution profile, and the range is limited (e.g., to ±5% of target). Best performance is achieved with standards on every 3–5 track. This method is used to evaluate analyses with a known target value and close tolerances, e.g., in quality control. Make sure that on standard tracks, an amount is available for each peak present in a substance window after assigning the peaks.

The single level method of calibration is the one with the most favorable error propagation. The ratio of analyses to standard tracks is also highly advantageous. The mean measured value for all standard tracks, calculated as a line through the origin, is used as reference. Calibration is (theoretically) admissible only over a narrow range about the standard mean. The Permitted deviation specified in the Substances dialog is used as a delimiter for this range. The extent of this range must

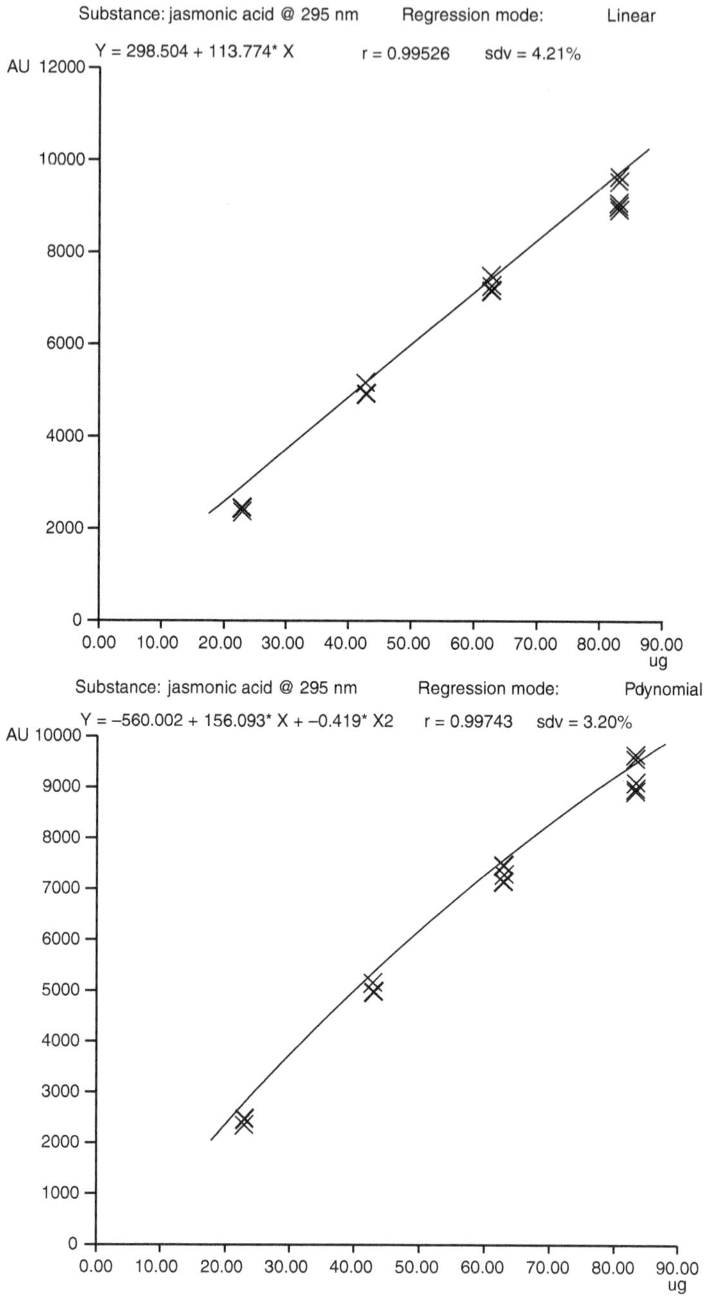

Fig. 12.2 Calibration curves (**a**) linear regression curve and (**b**) polynomial regression curve of jasmonic acid scanned at 295 nm at four different concentrations

be defined during method validation. The input is valid on both sides (+/−). Each substance to be calibrated should be present on at least two standard tracks. If you chromatograph the standard substances as a mixture of all substances, at least two standard tracks should thus be available. Single level calibration can be carried out with only one standard track.

The result for height/area is calculated from the single level calibration point and averaged. The relative standard deviation (RSD) CV of the result can be calculated provided at least two valid replicas (tracks) per unknown are available. Before a single level method can be developed, the "target amount" and the permitted deviation must first be determined by linear regression. The Permitted deviation is found from the difference between the linear regression and the single level line (through the origin). The closer to the origin the linear regression intersects the X-axis, the larger can be the permitted deviation range.

Multilevel Calibration

For multilevel calibration, select calibration functions, viz. linear regression, non-linear regression, polynomial regression, or Michaelis Menten regression best suited for our method validation (Ravi et al. 2006; Jagannath and Kanan 2008). Linear regression is normally used for fluorescent substances, for calibration in a limited concentration range (1:5–1:10), or for low amounts (within the linear range of the detector). Make sure that on standard tracks, an amount is available for each peak present in a substance window after assigning the peaks. A regression curve (in this case a line) is calculated by means of the calibration points. This line is (theoretically) defined only within the calibration range (i.e., between the measured values for the smallest and largest standard amounts). The Permitted deviation specified in the Substances dialog is used as a delimiter for this range. Enter a value larger than 0 only as an exception (e.g., to estimate an amount outside the calibration range). Outside the calibration range, error propagation can be undefined. Each substance to be calibrated should be present on at least three standard tracks but at two different levels. If you chromatograph the standards as a mixture of all substances, at least three standard tracks should thus be available. Linear regression can be carried out with only two standard tracks. The result for height/area is calculated from the calibration curve (line) and averaged. The RSD CV of the result can be calculated provided at least two valid replicas (tracks) per unknown are available (Fig. 12.2a).

Polynomial regression (second order) is used if the amounts of substance cover a wider range of concentrations (1:50–1:100) or if large amounts are calibrated (in the nonlinear range of the detector) (Fig. 12.2b). If the signal is too close to the detector saturation, the polynomial may reach a maximum. Calibration is terminated before this point. Make sure that on standard tracks, an amount is available for each peak present in a substance window after assigning the peaks. A second degree curve is calculated through the calibration points. This curve is (theoretically) defined only within the calibration range (i.e., between the measured values for the smallest and largest standard amounts). The Permitted deviation specified in the Substances dialog is used as a delimiter for this range. Enter a value larger than 0 as an exception only (e.g., to estimate an amount outside the calibration range). Even with a permitted deviation greater than 0, the curve is never calculated quite as far as the apex, but error propagation can become undefined earlier. Each substance to be calibrated should be present on at least four standard tracks but at three different levels. If you chromatograph the standards as a mixture of all substances, at least four standard tracks should thus be available. Polynomial regression can be performed with only three standard tracks. The result for height/area is calculated from the calibration curve and averaged. The RSD CV of the result can be calculated provided at least two valid replicas (tracks) per unknown are available.

Classical Internal Standard Method

It used to calculate unknown substances (e.g., decomposition products) with a known amount of a standard (e.g., parent substance) on the same track. A peak defined as internal standard on each track is used as standard for all other peaks on the same track. The different (relative) absorption capacities of the internal standard and of the unknown can be compensated mathematically by means of the response factor. This is a method to correct sample preparation, application, and chromatography errors. It can be used in combination with any one of the calibration modes, the single level, linear, or polynomial regression, as long as the amount of the internal standard is the same on all tracks (Prosek et al. 2002). The different (relative) absorption capacities of the internal standard and of the unknown does not need to be compensated mathematically by means of the response factor because a "track correction factor" is calculated by comparing the results of the internal standard between all tracks. The target application amount of the internal standard substance must definitely be the same on all tracks. If sample preparation errors are to be compensated for, you have to prepare the standards using the identical routine as for the unknowns. The same results as for the calibration mode used are calculated. There will be no results for the substance used as internal standard. Concentration of analyte in the sample is calculated by considering the weight of the sample initially taken and dilution factors. Since the contents of the active ingredients in the pharmaceutical formulations are well defined, part of linearity curve falling in the range of 75–125% of the labeled/declared amount is adequate. It has been observed that a straight line fitting through 75–125% in a five point curve yields assay values with ±1% accuracy than those obtained with five point curve usually recommended in the analysis; thus three level calibration curve usually serves the purpose for quantitative analysis (Sethi 1996). This advantage is significant in the view of additional space available in the plate without sacrificing the quality of the analytical results. It may be emphasized that without reference compounds, identification is rather difficult and ultimate quantification is always questionable. Spot/band intensity on the developed chromatogram is likely to change as time passes. Developed spots/bands should either be stabilized or scanned as early as possible, though it is the inherent advantage of HPTLC that all fractions remain stored on the plate. They can be scanned at different wavelengths consecutively.

Validation of the Method

Validation is a systemic approach to identifying, measuring, evaluating, documenting, and reevaluating all the critical steps responsible before establishing the validity of the method. The developed method is validated in terms of linearity, precision, limit of detection (LOD), limit of quantification, accuracy, robustness, and ruggedness.

Principles of Method Validation

- Concentration range over which analyte can be estimated should be defined in the method. The relationship between response and concentration should be demonstrated to be continuous and linear with at least 5–6 points; extrapolation of the plot on either end is not desirable.
- A standard curve should be maintained.
- Accuracy and precision of the method over the range of the standard curve should be clearly established.
- Stability of the analyte during purification/extraction procedure needs to be established.
- Lowest concentration of analyte in the matrix for which a reliable and reproducible estimation can be obtained should be established (LOQ).
- Accuracy and precision of the method for estimation of sample of the known concentration should be established.
- A minimum of three concentrations, one near but more than LOQ, one near center, and one near upper portion of the curve, should be taken for validation.
- Recovery of the analyte should be evaluated and documented and should be constant.
- Available, they should be chromatographed.

Selectivity

Ability of the developed analytical procedure to detect analyte quantitatively in the presence of expected or related substances in the sample matrix is known as selectivity. Results are represented as resolution. If expected impurities or related substances are available, they should be chromatographed along with the standard analyte to check the system stebility, retention factor, tailing factor, and resolution, the main parameters to be observed (Sethi 1996).

Specificity

The specificity of the method was determined by analyzing standard analyte and test samples. The spot for analyte in the samples has to be confirmed by comparing the R_f and spectrum of the spot with that of a standard (Fig. 12.3). The peak purity of analyte has to be assessed by comparing their respective spectra at the peak start (S), apex (M), and peak end (E) positions of that between the standard and sample spectra of analyte.

These studies are carried out to ascertain how accurately and specifically the analyte of interest is estimated in the presence of other components (e.g., impurities, degradation products, etc.) by exposing the sample to different stress conditions such as light, heat, oxidation, acids, and alkali and then analyzing them by proposed method. Estimation of bisoprolol fumarate in acid, alkali, UV light, and heat-exposed samples shows no significant difference with assay results, which indicate that there is no degradation (Kakde et al. 2008).

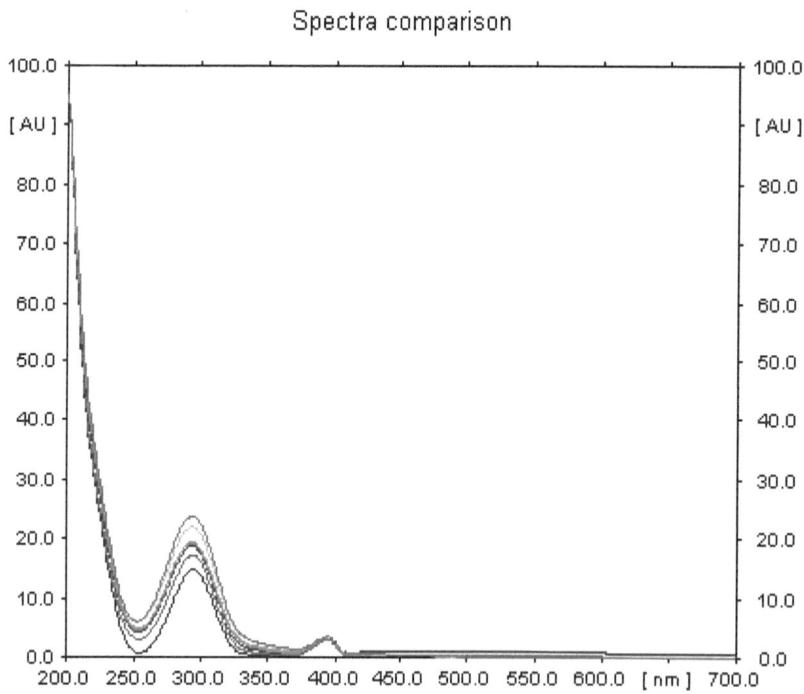

Fig. 12.3 Standard JA and JA separated from sample on TLC foils were scanned in the range of 200–700 nm. The spectra of JA separated from sample showed same absorption pattern as standard JA

Linearity and Range

It is the ability of the method, with in given range, to obtain test results in direct proportion to the concentrations of the analyte in the sample-calibration curve for the analyte. Misra et al. (2009) reported that the caffeine estimation response was found to be linear over the range of 2–14 μg/band. Dhandhukia and Thakkar (2008) reported linearity over the range of 6–90 μg/band for jasmonic acid isolated from the broth of *Lasiodiplodia theobromae* with linearity coefficient of 0.9956.

Precision

It provides introduction of the random error; results should be expressed as RSD or coefficient of variation (COV).

Repeatability (Precision on Replication)

Precision on replication is determined under same conditions, e.g., same analyte, same apparatus, short interval of time, and identical reagents using the same sample, in terms of repeatability of the measurement of peak area where RSD is not more than (NMT) 2%, based on seven times measurement of the same spot.

Replicate estimation of bisoprolol fumarate in the multiple samples of homogenous sample yielded quite concurrent results showing %RSD value well below 1.5%, which speak repeatability of the method (Kakde et al. 2008). Repeatability of position can be determined with RSD NMT 2%, based on seven times repositioning the instrument after each measurement. Moreover, Repeatability of sample application can be determined by application of six spots with equal volume. From the peak areas, the % RSD is determined, which should NMT 3% (Ravi et al. 2006).

Reproducibility (Precision of Comparison)

Precision of comparison is determined under different conditions, e.g., different analyte, different laboratory, different days, and different reagents from different sources using the same sample. Intraday assay precision is determined by analysis of standard analyte three times on the same day. Interday assay precision is carried out using the standard analyte on three different days and %RSD is calculated. Limit of acceptable RSD should NMT 10% within laboratory reproducibility (Ravi et al. 2006).

The precision of the method is verified by repeatability and intermediate precision studies. Repeatability studies are performed by analysis of at least three different concentrations of the analyte six times on the same day. The intermediate precision of the method is checked by repeating studies on three different days. Additionally, the developed HPTLC method is checked by separation studies on the mixture of reaction solutions on a different chromatographic system on a different day (Gbaguide et al. 2005).

Detection and Limit of Quantitation

It is calculated by standard deviation of response and slope of calibration curve. LOD and limit of quantitation (LOQ) of the methods are established according to ICH definitions (DL $= 3.3\sigma/S$) (QL $= 10\sigma/S$), where DL – detection limit, QL – quantitation limit, σ – standard deviation of regression line, and S – slope of the calibration curve (Kakde et al. 2008).

LOD is the lowest amount of analyte that can be detected NMT 10% of the individual Impurity limit. If this is not possible, then the amount of analyte to be applied has to be increased. LOD is determined on the basis of signal and noise ratio. Mean of 15 noise peak areas and their absolute SD values are determined. LOD is the amount of applied sample producing a peak area that is equal to the sum of mean blank area and three standard deviations.

LOQ is the lowest amount of analyte that can be quantitatively determined in sample with defined precision and accuracy under standard conditions. LOQ should NMT 20% of individual impurity. LOQ precisely is the amount of loaded sample producing a peak area that is equal to the sum of mean blank area and ten times its SD. Alternatively, decreasing amount of substance are applied in triplicate on the sample plate and RSD calculated from peal area of triplicate sample.

LOQ is smallest quantity of the substance for which the peak area of triplicate sample is either NMT 5% or when RSD value for repeatability of the method does not exeed 5%, not greater than this repeatibility value. LOQ determined, depending on the PHA, was 0.08–0.44 ng/band with linearity coefficients of correlation $\geq$0.9920 (Morlock and Kopacz 2008).

In order to estimate the LOD and LOQ, blank methanol alone is spotted six times to determine the signal-to-noise ratio. The LOD and LOQ are experimentally verified by diluting known concentrations of a standard solution of analyte until the average responses are approximately 3 or 10 times the standard deviation of the responses for the six replicate determinations.

Robustness

This is one of the most important parameters for the validation of any analytical method and therefore is the case for procedure involving application. Following experiments usually recommended evaluating ruggedness of a HPTLC method.

Sample preparation: Parameters to be checked are composition, quantity of solvent, pH, shaking time, temperature, and number of extractions. *Sample application:* Ruggedness is tested in terms of volume applied, spot/band, and spot stability (as describe earlier). *Separation:* This can be performed on at least three different plates from the same or different manufacturers. *Chromatographic conditions:* Evaluated in terms of chamber saturation, elution composition, eluent volume, temperature, humidity, and development distance. *Spot visualization:* this can be assayed in terms of postchromatographic derivatization, spraying, dipping, reaction temperature, and time. *Quantitative evaluation:* This can be performed by conditions for drying plate and detection wavelength used.

Once the ananlytical method is developed the parameters, which can be studied to ascertain the ruggedness of the method, are analyzed independently by three different analysts, well conserved with practicle aspects of the technique, for sample on same day in short time interval as well as same sample for three different days under same experimental conditions. If a validated method can be satisfactorily performed by an analyst not familiar with the specific procedure, ruggedness of the method can undoubtedly be relied upon (Sethi 1996; Kakde et al. 2008).

Accuracy

Accuracy of the analysis is determined by systemic error involved. It is defined as closeness of agreement between the actual value and mean analytical value obtained by applying the test method a number of times. Recoveries of the analyte observed to be very close to 100% representing the accuracy of the method and also noninterference of excipients (Kakde et al. 2008). The accuracy of the developed method is determined using at least three concentrations of analyte and six determinations performed at each concentration. The recovery value near 100% with RSD NMT 5% gives adequate accuracy.

Examples for the Successful Usage of HPTLC Method for Quantification

The HPTLC technique is now being used in toxicology/forensic laboratory to identify heroin, morphine, codeine, thebaine, papaverine, narcotine, paracetamol, etc. (Dongre and Kamble 2003). Identification of new/unknown substances is also possible with the HPTLC–FTIR coupled system, and on this basis, new pthalazine derivatives have been identified (Linares et al. 1998; Cimpoui et al. 2003).

An HPTLC-densitometry method was developed for the quality control and stability analysis of commercial emulsifiable concentrate (EC) formulations of the synthetic pyrethroids cypermethrin, α-cypermethrin, and λ-cyhalothrin. Reference standards and the EC formulation are applied with a Linomat IV band applicator and were chromatographed on an aluminum backed silica gel 60 F_{254} layer in a vapor presaturated twin-trough chamber with hexane: toluene (1:1 v/v) mobile phase. Quantification was carried out by single wavelength reflectance scanning at 220 nm using a TLC Scanner II. Calibration plots were linear in the range 8–24 μg, and the linearity correlation coefficients ranged between 0.97 and 0.99. Recoveries from laboratory-prepared test samples of the EC formulations were in the range 95–99% (Sharma 2002). Similar methods were described for analysis of formulations of the pyrethroids fenvalerate and deltamethrin (Hamada and Wintersteiger 2002). Calibration plots for these pesticides were linear in the range 3–23 μg, and recoveries from laboratory prepared EC formulations were 96–100%. In both cases, HPTLC was comparable to those obtained using a more complex, slower, and more costly GC-flame ionization detection method.

The six triazine herbicides terbutryn, ametryn, atrazine, propazine, terbuthylazine, and simazine were determined in drinking water after C_{18} SPE; separation on silica gel 60 F_{254} (*n*-heptane:ethyl acetate (1:1 v/v) mobile phase), C_{18} F_{254} (acetonitrile:water, 7:1 v/v), or diol F_{254} (acetone:water:tetrahydrofuran, 4:6:1 by vol) layers; and reflectance dual-wavelength densitometry at 222 nm. Recoveries between 88 and 99% were achieved, and the LOD was 100 ng/L. A fluorescence densitometry screening method was also described involving SPE recovery, derivatization with dansyl chloride, and scanning with an excitation wavelength of 280 nm; the linearity range was 20–1,200 pg/spot and detection limit 2 pg/spot (Tellier et al. 2002).

The metabolism of the cyanooxime fungicide cymoxanil and analogs was studied on several strains of *Botrytis cinera* owing to their difference in sensitivity towards cymoxanil. It was found that HPLC was suitable for analysis of cymoxanil but not its ionizable metabolites; so an ion-pairing HPTLC method with $C_{18}W$ F_{254} plates was devised to monitor these metabolites in unextracted culture media for substrates that were demonstrated to decompose most rapidly. Prior to chromatography, plates were impregnated by dipping into 50 or 70 mM methanolic TBAB (tetra-butylammonium bromide); initial zones were applied with an Automatic TLC Sampler. Plates were developed with phosphate buffer (0.01 M, pH 6): methanol mobile phase in a horizontal chamber, and quantification done by UV reflectance densitometry at 243 nm (Zhang et al. 2003).

The sulfonylurea herbicides metasulfuron-methyl, chlorsulfuron, bensulfuron-methyl, tribenuron, and chlorimuron-methyl were separated on aluminum-backed silica gel F_{254} sheets developed with chloroform: acetone: acetic acid (90:10:0.75 by vol) (mobile phase A) in a twin-trough chamber. The optimum humidity range was 18–42%, as determined using a Vario-TLC chamber and various water–sulfuric acid ratios for vapor phase conditioning. Quantification of bensulfuron-methyl added to tap water at levels of 5–20 μg/kg was carried out using extraction by C_{18} SPE, separation on alumina with double development by mobile phase A and then toluene:ethyl acetate (1:1 v/v), and densitometric scanning at 201 nm. Detection limits ranged from 2 to 8 ng/zone for the herbicides (Tripathi et al. 2006). A HPTLC method, suitable for the rapid screening for germplasm of *Phyllanthus* species for the determination of chemical profiles and quantification of the major lignans, phyllanthin and hypophyllanthin, was developed. The assay was applied without any special pretreatment of the sample, and a large number of samples can be analyzed without compromising accuracy (Tripathi et al. 2008).

Sukumar et al. also reported a simple, rapid, and sensitive HPTLC method to identify and quantify sesamin and sesamolin, the major lignans of the sesamum oil, and the method was applied to polyherbal formulations containing the oil for their quantitative estimation. Kaur et al. (2008) developed a new, simple, sensitive, precise, and robust high-performance thin layer chromatographic (HPTLC) method for the estimation of conessine in herbal extracts and pharmaceutical dosage forms. After derivatizing the plate with modified Dragendroff's reagent, Camag TLC scanner III was used for spectrodensitometric scanning and analysis of the plate in absorbance mode at 520 nm. The system was found to give compact spots for conessine (R_f value of 0.82). The data for calibration plots showed good linear relationship with $r^2 = 0.9998$ in the concentration range of 1–10 μg with respect to peak area. Arbinda et al. described HPTLC method for estimation of tetrahydroamentoflavone (THA), a major bioactive biflavonoid as a chemical marker of *Semecarpus anacardium* (Arbinda et al. 2008). The method was validated in terms of their linearity, LOD, LOQ, precision, and accuracy and compared with RP-HPLC-DAD method. The methods were able to identify and quantify tetrahydroamentoflavone from complex mixtures of phytochemicals and could be extended to the marker-based standardization of polyherbal formulations, containing *S. anacardium*.

HPTLC method was found to be useful in determination of the poor UV absorbing compounds such as jasmonic acid. Dhandhukia and Thakkar (2008) reported the robust HPTLC method for quantitative determination of jasmonic acid that was extracted from the broth of *L. theobromae* with a good regression coefficient ($r = 0.9956$) for curve in the range of 6–90 μg without derivatization of the sample. The method was found to be sensitive for use in quantification of JA.

Sharma et al. (2009) reported that the quantitative densitometric evaluation was performed in the absorbance/reflectance mode at 530 nm for two tropane alkaloids, hyposcyamine and scopolamine. The average recovery of hyposcyamine and scopolamine was 97.4 and 98.6, respectively. The calibration curves were linear in the range of 1,000–4,000 ng and 500–2,000 ng for hyposcyamine and scopolamine, respectively.

Notes

HPTLC provides an efficient, fast, and reliable alternative for quantitative determination of natural products. Advantages of HPTLC over other methods are that once the separation is achieved, sample can be reextracted from the solid support or each band can be scanned with absorbance. Accurate application of the sample can be performed using automated TLC applicator under flow of N_2, e.g., Linomate applicator. Densitometry is the in situ instrumental measurement of visible, UV absorbance, fluorescence, and fluorescent quenching directly on the layer without resorting to scrapping and elution. Depending upon the analytical problem, one can resort the single level or multilevel calibration. Validation is a systemic approach to identifying, measuring, evaluating, documenting, and reevaluating all the critical steps responsible before establishing the validity of the method. The developed method is validated in terms of linearity, precision, LOD, limit of quantification, accuracy, robustness, and ruggedness. HPTLC for fingerprinting of herbal extracts, characterization, and quantification of secondary metabolites isolated in small quantities from a large number of plants affords a number of advantages over conventional methods in use for this purpose.

References

Anderson JM. Simultaneous determination of abscisic acid and jasmonic acid in plant extracts using high performance liquid chromatography. J Chromatogr 1985; 330, 347–355.

Anderson JM. Fluorescent hydrazides for the high performance liquid chromatographic determination of biological carbonyls. Anal Biochem 1986; 152, 146–153.

Aravind SG, Arimboor R, Rangan M, Madhavan SN, Arumughan C. Semi-preparative HPLC preparation and HPTLC quantification of tetrahydroamentoflavone as marker in *Semecarpus anacardium* and its polyherbal formulations. J Pharm Biomed Anal 2008; 48(3), 808–813.

Cimpoui C, Miclaus V, Damain G, Puia M, Casoni D, Bale C, Hodison T. Identification of new pthalazine derivatives by HPTLC-FTIR and characterization of their separation using some molecular properties. J Liq Chromatogr Rel Technol 2003; 26, 2687–2696.

Dhandhukia PC, Thakkar VR. Separation and quantification of jasmonic acid using HPTLC. J Chromatogr Sci 2008; 46, 320–325.

Dongre VG, Kamble VW. HPTLC detection and identification of heroin (Diacetylmorphine) in forensic samples. J Planar Chromatogr 2003; 16, 458–460.

Gbaguide F, Muccioli G, Accrombessi G, Moudachirou M, Quetin-Leclercq J. Densitometric HPTLC quantification of 2-azaanthraquinone isolated from *Mitracarpus scaber* and antimicrobial activity against *Dermatophilus congolensis*. J Planar Chromatogr-Mod TLC 2005; 18 (5), 377–379.

Hahn-Deinstrop E. The documentation of thin layer chromatograms by non densitometric methods. J Planar Chromatogr 1991; 4, 154–157.

Hamada M, Wintersteiger R. Rapid screening of triazines and quantitative determination in drinking water. J Biochem Biophys Methods 2002; 53, 229–239.

Hans-Deinstrop E (2000). Applied Thin Layer Chromatography: Best Practice and avoidance of Mistakes, 2nd edition, published by Willy-VCH Verlag GmbH and Co, Weinheim.

Jagannath NS, Kanan K. HPTLC method for estimation of ellagic acid and gallic acid in Triphala churanam formulations. Res J Phytochem 2008; 2(1), 1–9.

Kakde RB, Kotak VH, Kale DL. Estimation of bisoprolol fumarate in pharmaceutical preparations by HPTLC. Asian J Res Chem 2008; 1(2), 70–73.

Kaur AD, Ravichandran V, Jain PK, Agrawal RK. High-performance thin layer chromatography method for estimation of conessine in herbal extract and pharmaceutical dosage formulations. J Pharm Biomed Anal 2008; 46(2), 391–394.

Khoobdel M, Nematollah J, Babak S. Quantitative and qualitative determination of dimethyl phthalate and N,N-diethyl-m-toluamide in repellents commercial formulations by high performance thin layer chromatography. Pak J Biol Sci 2007; 10(20), 3678–3682.

Linares RM, Ayala JH, Afonse AM, Gonzalez V. Quantitative analysis of biogenic amines by high performance thin layer chromatography utilizing fibre optic fluorescence detector. Anal Lett 1998; 31, 475–478.

Misra H, Mehta D, Mehta BK, Soni M, Jain DC. Study of extraction and HPTLC-UV method for estimation of caffeine in marketed tea (*Camellia sinensis*) granules. Int J Green Pharm 2009; 3 (1), 47–51.

Morlock G, Kopacz S. Fast and precise SBSE-HPTLC/FLD method for quantification of six polycyclic aromatic hydrocarbons frequently found in water. J Liq Chromatogr Relat Technol 2008; 31(13), 1925–1942.

Mueller MJ, Brodschelm W. Quantification of jasmonic acid by capillary gas chromatography-negative chemical ionization-mass spectrometry. Anal Biochem 1994; 218, 425–435.

Pawar PK, Borse TP, Pinjari RZ, Maheshwari VL. A simple technique for rapid quantitative determination of solasodine from cultured hairy roots of *Solanum surattense*. J Herb Med Toxicol 2008; 2(1), 7–10.

Prosek M, Golc-Wondra A, Vovk I. Uncertainty in quantitative thin-layer chromatography. J Chromatogr Sci 2002; 40(10), 598–602.

Raverkar SR. High performance thin layer chromatographic analysis of nitrobenzene process streams. Anal Sci 2001; 17, 287–289.

Ravi TK, Gandhimathi M, Sireesha KR, Jacob S. HPTLC method for the simultaneous estimation of Tizanidine and Rofecoxib in tablets. Indian J Pharm Sci 2006; 68, 234–236.

Sethi PD (1996). HPTLC Quantitative Analysis of Pharmaceutical Formulations, 1st edition, published by CBS Publisher and Distributor, New Delhi.

Sharma KK. Determination of active ingredient in synthetic pyrethroid formulations by high-performance thin-layer chromatography/densitometry. J AOAC Int 2002; 85 (6), 1420–1424.

Sharma V, Sharma K, Singh B, and Gupta RC. Cytomorphological studies and HPTLC finger-printing in different plant parts of three wild morphotypes of *Datura metel* L. "Thorn Apple" from north India. Int J Green Pharm 2009; 3(1), 40–46.

Sherma J. Thin-layer chromatography of pesticides – a review of applications for 2002–2004. ACTA Chromatographica 2005; 15, 5–30.

Sukumar D, Ranjith A, Arumughan C. HPTLC fingerprinting and quantification of lignans as markers in sesame oil and its polyherbal formulations. J Pharm Biomed Anal 2008; 47(4–5), 795–801.

Tellier F, Fritz R, Leroux P, Carlin-Sinclair A, Cherton, JJC. Analytical technologies in the biomedical and life sciences. J Chromatogr B 2002; 769(1), 35–46.

Tripathi AK, Verma RK, Gupta AK, Gupta MM, Khanuja SP. Quantitative determination of phyllanthin and hypophyllanthin in *Phyllanthus* species by high performance thin layer chromatography. Phytochem Anal 2006; 17, 394–397.

Ueda J, Miyamoto K. Separation of a new type of plant growth regulator, jasmonates, by chromatographic procedures. J Chromatogr A 1994; 658, 129–142.

Wagner SD and Sherma J. Analysis of an active ingredient Cimetidine in acid reduction tablets by high performance thin layer chromatography with ultraviolet absorption densitometry. Chromatography 2001; 22(2), 97 99.

Zhang R, Yue Y, Hua R, Tang F, Cao H, Ge S. J Planar Chromatogr 2003; 16, 127–130.

Chapter 13
Quantification of Low Molecular Mass Compounds Using Thermostated Planar Chromatography

Paweł K. Zarzycki

Abstract Separation, detection, and quantification of selected low-molecular mass compounds involving planar chromatography working under temperature-controlled conditions has been described, focusing mainly on the author and co-workers approach. Particularly, basic theory concerning temperature effects in chromatography as well as construction of simple vertical and horizontal chamber units for thermostated thin-layer chromatography (TLC) are discussed from a practical point of view. Moreover, an advantage of microplanar chromatography for fast one- and two-dimensional separation of multicomponent samples, including plant extracts, over separation protocols achieved via regular size TLC as well as its column counterpart, namely high-performance liquid chromatography (HPLC), is also demonstrated. Additionally, temperature effect that may significantly improve sensitivity, selectivity, and robustness of spots detection process performed by postrun derivatization with phosphomolybdic acid (PMA) is reported along with selected applications.

Temperature Effects

One of the important parameters that may significantly affect chromatographic retention, separation efficiency, and in particular cases, detection of components of interest is temperature. It is easy to observe that in elevated temperature, numbers of solutes are less retained on the stationary phase in comparison to room or subambient temperature regions, using either gas or liquid mobile phases. Generally, it has been found that solutes' retention is inversely related to temperature

---

P.K. Zarzycki
Section of Toxicology and Bioanalytics, Koszalin University of Technology, Śniadeckich 2, 75-453 Koszalin, Poland
e-mail: pkzarz@wp.pl

MM. Srivastava (ed.), *High-Performance Thin-Layer Chromatography (HPTLC)*, DOI 10.1007/978-3-642-14025-9_13,

(Snyder 1970; Laub and Purnell 1978; Chmielowiec and Sawatzky 1979). When the retention mechanism is the same over the temperature range investigated, the resulting retention profile yields a straight line as it is presented in Fig. 13.1. Under such conditions, the slope and intercept of linear plot $y = ax + b$ correspond to the enthalpy and entropy changes, according to the (13.1) expressed as:

$$\ln k = -\Delta H^{\mathrm{o}}/RT + -\Delta S^{\mathrm{o}}/RT/\ln \varphi, \tag{13.1}$$

where k denotes chromatographic retention factor, ΔH^{o} and ΔS^{o} are related to enthalpy and entropy changes of solute transfer from the mobile phase to stationary phase, respectively, and T expresses absolute temperature measured in Kelvin scale. Gas constant and phase ratio of the column are designated by R and φ symbols, respectively. The graphical representation of the above equation, where X axis is related to reversed T values and Y axis represents logarithmic form of the retention factor, is commonly described as Van't Hoff plot. In gas chromatography (GC), temperature is considered as one of the critical and most useful parameters affecting the retention of solutes, peaks separation, and total analysis time. However, it should be noted that in reality, the temperature effect on retention in classical chromatography is relatively week. Small slope values of the lines on the Van't Hoff plots are typical for number of low molecular mass compounds, including, e.g., n-alkanes or polycyclic aromatic hydrocarbons. Therefore, in GC, wide range of temperature from 100 to 300°C must be considered to get a significant effect on the analytes retention and subsequently improve mixture separation. Therefore, this approach may be applied for separation of temperature-stable components of

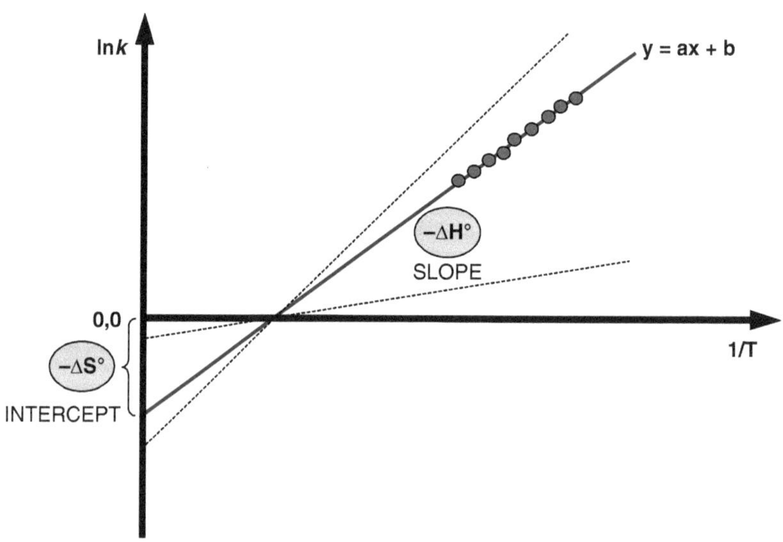

Fig. 13.1 Graphical representation of Van't Hoff plot based on chromatographic retention

interest only, which is the main disadvantage of the use of GC technique for a number of thermally nonstable bioactive substances.

One of the consequences of the weak retention response for temperature alternated within room condition (with the common range of 7–32°C) is that the temperature was almost neglected as an important factor controlling the retention in non-GC techniques. Particularly, chromatographic systems driven by liquid mobile phases included capillary electrophoresis (CE), high-performance liquid chromatography (HPLC), and thin-layer chromatography (TLC). Nevertheless, any reversible process that may alter the enthalpy or entropy of solutes' adsorption in principle gives rise to nonlinear Van't Hoff plots. This may significantly change the retention and separation response even within narrow temperature range. Among others, the changes in solutes' conformation, presence of multiple types of retention mechanism, or multiple types of binding sites may lead to nonlinearity of the Vant't Hoff plots (Melander et al. 1979; Horvath et al. 1976; Pirkle 1991; Boehm et al. 1988; Dimov and Simova 1999; Lou et al. 2006; Saito et al. 2003).

From a practical point of view, strong nonlinear temperature effect within narrow temperature range close to room condition can be observed if the retention of analytes is modulated through mobile phase inclusion additives, e.g., cyclodextrins (CD). Such a group of toroidal-shaped oligosaccharides may efficiently contribute to several guest-associated phenomena in solutions. The host–guest complexation with CDs is highly selective and is the method of choice for resolution of various isomers including enantiomers. Particularly, in planar chromatography, cyclodextrins were frequently used as very efficient chiral and nonchiral selectors (Hinze and Armstrong 1980; Armstrong 1980; Dębowski et al. 1985; Lepri et al. 1994). It should be noted that the CD complexation process is extremely temperature-dependent and, under particular conditions, the association constant of host–guest complexes can change more than 200% per 1°C (Zarzycki and Lamparczyk 1996, 1998a). The experimental data indicates that retention of inclusion complexes can be varied between two lines formed by the Van't Hoff plot of the cyclodextrin and the Van't Hoff plot of uncomplexed solute (Fig. 13.2) (Zarzycki and Smith 2001). Noteworthy, if the retention of cyclodextrin is considerably lower (preferably close to the retention time of nonretarded marker, t_0) than the retention of solutes chromatographed, then the temperature can produce a massive change in separation power of chromatographic system (Lamparczyk et al. 1994; Zarzycki and Lamparczyk 1998b; Zarzycki et al. 2009a). Importantly, in low temperatures, the inclusion modifier action is more efficient due to high values of the binding constant of the complexes created and large differences in retention of free CD and uncomplexed solutes. It has been observed that small changes in the temperature of the mobile phase modified with macrocyclic additives may produce substantial differences in the separation power of planar chromatographic system designed for separation of enantiomeric pair mixtures or structural isomers of low molecular mass compounds, for example, amino acids including tryptophan and its derivatives (Lepri et al. 1990; Xuan and Lederer 1994).

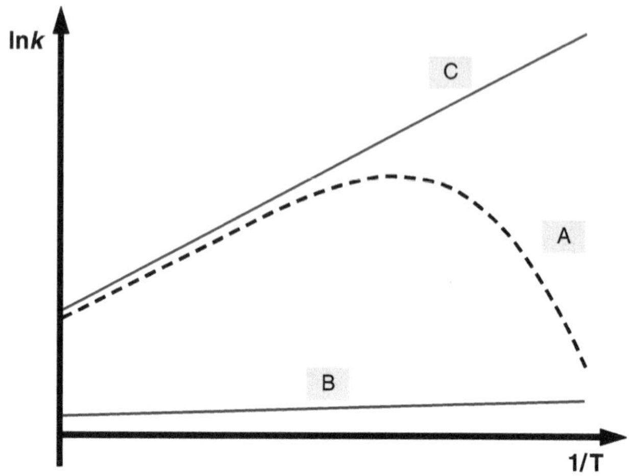

Fig. 13.2 Deviation from linear Van't Hoff plot (**a**) as the result of host–guest interaction existing between mobile phase macrocyclic additive (which should be nonretarded by the stationary phase components); (**b**) and chromatographed substance of interest (**c**)

Approach to Temperature-Controlled Planar Chromatography

Despite the phenomenon described above, planar chromatographic process may be significantly affected by the heat evolved during solvent adsorption. Contrary to column chromatography in TLC, stationary phase is not immersed with liquid mobile phase before chromatographic run. Therefore, due to the heat evolved during solvent adsorption, the temperature of the stationary phase usually increases near solvent front. When low heat transfer glass plates are used, this phenomenon makes it difficult to control the temperature of the chromatographic process. This process can increase the stationary phase temperature as much as 3–4°C and the basic separation conditions can be considerably different for spots migrating far from or near to of the mobile phase front. The simplest way to avoid such phenomenon is to replace glass-based plates into aluminum-backed sheets and/or to perform the chromatographic run using small internal volume thermostated chamber (Zarzycki 2001a) (Table 13.1).

Despite the mobile/stationary phase problems, each classical TLC run involves gas phase containing mobile phase vapor. Therefore, any irregularity in the temperature distribution inside chromatographic chamber may lead to substantial changes in the spots' migration distance because some of the mobile phase might evaporate from the plate. Such distillation process, which especially may be observed at an elevated temperature using improperly thermostated and isolated TLC chambers give rise to strong deviations from expected linear Van't Hoff plots (Zarzycki 2001a). It is noteworthy that this phenomenon may not only improve the spots' separation but also make such chromatographic systems nonrobust and

Table 13.1 Examples of planar chromatographic separations performed at elevated and subambient temperature using classical TLC chambers working inside temperature controlling devices

Analytes	Type of temperature controlling device	Temperature range (°C)	Authors and publication year
Dyes	Climatic cabinets	+5 to +60	Geiss and Schlitt (1968)
Steroids, PAHs, fatty acids	Water bath	+4 to +40	Singh and Gershbein (1970)
Dyes	Circulating-air gas chromatograph oven	+20 to +160	Berezkin and Bolotov (1981, 1983)
Polar lipids	Thermostated water bath	+20 to +32.5	Kolarovic et al. (1985), Kolarovic and Traitler (1985)
Polyethylene	Heated, insulated external chamber	+100 to +140	Armstrong and Yang (1988)
Fatty acids, biphenyls, cholesteryl esters	Water bath	+4	Touchstone et al. (1989)
Amino acid enantiomers	Heating oven and cooling cabinet	−10 to +63	Kuhn et al. (1989)
Thiazine dyes	Laboratory drying oven	+5 to +40	Liedekerke et al. (1991)
Bile acids	Freezer, refrigerator, and incubator	−20 to +50	Rivas-Nass and Müllner (1994)
Bile acids	Incubator	+18 to +40	Pyka et al. (2005)

virtually useless for any thermodynamic data collection and physicochemical investigations.

Under proper experimental conditions, in which the heat evolved during solvent adsorption and "distillation process" is minimized, the planar chromatographic systems can be very suitable for separation at elevated and subambient temperatures. This is mainly because of the low flow rate and small amount of the mobile phase that is necessary to perform the separation process. Moreover, there is no Joule's heat evolved due to electric current flow, as in, for example, planar electrophoresis systems (Novotny et al. 2006). Hence, to ensure proper plate temperature, the thermostating devices can work with great precision and consume much less energy. Unfortunately, in contrast to great progress in the control of humidity and/or use of gas-phase-controlled TLC chambers, the problem of high precision and reproducibility of plate temperature in commercially available devices has not been successfully resolved, and this problem is still poorly recognized (Camag Automatic Developing Chamber ADC 2; Dzido 1996, 2001; Zarzycki and Lamparczyk 2001; Dzido et al. 2002; Berezkin et al. 2006). The simplest way to perform an isothermal separation at subambient or elevated temperature is the use of a standard TLC chamber working within a thermostatically controlled container, including a freezer, refrigerator, incubator, water bath, laboratory drying oven, or circulating-air gas chromatograph oven. As it is listed in Table 13.2, this approach has been successfully implemented for planar chromatographic separations of different classes of analytes within wide range of temperatures from −20 to +160°C (Geiss and Schlitt 1968; Singh and Gershbein 1970; Berezkin and Bolotov

Table 13.2 List of chemical standards and biological extracts separated and/or quantified under thermostated TLC conditions[a] reported by the author and co-workers

Substance or biological extract	Developing device type and saturation mode	Temperature (°C)	Retention data	Quantitative data	Ref.
Azucalen herbs extract (*Chamomillae* and *Calendulae* extractum fluidum)	Horizontal microchamber; unsaturated	+20	No	Yes	Zarzycki (2008), Zarzycki et al. (2009b)
Bromocresol green	Horizontal microchamber; unsaturated	+20	No	Yes	Zarzycki and Zarzycka (2008a)
Calix[4]arene	Vertical Dewar-type; saturated	+30	Yes	No	Zarzycki (2001b)
Chenodeoxycholic acid (CDCA)	Horizontal; saturated	+5 to +60	Yes	No	Zarzycki et al. (1999a)
Chenodeoxycholic acid, sodium salt (NaCDC)	Horizontal; saturated	+5 to +60	Yes	No	Zarzycki et al. (1999a)
Cholecalciferol	Horizontal Chromdes DS-L; saturated	+30	Yes	Yes	Zarzycki et al. (2007c)
4-Cholesten-3-one	Horizontal Chromdes DS-L; saturated	+30	Yes	Yes	Zarzycki et al. (2007c)
Cholesterol	Vertical Dewar-type; saturated	+20	No	Yes	Zarzycki and Zarzycka (2008a), Modzelewski et al. (2008)
Cholesterol	Horizontal Chromdes DS-L; saturated	+5 to +39	Yes	No	Zarzycki et al. (2007a)
Cholesterol	Horizontal; saturated	+5 to +60	Yes	No	Zarzycki et al. (1999a)
Cholesterol acetate	Horizontal Chromdes DS-L; saturated	+30	Yes	Yes	Zarzycki et al. (2007c)
Cholic acid (CA)	Horizontal; saturated	+5 to +60	Yes	No	Zarzycki (2001a), Zarzycki et al. (1999a)
Cholic acid, sodium salt (NaCl)	Horizontal; saturated	+5 to +60	Yes	No	Zarzycki et al. (1999a)
Crystal violet	Horizontal microchamber; unsaturated	+20	No	Yes	Zarzycki and Zarzycka (2008a)
α-Cyclodextrin	Horizontal; saturated	+5 to +60	Yes	No	Zarzycki et al. (1997, 1999b)
α-Cyclodextrin		+30	Yes	No	Zarzycki et al. (1995, 1996)

	Vertical immersible; saturated				
β-Cyclodextrin	Horizontal microchamber; saturated	+20	No	Yes	Zarzycki (2008)
β-Cyclodextrin	Horizontal; saturated	+5 to +60	Yes	No	Zarzycki et al. 1997, 1999b
β-Cyclodextrin	Vertical immersible; saturated	+30	Yes	No	Zarzycki et al. 1995, 1996
γ-Cyclodextrin	Horizontal; saturated	+5 to +60	Yes	No	Zarzycki et al. 1997, 1999b
γ-Cyclodextrin	Vertical immersible; saturated	+30	Yes	No	Zarzycki et al. (1995, 1996)
Deoxycholic acid (DCA)	Horizontal; saturated	+5 to +60	Yes	No	Zarzycki et al. (1999a)
Deoxycholic acid, sodium salt (NaDC)	Horizontal; saturated	+5 to +60	Yes	No	Zarzycki et al. (1999a)
Dihydrocholesterol	Horizontal Chromdes DS-L; saturated	+30	Yes	Yes	Zarzycki et al. (2007c)
13,14-Dihydro-15-keto-prostaglandin E_2 (PGEM)	Vertical Dewar-type; saturated	−20 to +60	Yes	No	Welsh et al. (2003)
13,14-Dihydro-15-keto-prostaglandin $F_{2\alpha}$ (PGFM)	Vertical Dewar-type; saturated	−20 to +60	Yes	No	Welsh et al. (2003)
Dimethyl-β-cyclodextrin	Horizontal microchamber; saturated	+20	No	Yes	Zarzycki (2008)
Equilin	Horizontal microchamber; saturated	+20	Yes	No	Zarzycki et al. (2009c)
Ergosterol	Horizontal Chromdes DS-L; saturated	+5 to +39	Yes	No	Zarzycki et al. (2007a)
Ergosterol	Horizontal Chromdes DS-L; saturated	+30	Yes	Yes	Zarzycki et al. (2007c)
Estetrol	Horizontal micro-chamber; saturated	+20	Yes	No	Zarzycki (2008), Zarzycki et al. (2009c)
Estetrol	Horizontal microchamber; unsaturated	+20	No	Yes	Zarzycki (2008)
Estetrol		−20 to +60	Yes	No	Zarzycki (2002), Zarzycki et al. (2005)

(continued)

Table 13.2 (continued)

Substance or biological extract	Developing device type and saturation mode	Temperature (°C)	Retention data	Quantitative data	Ref.
	Vertical Dewar-type; saturated				
17α-Estradiol	Horizontal microchamber; saturated	+20	Yes	No	Zarzycki et al. (2009c)
17β-Estradiol	Horizontal microchamber; saturated	+20	Yes	No	Zarzycki et al. (2009c)
17β-Estradiol	Horizontal microchamber; unsaturated	+20	No	Yes	Zarzycki (2008)
17β-Estradiol	Vertical Dewar-type; saturated	−20 to +60	Yes	No	Zarzycki (2002), Zarzycki et al. (2005)
Estriol	Horizontal microchamber; saturated	+20	Yes	No	Zarzycki et al. (2009c)
Estriol	Horizontal microchamber; unsaturated	+20	No	Yes	Zarzycki (2008), Zarzycki and Zarzycka (2008b)
Estriol	Vertical Dewar-type; saturated	−20 to +60	Yes	No	Zarzycki (2002), Zarzycki et al. (2005)
Estrone	Horizontal microchamber; saturated	+20	Yes	No	Zarzycki et al. (2009c)
Estrone	Horizontal microchamber; unsaturated	+20	No	Yes	Zarzycki (2008)
Estrone	Vertical Dewar-type; saturated	−20 to +60	Yes	No	Zarzycki (2002), Zarzycki et al. (2005)
Fluorescein	Horizontal microchamber; unsaturated	+20	No	Yes	Zarzycki and Zarzycka (2008a)
Fullerene C60	Horizontal microchamber; unsaturated	+20	Yes	Yes	Zarzycki et al. (2009b)
Fullerene C60	Horizontal microchamber; unsaturated	−20 and +20	Yes	Yes	Zarzycki (2008)
Fullerene C60	Horizontal microchamber; saturated; unsaturated	+20	Yes	Yes	Zarzycki et al. (2007b)

Fullerene C70	Horizontal microchamber; unsaturated	+20	Yes	Yes	Zarzycki et al. (2009b)
Fullerene C70	Horizontal microchamber; unsaturated	−20 and +20	Yes	Yes	Zarzycki (2008)
Fullerene C70	Horizontal microchamber; saturated; unsaturated	+20	Yes	Yes	Zarzycki et al. (2007b)
Glycocholic acid (GCA)	Horizontal; saturated	+5 to +60	Yes	No	Zarzycki et al. (1999a)
Glycocholic acid, sodium salt (NaGC)	Horizontal; saturated	+5 to +60	Yes	No	Zarzycki et al. (1999a)
Glycodeoxycholic acid (GDCA)	Horizontal; saturated	+5 to +60	Yes	No	Zarzycki (2001a), Zarzycki et al. (1999a)
Glycodeoxycholic acid, sodium salt (NaGDC)	Horizontal; saturated	+5 to +60	Yes	No	Zarzycki et al. (1999a)
17α-Hydroxyprogesterone	Horizontal microchamber; saturated	+20	Yes	No	Zarzycki et al. (2009c)
20α-Hydroxyprogesterone	Horizontal microchamber; saturated	+20	Yes	No	Zarzycki et al. (2009c)
15-Keto-prostaglandin E_2 (15-keto-PGE_2)	Vertical Dewar-type; saturated	−20 to +60	Yes	No	Welsh et al. (2003)
15-Keto-prostaglandin $F_{2\alpha}$ (15-keto-$PGF_{2\alpha}$)	Vertical Dewar-type; saturated	−20 to +60	Yes	No	Welsh et al. (2003)
Lithocholic acid (LCA)	Horizontal; saturated	+5 to +60	Yes	No	Zarzycki et al. (1999a)
Methyl-β-cyclodextrin	Horizontal microchamber; saturated	+20	No	Yes	Zarzycki (2008)
Methyl red	Horizontal microchamber; unsaturated	+40	No	Yes	Zarzycki and Zarzycka (2008c)
Methyl red	Horizontal microchamber; unsaturated	+20	No	Yes	Zarzycki and Zarzycka (2008a)
Methyltestosterone	Horizontal microchamber; saturated; unsaturated	−20 to +60	Yes	Yes	Zarzycki and Zarzycka (2008b)
Phenolphthalein	Horizontal microchamber; unsaturated	+20	No	Yes	Zarzycki and Zarzycka (2008a)
Progesterone		+20	Yes	No	Zarzycki et al. (2009c)

(continued)

Table 13.2 (continued)

Substance or biological extract	Developing device type and saturation mode	Temperature (°C)	Retention data	Quantitative data	Ref.
	Horizontal microchamber; saturated				
Prostaglandin E_2 (PGE_2)	Vertical Dewar-type; saturated	+30	No	Yes	Zarzycki and Bartoszuk (2008)
Prostaglandin E_2 (PGE_2)	Vertical Dewar-type; saturated	−20 to +60	Yes	No	Welsh et al. (2003)
Prostaglandin $F_{2\alpha}$($PGF_{2\alpha}$)	Vertical Dewar-type; saturated	+30	No	Yes	Zarzycki and Bartoszuk (2008)
Prostaglandin $F_{2\alpha}$($PGF_{2\alpha}$)	Vertical Dewar-type; saturated	−20 to +60	Yes	No	Welsh et al. (2003)
Pyrene	Vertical Dewar-type; saturated	+30	Yes	No	Zarzycki (2001b)
Rifampicin	Horizontal; saturated	+5 to +60	Yes	No	Zarzycki et al. (1997)
Rifamycin B	Horizontal; saturated	+5 to +60	Yes	No	Zarzycki et al. (1997)
Spirulina maxima dyes from pharmaceutical formulations	Horizontal microchamber; unsaturated	+40	Yes	Yes	Zarzycki (2008), Zarzycki et al. (2009b), Zarzycki and Zarzycka (2008a, 2008c)
Stigmasterol	Horizontal Chromdes DS-L; saturated	+5 to +39	Yes	No	Zarzycki et al. (2007a)
Stigmasterol	Horizontal Chromdes DS-L; saturated	+30	Yes	Yes	Zarzycki et al. (2007c)
Taurodeoxycholic acid, sodium salt (NaTDC)	Horizontal; saturated	+5 to +60	Yes	No	Zarzycki et al. (1999a)
Testosterone	Horizontal microchamber; unsaturated	+20	Yes	Yes	Zarzycki et al. (2009b)
Testosterone	Horizontal microchamber; saturated; unsaturated	−20 to +60	Yes	Yes	Zarzycki (2008), Zarzycki and Zarzycka (2008b)
Testosterone caprate	Horizontal microchamber; saturated; unsaturated	−20 to +60	Yes	Yes	Zarzycki (2008), Zarzycki and Zarzycka (2008b)
Testosterone enanthate		−20 to +60	Yes	Yes	

	Horizontal microchamber; saturated; unsaturated				Zarzycki (2008), Zarzycki and Zarzycka (2008b)
Testosterone isobutyrate	Horizontal microchamber; saturated; unsaturated	−20 to +60	Yes	Yes	Zarzycki (2008), Zarzycki and Zarzycka (2008b)
Testosterone isocaproate	Horizontal microchamber; saturated; unsaturated	−20 to +60	Yes	Yes	Zarzycki (2008), Zarzycki and Zarzycka (2008b)
Testosterone phenylpropionate	Horizontal microchamber; saturated; unsaturated	−20 to +60	Yes	Yes	Zarzycki (2008), Zarzycki and Zarzycka (2008b)
Testosterone propionate	Horizontal micro-chamber; saturated; unsaturated	−20 to +60	Yes	Yes	Zarzycki (2008), Zarzycki and Zarzycka (2008b)
4-*tert*-Butylcalix[4]arene tribenzoate	Vertical Dewar-type; saturated	+30	Yes	No	Zarzycki (2001b)
4-*tert*-Butylcalix[5]arene	Vertical Dewar-type; saturated	+30	Yes	No	Zarzycki (2001b)
Trimethyl-β-cyclodextrin	Horizontal microchamber; saturated	+20	No	Yes	Zarzycki (2008)

[a]Chamber body material as well as developing unit internal dimensions (W, H ,D, ϕ; mm) and volumes (V; cm^3):
Horizontal chromdes DS-L PTFE: $W = 250$, $H = 15$, $D = 130$, $V = 488$
Horizontal chamber aluminum: $W = 60$, $H = 15$, $D = 110$, $V = 99$
Horizontal microchamber PTFE or brass: $W = 56$, $H = 8$, $D = 52$, $V = 23$
Vertical immersible glass: $W = 40$, $H = 120$, $D = 15$, $V = 72$
Vertical Dewar-type glass: $\phi = 36$, $H = 200$, $V = 204$

1981, 1983; Kolarovic and Traitler 1985; Kolarovic et al. 1985; Armstrong and Yang 1988; Touchstone et al. 1989; Kuhn et al. 1989; Liedekerke et al. 1991; Rivas-Nass and Müllner 1994; Pyka et al. 2005). From the experiences of the author and co-workers, a constant and reproducible plate temperature conditions may be easy obtained when the developing unit is placed directly into thermostatically controlled water bath or air-circulating oven (Figs. 13.3 and 13.4) (Zarzycki 2001a; Zarzycki et al. 1995, 2007a). Despite equipment simplicity, the main disadvantage of such equipment configuration is the relatively low rate of temperature equilibration because of the chamber body material, chamber weight, and type of heat-exchange medium. In case of water bath, there are some additional problems with humidity control inside the developing unit. For typical PTFE chamber such as Chromdes DS-L (Dzido 2001; Dzido et al. 2002) to achieve desired temperature

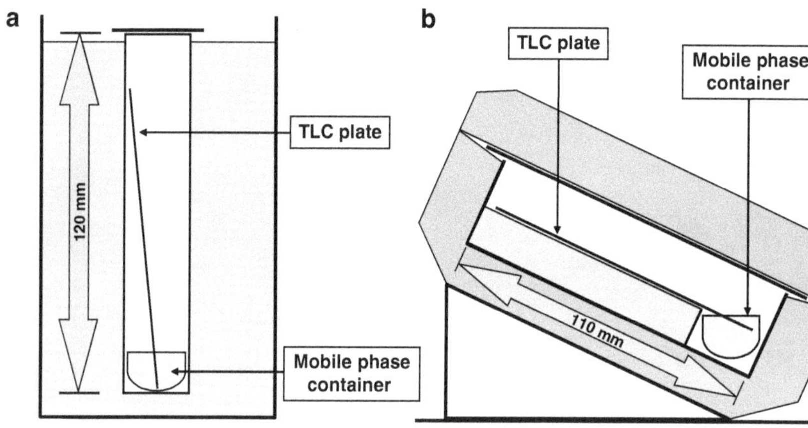

Fig. 13.3 Section drawings of simple temperature-controlled TLC devices: submersible vertical glass chamber unit working in water bath (**a**) (Zarzycki et al. 1995) and horizontal aluminum made and foam insulated chamber with internal heat exchange compartment, connected to the external liquid circulating thermostat (**b**) (Zarzycki 2001a)

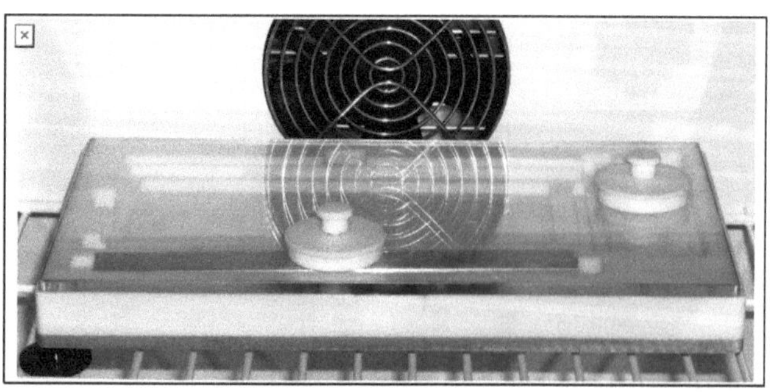

Fig. 13.4 Commercial horizontal TLC chamber with eluent distributor (Chromdes DS-L, Lublin, Poland), working inside thermostated air-circulating incubator (Zarzycki et al. 2007a)

within relatively narrow range from 5 to 40°C, developing unit must be thermostated for at least 2 h before chromatographic run (Zarzycki et al. 2007a). However, after temperature equilibration of the developing unit, a typical aluminum-backed TLC plate reaches the given temperature within just few minutes, without alteration of the chamber body temperature. Therefore, the main applications of such equipment setup are parallel chromatographic runs at a fixed temperature rather than for temperature screening for optimization of chromatographic separation or physicochemical investigations.

From practical point of view, especially considering the temperature range available as well as reproducibility and temperature equilibration speed, two different but complementary approaches have been used: Glass Dewar flask-based chamber with removable TLC plate support (Fig. 13.5) (Zarzycki 2002) and PTFE or chromium-coated brass removable micro-TLC unit (Figs. 13.6 and 13.7) (Zarzycki 2008), for vertical and horizontal separations modes, respectively. Both developing systems were connected to an external liquid circulating thermostat, providing a constant TLC plate temperature ranging from −20 to +80°C with

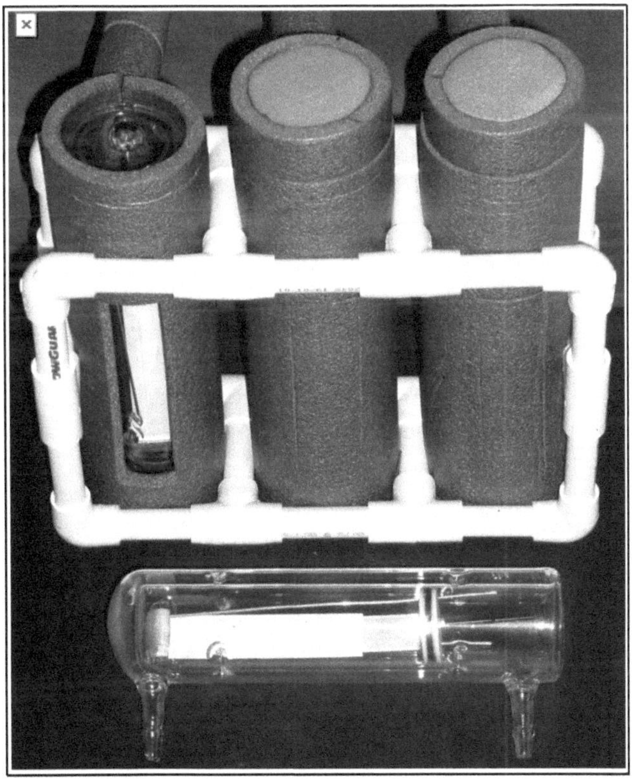

Fig. 13.5 The battery of the thermostated vertical developing chambers based on Dewar-type glass flask module (Zarzycki 2002)

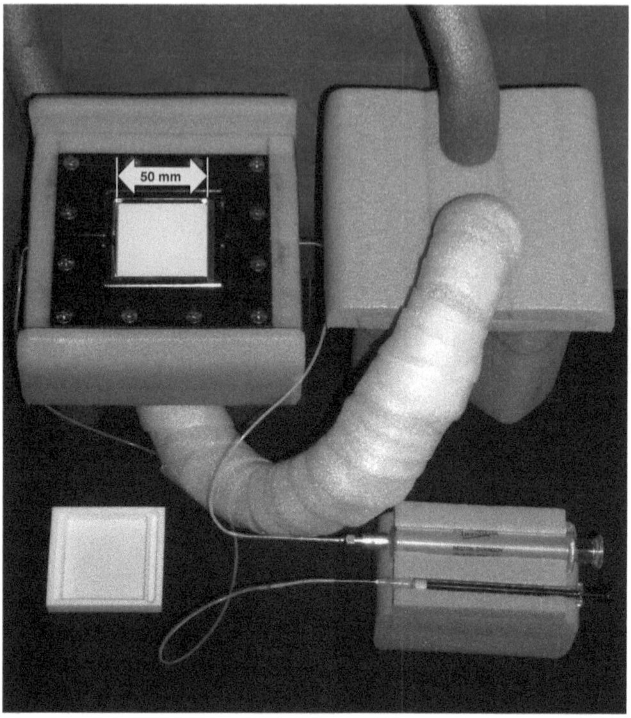

Fig. 13.6 Perspective view of the horizontal microchamber unit working inside temperature controlled metal oven (Zarzycki 2008). The cover and bottom part of the metal oven are both connected to the external liquid circulating thermostat using flexible foam insulated silica pipes

an accuracy at least of $\pm 0.1°C$. Vertical system allows developing up to 10 cm long plates or more, depending on the Dewar flask heights. Horizontal module was designed to develop up to 50 mm long/wide micro-TLC plates, allowing a maximum mobile phase front migration distance of 45 mm. It should be noted that using modern high-performance TLC plates (HPTLC), the mobile phase developing distance can be significantly reduced from 10 to 20 cm to below 5 cm. This conclusion is based on the observation that minimum values of the plate height (*H*) considering nonforced flow rate developing mode can be achieved if the solvent migration distance along the HPTLC plate ranges from 30 to 40 mm (Wall 2006). Therefore, it can be expected that within this migration distance range, more dense spots can be obtained, and sensitivity of quantification protocols will be improved. In practice, under such conditions, efficient separations for a number of types of substances were described, and total analysis time was dramatically reduced in comparison to chromatographic separations performed on common 20 or 10 cm long TLC plates (Fig. 13.8) (Wall 2006; Reich et al. 2003).

The main advantages of micro-TLC approach for separation in subambient and elevated temperatures include the temperature homogeneity inside chromatographic

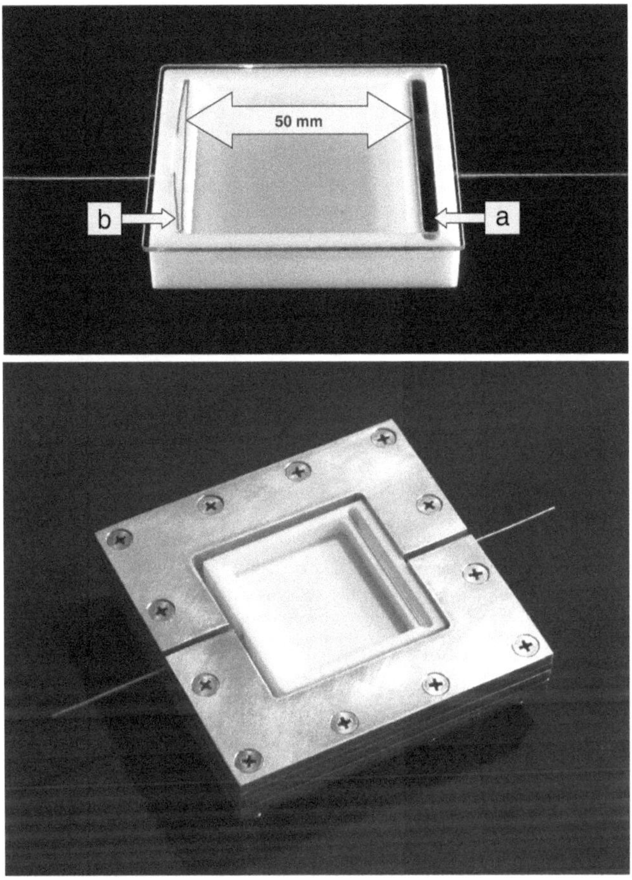

Fig. 13.7 Construction details of PTFE microchamber removable unit (*top picture*) and underside part of the temperature-controlled metal oven (*bottom picture*); **a** – eluent application bar, **b** – metal spring for the TLC plate positioning

chamber unit and minimizing of the mobile phase "distillation process," which can give rise to pseudo-nonlinear Van't Hoff plots. Moreover, because of the chamber material, weight, and volume, the high rate of the temperature equilibration can be obtained. In addition, due to chromatographic plate size, separation and detection processes require low amount of the mobile phases and visualization reagents. Typical working protocol to obtain microplanar chromatograms using horizontal TLC is fast and simple (Zarzycki 2008). Principally, TLC plate with components of interest spotted on the starting line is positioned horizontally inside the chamber module with the stationary phase layer placed down. Afterward, the chamber module is transferred into the thermostating oven cavity and sealed using 1 mm thin glass cover. Then, the movable thermostated cover of the oven is slid to reach the position above the TLC chamber module, and the temperature equilibration step is performed for few minutes, if necessary. The main chromatographic process is started after

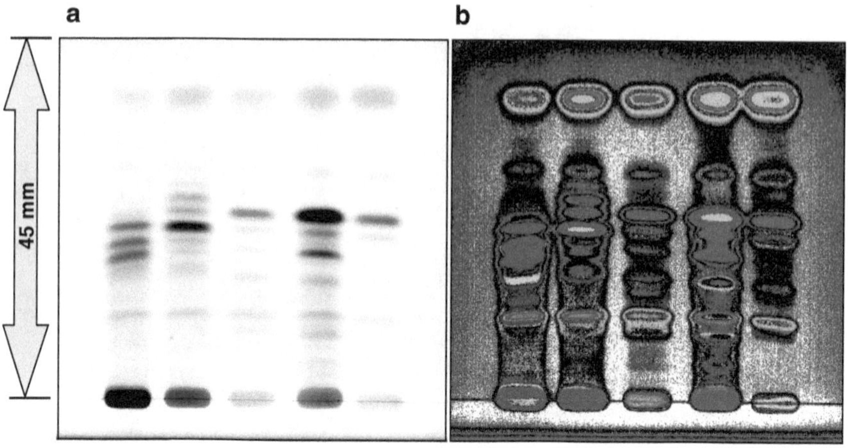

Fig. 13.8 Micro-TLC chromatogram of the *Spirulina maxima* dyes from the pharmaceutical formulation (visible light scan; (**a**) obtained at elevated temperature 40°C using thermostated horizontal microchamber unit (presented on Figs. 13.6 and 13.7). Corresponding artificial color representation of the original scan; (**b**) was prepared using Scion Image freeware (Rainbow Color Table Filter). Chromatographic lanes from the left: methanol, 1-propanol, acetonitrile, acetone, and *n*-hexane. Chromatographic conditions: stationary phase: HPTLC RP18W (Merck); mobile phase: 30% (v/v) acetone/*n*-hexane; chamber saturation mode: unsaturated; sample application mode: manual using home-made microsprayer (2 × 6 mm bands); sample volume: 30 μl; plate temperature equilibrium before developing: 15 min; developing time: 5 min; spots visualization and data acquisition method: direct digital scan via Plustek OpticPro S12 USB office scanner

injecting of given eluent in the volume ranging from 0.3 to 1 ml through the injection pipe into the mobile phase application bar. The TLC plate should be removed from the chamber module immediately after the mobile phase front reached to the plate edge located opposite to the application bar. Depending on the TLC plate supporting material (typically plastic, glass or aluminum), proper plate temperature equilibration can be obtained within 5–12 min (Zarzycki 2008), and a typical nonforced flow run can be finished within a short period of time ranging from 5 to 20 min (Zarzycki 2002, 2008); even separation process is performed under subambient temperature and below 0°C (Zarzycki 2002). Our experimental data revealed that micro-TLC plate (50 × 50 mm size coated with RP18 HPTLC stationary phase) working under two-dimensional developing mode is capable of separating more than 240 spots, which makes such methodology useful for nonexpensive, fast, and high throughput fingerprinting of complex biological samples including cyanobacterias or plant extracts from pharmaceutical formulations (Zarzycki et al. 2009b). In particular cases, e.g., for fullerenes C60/70, microplanar separation performed on HPTLC plates with a developing distance of 45 mm is better for those observed on a 25 cm length analytical HPLC column under similar conditions including stationary phase carbon coverage, mobile phase composition, and temperature (Zarzycki et al. 2007b).

It should be noted that from its nature, isocratic planar chromatography has a great capability to separate mixtures composed of low retarded components of interest. This problem can be simply explained taking into account nonlinear relationship that exists between column (k) and planar (R_F) retention parameters and is described by (13.2):

$$\log k = R_M, \tag{13.2}$$

where k describes retention factor derived from column chromatography and R_M corresponds to mobility factor calculated from planar chromatographic retention parameter R_F using common (13.3) (Gasparic et al. 1966; Soczewiński 2002)

$$R_M = \log[1/(R_F - 1)]. \tag{13.3}$$

One of the interesting consequences of such nonlinear interdependence between raw k and R_F retention parameters is the high capability of planar systems to separate complex mixtures composed of relatively weakly retarded analytes ($k < 10$). This can be easily demonstrated through, e.g., graphical representation of $\log k = R_M$ equation (Zarzycki et al. 2005). Particularly, within low retarded group of analytes chromatographed under similar experimental conditions (the same type of stationary phase and mobile phase composition), a regular distribution of spots on the plate corresponds to strong irregular dispersion of peaks on the chromatogram generated by the column method. Such approach under temperature-controlled conditions involving both vertical and horizontal separations methodology was used for the efficient separation and quantification of number of low molecular mass analytes including dyes, fullerenes, cyclodextrins, calixarenes, macrocyclic antibiotics, prostaglandins, steroids, as well as complex biological mixtures as it is listed in Table 13.2 (Zarzycki 2001a, b, 2002, 2008; Zarzycki et al. 1995, 1996, 1997, 1999a, b, 2005, 2007a, b, c, 2009b, c; Zarzycki and Zarzycka 2008a, b, c; Modzelewski et al. 2008; Welsh et al. 2003; Zarzycki and Bartoszuk 2008).

Temperature-Sensitive Quantification of Target Compounds

Planar chromatography allows fast visualization and detection of a number of UV–Vis-transparent compounds by postrun proceedings of the developed plates involving temperature alteration. For example, temperature-controlled devices for mapping of the analyte spots on the TLC plates based on the liquid crystal detection approach was invented and described by Witkiewicz (1989) and Błądek (1988). The temperature, in which maximum sensitivity of such detection system was observed, depends mainly on the clearing point of the liquid crystals as well as the chemical structure of substances detected. They described a quantitative relationship between temperature and the detection sensitivity of the air pollutants investigated. In practice, however, a number of low molecular mass compounds may be sensitively detected without use of any sophisticated detectors, by simply using common staining reagents. In particular, phosphomolybdic acid (PMA) can give intensely blue mixed oxides, in which the initial Mo (VI) is reduced to Mo (IV), with a number of substances, especially for those that are transparent to UV–Visible light. This reagent is frequently referred to as the most commonly employed stain for thin-layer chromatographic detection of bioactive compounds including lipids, saponins, prostaglandins, terpenes, aminophospholipids, and common sterols (Wall 2006; Sherma and Fried 2003; Lamparczyk 1992). Moreover, visualization protocols based on the color reaction of transparent analytes with the PMA are frequently applied for purity confirmation and substance identification from complex pharmaceutical formulations, according to the European Pharmacopoeia qualitative and quantitative analytical procedures involving planar chromatography (European Pharmacopoeia). The main advantages of staining procedures utilizing PMA reagent are simplicity and sensitivity as well as the stability of the colored reaction products and the reagent itself. For TLC purpose, PMA is regarded as a multipurpose stain, which is sensitive to small amounts of analytes within nanograms per spot range (Wall 2006; Sherma and Fried 2003). PMA may be successfully employed on the TLC plates coated with silica gel, aluminum oxide, polyamide, cellulose, and even *n*-alkanes bonded adsorbents. It has recently been found that the quantitative effect of PMA dyeing is strongly time- and temperature-dependent (Zarzycki et al. 2006). According to the data presented in the literature, typical conditions for analytes' visualization with PMA reagent involve fast heating of TLC plates, usually less than 10 min at a elevated temperature more than 100°C (Wall 2006; Sherma and Fried 2003; Lamparczyk 1992; European Pharmacopoeia). Unfortunately, robust spot detection under such conditions cannot be obtained because within the first minutes of the plate heating, the temperature changes very quickly and the average surface temperature is substantially different from the desired destination temperature. Such unsteady, nonlinear, and uncontrolled conditions can be a significant source of the poor precision and uncertainty in quantitative TLC, especially performed on the common glass-based plates. Moreover, it has been found that if the PMA detection is performed for a short time and under high temperature conditions, the signal/noise ratio and spots stability are usually low (Modzelewski et al. 2008; Pyka 2008). Our recent experimental work revealed that the best conditions for robust detection and quantification of analytes including wide range of steroids and prostaglandins can be obtained if the silica or octadecylsilane (C18) coated plates are heated within relatively low temperature region, between 40 and 80°C and for more than 20 min (Zarzycki et al. 2006, 2007c; Modzelewski et al. 2008; Zarzycki and Bartoszuk 2008). It has been found that under such conditions, background noise level can be substantially reduced and analytes' signal intensity greatly improved, particularly on glass plates coated with C18 stationary phase. In general, optimization of the temperature and time of the plate heating is necessary to minimize uncertainty and maximize reproducibility of TLC detection and quantification based on the analytes' spot staining with PMA reagent.

Notes

The main advantages of nonforced flow rate planar chromatography result from its simplicity, ease of operation, and the inexpensive equipment needed. In contrary to great progress in the construction of TLC devices allowing automatic chromatograms development under optimal saturation and humidity, the problem of high precision and reproducibility of the plates temperature in commercially available chambers was not yet successfully resolved. As it was argued, the selectivity and efficiency of TLC separation as well as precision and reproducibility of the analysis are strongly affected by temperature. Therefore, temperature should be carefully controlled for both qualitative and quantitative data collection. Particularly, for physicochemical measurements, the problem of pseudo-nonlinear Van't Hoff plots should be minimized by secure and equal temperature within working volume of the chamber body. From the experiences of the authors and co-workers, such conditions can be provided using simple Dewar-type glass flask or horizontal microunits working inside metal oven connected to the external liquid circulating thermostat. Despite the temperature effect on the separation process, this parameter can strongly affect quantification of components of interest, if they are detected by postrun derivatization with staining reagents such as PMA. It has been reported that optimization of the plates' heating temperature is necessary to minimize uncertainty and maximize sensitivity as well as reproducibility of the spots detected on TLC plates.

References

Armstrong DW. Pseudophase liquid chromatography. Applications to TLC. J Liq Chromatogr 1980; 3, 895–900.

Armstrong DW and Yang XF. Hot thin-layer chromatographic fractionation of polyethylene. J Chromatogr 1988; 456, 440–443.

Berezkin VG and Bolotov SL. Thin-layer chromatography with high-boiling mobile phases. Part 1. Investigation of the flow of the high-boiling mobile phases. J High Resolut Chromatogr Commun 1981; 4, 398–401.

Berezkin VG and Bolotov SL. Thin-layer chromatography with high-boiling mobile phases. Part 2: effect of temperature on retention characteristics in TLC with high-boiling mobile phases. J High Resolut Chromatogr Commun 1983; 6, 203–208.

Berezkin VG, Sumina EG, Shtykov SN, Atayan VZ, Zagniboroda DA and Nekhoroshev GA. Effect of chamber gas phase on mobile phase pH and on separation efficiency in TLC. A new mode of chromatography. Chromatographia 2006; 64, 105–108.

Błądek J. Parameters of the liquid crystal method of visualizing thin-layer chromatograms. J Chromatogr 1988; 437, 131–137.

Boehm RE, Martire DE and Armstrong DW. Theoretical considerations concerning the separation of enantiomeric solutes by liquid chromatography. Anal Chem 1988; 60, 522–528.

Camag Automatic Developing Chamber ADC 2; Webpage address: http://www.camag.com/v/products/development/adc2.html.

Chmielowiec J and Sawatzky H. Entropy dominated high performance liquid chromatographic separations of polynuclear aromatic hydrocarbons. Temperature as separation parameter. J Chromatogr Sci 1979; 17, 245–252.

Dębowski J, Grassini-Strazza G and Sybilska D. Alpha- and beta-cyclodextrins as selective agents for the separation of isomers by reversed-phase high-performance thin-layer and column liquid chromatography. J Chromatogr 1985; 349, 131–136.
Dimov N and Simova M. Unexpected behavior of digoxin at elevated temperatures in RP LC. Chromatographia 1999; 49, 306–308.
Dzido TH (1996) Polish Patent Appl. No 313053.
Dzido TH. Effect of temperature on the retention of aromatic hydrocarbons with polar groups in binary reversed-phase TLC. J Planar Chromatogr 2001; 14, 237–245.
Dzido TH, Gołkiewicz W and Piłat JK. The effect of temperature on the separation of some test solutes in preparative thin-layer chromatography. J Planar Chromatogr 2002; 15, 258–262.
European Pharmacopoeia 5th Edition (2004), General Monographs, European Directorate for the Quality of Medicines, France, Strasbourg.
Gasparic J, Gemzova I and Snobl D. Investigations of the relationship between structure and chromatographic retention of water azo dyes. Czech Chem Commun 1966; 31, 1712–1741.
Geiss F and Schlitt H. Temperatureffekte in der Dünnschichtchromatographie. J Chromatogr 1968; 33, 208–216.
Hinze WL and Armstrong DW. Thin layer chromatographic separation of ortho, meta, and para substituted benzoic acids and phenols with aqueous solutions of alpha-cyclodextrin. Anal Lett 1980; 13, 1093–1104.
Horvath C, Melander W and Molnar I. Solvophobic interactions in liquid chromatography with nonpolar stationary phases. J Chromatogr 1976; 125, 129–156.
Kolarovic L and Traitler H. The application of gradient saturation in unidimensional planar chromatography of polar lipids. Part 1. Presentation of the HPTLC system. J High Resolut Chromatogr Commun 1985; 8, 341–346.
Kolarovic L, Cousin H and Traitler H. The application of gradient saturation in unidimensional planar chromatography of polar lipids. Part 2. Optimization of the system. J High Resolut Chromatogr Commun 1985; 8, 838–842.
Kuhn AO, Lederer M and Sinibaldi M. Adsorption chromatography on cellulose. IV. Separation of D- and L-tryptophan and D- and L-methyltryptophan on cellulose with aqueous solvents. J Chromatogr 1989; 469, 253–260.
Lamparczyk H (1992). Analysis and Characterization of Steroids. CRC Press, Boca Raton, FL.
Lamparczyk H, Zarzycki PK and Nowakowska J. Effect of temperature on separation of norgestrel enantiomers by high-performance liquid chromatography. J Chromatogr A 1994; 668, 413–417.
Laub RJ and Purnell JH. Quantitative optimization of system variables for chromatographic separations. The column temperature. J Chromatogr 1978; 161, 49–57.
Lepri L, Coas V, Desideri PG and Checchini L. Separation of optical and structural isomers by planar chromatography with development by beta-cyclodextrin. J Planar Chromatogr 1990; 3, 311–316.
Lepri L, Coas V and Desideri PG. Planar chromatography of optical and structural isomers with eluents containing modified beta-cyclodextrins. J Planar Chromatogr 1994; 7, 322–326.
Liedekerke BM, Leenheer AP and Spiegeleer BM. High-performance thin-layer chromatographic analysis of thiazine dyes on silica, cyano, and reversed-phase C18 layers. J Chromatogr Sci 1991; 29, 49–53.
Lou DW, Saito Y, Zarzycki PK, Ogawa M and Jinno K. Isocratic separation of ginsenosides by high-performance liquid chromatography on a diol column at subambient temperatures. Anal Bioanal Chem 2006; 385(1), 96–104.
Melander WR, Nahum A and Horváth C. Mobile phase effects in reversed-phase chromatography. III. Changes in conformation and retention of oligo (ethylene glycol) derivatives with temperature and eluent composition. J Chromatogr 1979; 185, 129–152.
Modzelewski T, Włodarczyk E, Harasimiuk FB, Zarzycka MB, Baran M and Zarzycki PK. Simple device for TLC spots visualisation at elevated temperatures. Pomiary Automatyka Kontrola 2008; 54(4), 184–186.

Novotny AL, Nurok D, Replogle RW, Hawkins GL and Santini RE. Results with an apparatus for pressurized planar electrochromatography. Anal Chem 2006; 78, 2823–2831.

Pirkle WH. Unusual effect of temperature on the retention of enantiomers on a chiral column. J Chromatogr 1991; 558, 1–6.

Pyka A. TLC of selected bile acids: detection and separation. J Liq Chromatogr 2008; 31, 1373–1385.

Pyka A, Dołowy M and Górak D. Separation of selected bile acids by TLC. V. Influence of temperature on the separation. J Liq Chromatogr 2005; 28, 631–640.

Reich E, Blatter A and Meier B (2003) TLC for the analysis of herbal drugs, Camag Scientific Note, http://www.camag.ch/index.php.

Rivas-Nass A and Müllner S. The influence of critical parameters on the TLC separation of bile acids. J Planar Chromatogr 1994; 7, 278–285.

Saito Y, Ohta H and Jinno K. Design and characterization of novel stationary phases based on retention behavior studies with various aromatic compounds. J Sep Sci 2003; 26, 225–241.

Sherma J, Fried B (2003). Handbook of thin-layer chromatography, 3rd edition, Marcel Dekker, New York.

Singh EJ and Gershbein LL. TLC separation of organic mixtures at various temperatures. J Chromatogr Sci 1970; 8, 162–163.

Snyder LR. Comparisons of normal elution, coupled-columns, and solvents, flow or temperature programming in liquid chromatography. J Chromatogr Sci 1970; 8, 692–706.

Soczewiński E (2002) In: H.J. Issaq (ed.), A Century of Separation Science: Quantitative Retention–Eluent Composition Relationships in Partition and Adsorption Chromatography, Marcel Dekker, New York.

Touchstone JC, Sheikh SU, Fang XG, Alvarez JG and Levin SS. Subambient temperature in planar chromatography. J Liq Chromatogr 1989; 12, 3193–3199.

Wall PE (2006) Thin-Layer Chromatography: A Modern Practical Approach, Springer/Royal Society of Chemistry, Cambridge, UK.

Welsh T, Zakar T, Mesiano S and Zarzycki PK. Separation of bioactive prostaglandins and their metabolites by reversed-phase thin-layer chromatography. J Planar Chromatogr 2003; 16, 95–101.

Witkiewicz Z. Application of liquid crystals in chromatography. J Chromatogr 1989; 466, 37–87.

Xuan HTK and Lederer M. Adsorption chromatography on cellulose. XI. Chiral separations with aqueous solutions of cyclodextrins as eluents. J Chromatogr A 1994; 659, 191–197.

Zarzycki PK. Some technical problems associated with temperature-controlled thin-layer chromatography. J Planar Chromatogr 2001; 14, 63–65.

Zarzycki PK. Retention and separation studies of selected calixarenes and pyrene using NP and RP thin-layer chromatography. J Planar Chromatogr 2001; 14, 113–118.

Zarzycki PK. Simple chamber for temperature-controlled planar chromatography. J Chromatog A 2002; 971, 193–197.

Zarzycki PK. Simple horizontal chamber for thermostated micro-thin layer chromatography. J Chromatogr A 2008; 1187, 250–259.

Zarzycki K and Bartoszuk MA. Improved TLC detection of prostaglandins by post-run derivatization with phosphomolybdic acid. J Planar Chromatogr 2008; 21(5), 387–390.

Zarzycki PK and Lamparczyk H. A simple experiment demonstrating the temperature effect in supramolecular chemistry. J Chem Educ 1996; 73, 459–460.

Zarzycki PK and Lamparczyk H. The equilibrium constant of beta-cyclodextrin-phenolphthalein complex; Influence of temperature and tetrahydrofuran addition. J Pharm Biomed Anal 1998; 18, 165–170.

Zarzycki PK and Lamparczyk H. Evidences for temperature-dependent mechanism of host–guest complexation. Chromatographia 1998; 48, 377–382.

Zarzycki PK and Lamparczyk H. Temperature controlled thin-layer chromatography. Chem Anal 2001; 46, 469–475.

Zarzycki PK and Smith R. Separation of steroids using temperature-dependent inclusion chromatography. J Chromatogr A 2001; 912, 45–52.

Zarzycki PK and Zarzycka MB. Low cost, efficient TLC sprayer. J Planar Chromatogr 2008; 21(3) 221–223.

Zarzycki PK and Zarzycka MB. Application of temperature-controlled micro planar chromatography for separation and quantification of testosterone and its derivatives. Anal Bioanal Chem 2008; 391(6), 2219–2225.

Zarzycki PK and Zarzycka MB. Evaluation of the water and organic liquids extraction efficiency of the *Spirulina maxima* dyes using thermostated micro-thin-layer chromatography. J AOAC Int 2008; 91(5), 1196–1202.

Zarzycki PK, Nowakowska J, Chmielewska A and Lamparczyk H. Retention properties of cyclodextrins in RP-HPTLC. J Planar Chromatogr 1995; 8, 227–231.

Zarzycki PK, Nowakowska J, Chmielewska A and Lamparczyk H. Retention properties of cyclodextrins on polyamide TLC plates. J Planar Chromatogr 1996; 9, 260–263.

Zarzycki PK, Nowakowska J, Chmielewska A, Wierzbowska M and Lamparczyk H. Thermodynamic study of retention of selected macrocycles using RP-HPTLC plates and methanol/water mobile phases. J Chromatogr A 1997; 787, 227–233.

Zarzycki PK, Wierzbowska M and Lamparczyk H. Retention and separation studies of cholesterol and bile acids using thermostated thin-layer chromatography. J Chromatogr A 1999; 857, 255–262.

Zarzycki PK, Wierzbowska M, Nowakowska J, Chmielewska A and Lamparczyk H. Interactions between native cyclodextrins and n-alcohols studied using thermostated thin-layer chromatography. J Chromatogr A 1999; 839, 149–156.

Zarzycki PK, Kulhanek KM, Smith R, Bartoszuk MA and Lamparczyk H. Planar chromatography versus column chromatography: a performance comparison. LCGC North America 2005; 23, 286–300.

Zarzycki PK, Bartoszuk MA and Radziwon AI. Optimization of TLC detection by phosphomolybdic acid staining for robust quantification of cholesterol and bile acids. J Planar Chromatogr 2006; 19, 52–57.

Zarzycki PK, Baran M and Włodarczyk E. Suitability of the CHROMDES DS-L horizontal TLC chamber for temperature-controlled planar chromatography. Acta Chromatogr 2007; 18, 249–259.

Zarzycki PK, Ohta H, Harasimiuk FB and Jinno K. Fast separation and quantification of C60 and C70 fullerenes using thermostated micro thin-layer chromatography. Anal Sci 2007; 23, 1391–1396.

Zarzycki PK, Baran M, Włodarczyk E and Bartoszuk MA. Improved detection of ergosterol, stigmasterol, and selected steroids on silica coated TLC plates using phosphomolybdic acid staining. J Liquid Chromatogr 2007; 30, 2629–2634.

Zarzycki PK, Włodarczyk E and Baran MJ. Determination of endocrine disrupting compounds using temperature-dependent inclusion chromatography I. Optimization of separation protocol. J Chromatogr A 2009; 1216, 7602–7611.

Zarzycki PK, Zarzycka MB and Głód BK. Estimation of micro-TLC plate peak capacity for one and two dimensional multiple sample separation. Pomiary Automatyka Kontrola 2009; 55, 276–279.

Zarzycki PK, Włodarczyk E, Zarzycka MB and Glod BK. Optimization of solid-phase extraction protocol for fractionation of selected steroids using retention data from micro thin-layer chromatography. Anal Sci 2009; 25, 935–939.

Part IV
HPTLC and Its Future to Combinatorial Approach

Chapter 14
Basic Principles of Planar Chromatography and Its Potential for Hyphenated Techniques

Tomasz Tuzimski

Abstract Sample preparation, detection, identification, and quantitative determination of biomolecules are presented in this chapter. Advantages of planar chromatography and the basic principles (chambers, sample application, and chromatogram development) are also described. Rapid detection of biomolecules plays a strategical role in their investigation. Hyphenated techniques such as planar chromatography coupled to UV diode array detection and to mass spectrometry provide on-line extensive structural information on the metabolites prior to their isolation. In this chapter, the combination of thin-layer chromatography (TLC) with biomolecules specific detection by diode array scanning (DAD), mass spectrometry (MS), nuclear magnetic resonance (NMR), and Fourier transform infrared spectroscopy (TLC-FTIR) is discussed. In the last part of the chapter, the reader will gain useful information about a recent method of planar chromatography – multidimensional planar chromatography (MD-PC) and information on application of different modes of multidimensional planar chromatography and combination of this technique with diode array detection (MD-PC–DAD) and HPLC–DAD for separation, detection, and qualitative and quantitative determination of biomolecules. Planar chromatography as a pilot technique for HPLC is also described.

Biomolecules are widespread throughout the world. It is an organic molecule produced by a living organism, including large polymeric molecules such as proteins, polysaccharides, and nucleic acids as well as small molecules such as a primary metabolites, secondary metabolites, and natural products. Many sample-preparation techniques are used in biomolecule analysis; the method selected depends on the

---

T. Tuzimski
Department of Physical Chemistry, Faculty of Pharmacy, Medical University of Lublin, Lublin, Poland

MM. Srivastava (ed.), *High-Performance Thin-Layer Chromatography (HPTLC)*,
DOI 10.1007/978-3-642-14025-9_14,

complexity of the sample, the nature of the matrix and the analytes, and the analytical techniques available. Planar chromatography is an important analytical method with other chromatographic techniques such as gas chromatography (GC), high performance liquid chromatography (HPLC), and supercritical fluid chromatography (SFC). Thin-layer chromatography (TLC), although less sensitive and efficient than some other separation methods, has many advantages. Planar chromatography is most effective for the low-cost analysis of samples requiring minimal sample clean-up. Planar chromatography is also selected for biomolecular step analysis because:

- Single use of stationary phase minimizes sample preparation requirements
- Parallel separation of numerous samples enhances high throughput
- Ease of postchromatographic derivatization improves method selectivity and specificity
- Detection and/or quantitation steps can easily be repeated under different conditions
- All chromatographic information is stored on the plate and can be (re-) evaluated if required
- Several screening protocols for different analytes can be carried out simultaneously
- Selective derivatizing reagents can be used for individual or group identification of the analytes
- Detection of the separated spots with specific and sensitive color reagents
- Visual detection of UV-absorbing compounds is possible in field analyses by use of a UV lamp
- Detection by contact with X-ray film, digital bio- and autoradiography, and even quantitative assay by use of enzymes is possible
- TLC plates can be documented by videoscans or photographs
- Planar chromatography combined with modern videoscanning and densitometry enables quantitative analysis
- Planar chromatography coupled with densitometry enables detection of the spots or zones through scanning of the chromatograms with UV–Vis light in the transmission, reflectance, or fluorescence mode
- With multiwavelength scanning of the chromatograms, spectral data of the analytes can be directly acquired from the TLC plates and can further be compared with the spectra of the analytes from software library
- Additional information for structural elucidation can be obtained by planar chromatography combined with MS (fast atom bombardment (FAB) and liquid secondary mass spectrometry (SIMS))
- The whole procedure of chromatographic development can be followed visually, so any distortion of the solvent front, etc., can be observed directly
- The chromatogram can be developed simply by dipping the plate into a mobile phase
- The possibility of two-dimensional development with a single adsorbent

- The possibility of two-dimensional development on, for example, silica – octadecyl silica coupled layers (Multi-K SC5 and CS5 dual phase)
- Planar chromatography is also the easiest technique, which performs multidimensional separation (e.g., by graft chromatography or multidimensional chromatography).

Summing up, planar chromatography is one of principal separation techniques, which plays the key role also in biomolecule analysis. In the first part of the chapter, the reader will gain useful information to avoid some problems about performing planar chromatography experiments, and in the second part, he will find useful information for application of planar chromatography for separation, detection, and qualitative and quantitative determination of biomolecules.

Principles of Planar Chromatography

The stages of planar chromatography procedure such as sample application, chromatogram development, registration of chromatogram, and its evaluation cannot be presently performed in one run using commercially available devices. It means that planar chromatography analysis cannot be completely automated at contemporary laboratory practice. Chromatographers have to separately optimize each of the mentioned stages using more or less sophisticated devices. There are various equipments for semi- or full-automatic operations at the mentioned stages of planar chromatography procedures. At each stage of the TLC procedure, the chromatographer should possess the basic skill, which substantially helps to accomplish TLC experiments correctly, to obtain reliable, repeatable, and reproducible results. He/she can meet many pitfalls during work with TLC mode. There are some fundamental books that help to overcome these problems (Geiss 1987; Nyiredy 2001; Hahn-Deinstrop 2000; Fried and Sherma 1999; Dzido and Tuzimski 2008). The present chapter gives some information, which can draw the reader's attention to the procedures and equipments mentioned, which are more often applied and proven in contemporary planar chromatography practice.

Modern Chambers for TLC

Various chambers have been used for the development of thin-layer chromatograms. The classification of the chromatographic chambers can be performed taking into account:

- Volume of vapor atmosphere inside the chamber – unsaturation or saturation with the vapor of the mobile phase system.
- Direction of mobile phase migration – linear development in which solvent migrates through a rectangular or square chromatographic plate from one of its edges to the opposite edge with constant width of front of the mobile phase

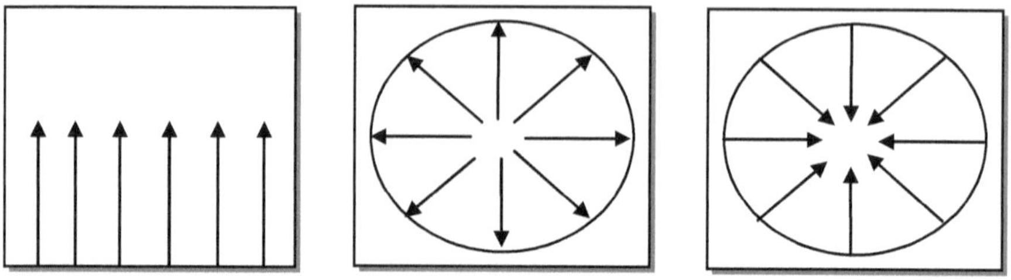

Fig. 14.1 Modes of development in planar chromatography [Adapted from Dzido and Tuzimski (2008)]

(Fig. 14.1a) or radial development (including circular (Fig. 14.1b) and anticircular (Fig. 14.1c) types of radial developments); in the circular type, the mobile phase is delivered at the center of the chromatographic plate and its front migrates toward the periphery of the adsorbent layer; meanwhile in the anticircular type, the mobile Phase migrates in the opposite direction and its front again is circular.

- Configuration of the chromatographic plate in the chamber (horizontal – then the chamber is named as horizontal chamber – or vertically – and then the chamber is named as vertical one).
- Degree of automation of chromatogram development (including temperature and humidity control, eluent and vapor phase delivery to the chromatographic chamber, and drying the chromatographic plate).

Regarding the volume of vapor atmosphere, two main types of the chamber can be distinguished: normal (conventional) chambers (N-chambers) and sandwich chambers (S-chambers). The above classification is not unequivocal because a chamber can belong to more than one chamber type.

Conventional Chambers (N-Chambers)

The N-chambers are typically made of glass as a vessel possessing cuboid or cylindrical form. Their dimensions are about 230 × 230 × 80 mm for the respective development of 200 × 200 mm TLC plates or 130 × 130 × 50 mm for the respective development of 100 × 100 mm TLC plates. This type of chamber can be very easily applied for conditioning with vapor phase by using saturation pads or an adequate size of filter paper (blotting paper) in the chromatographic chamber, which is very important especially when mixed mobile phase is used for chromatogram development. Then the repeatability of retention values is higher in comparison to development without vapor saturation.

Another type of N-chamber is the cuboid's twin-through chamber, which can be conveniently used for chromatography under different conditions of vapor saturation

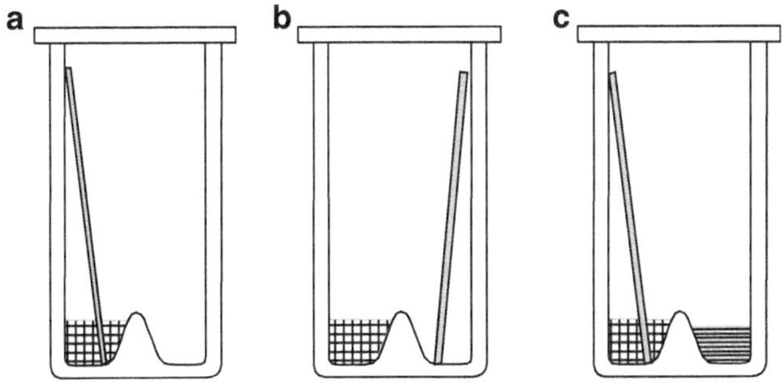

Fig. 14.2 The twin-trough chamber with various variants of chromatogram development [Adapted from Dzido and Tuzimski (2008)]

(Geiss 1987). Schematic view of the chamber is demonstrated in Fig. 14.2. The bottom of the chamber is divided by a ridge into two parallel troughs. This construction of the chamber enables to perform chromatogram development in three modes: without chamber saturation (Fig. 14.2a), with chamber saturation (Fig. 14.2b), and chamber saturation with one solvent followed by development with another one (Fig. 14.2c) (Dzido and Tuzimski 2008).

Horizontal Chambers for Linear Development

As it was mentioned above, the chromatographic plate is positioned horizontally in this chamber type. The most often applied in experiments, TLC horizontal chambers are produced by Camag (Muttenz, Switzerland), Desaga (Heidelberg, Germany), and Chromdes (Lublin, Poland). All have similar construction. The elements of the three types of horizontal chambers described below (horizontal developing chamber, horizontal DS (Dzido-Soczewiński) chamber, and H-separating chamber) are made of Teflon and glass; so they are very resistant to all solvents applied for chromatographic separations. The differences consist in the eluent delivery system. One example of horizontal chamber (Horizontal DS Chamber manufactured by Chromdes, Lublin, Poland) is presented as cross section before and during chromatogram development in Fig. 14.3a, b, respectively (http://www chromdes com; Dzido 2001a). The main feature of the chamber is formation of vertical meniscus of the solvent (dark area) between slanted bottom of the mobile phase reservoir (2) and glass strip (1). Chromatogram development is started by shifting the glass strip to the edge of the chromatographic plate (3) with adsorbent layer face down, which brings the solvent to contact with the chromatographic plate. During development, the meniscus of solvent moves in the direction of the chromatographic plate, which makes the chamber very economical (the solvent can be exhausted from the reservoir almost

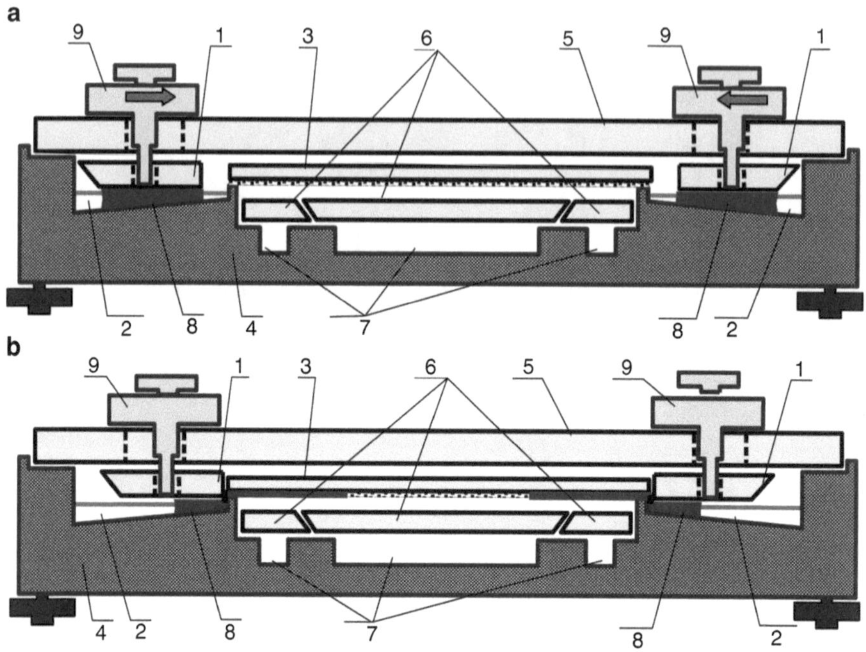

Fig. 14.3 Horizontal DS-II chamber (Chromdes): (**a**) before development, (**b**) during development; 1 – cover plate of the mobile-phase reservoir, 2 – mobile phase reservoir, 3 – chromatographic plate with layer face down, 4 – body of the chamber, 5 – main cover plate, 6 – cover plates (removable) of the troughs for vapor saturation, 7 – troughs for saturation solvent, 8 – mobile phase, 9 – mobile-phase distributor/injector. Adopted from http://www.chromdes.com

completely). Conditioning of the chamber atmosphere can be performed by pouring some drops of solvent onto the bottom of the chamber (lined with blotting paper) (7) after removing glass plates (6). All kinds of plates (foil and glass-backed, of dimension from 5 × 10 to 20 × 20 cm) can be developed in these chambers depending on the chamber type and size. The maximum distance of chromatogram development is equal to 20 cm. The consumption of solvents is very low, e.g., 3–5 ml for 100 × 200 mm plates.

Another example of horizontal chamber (horizontal developing chamber manufactured by Camag, Muttenz, Switzerland) is presented in Fig. 14.4 as a cross section (http://www camag com; Jaenchen 1996). The chromatographic plate (1) is positioned with adsorbent layer face down and is fed with the solvent from the reservoir (2). Chromatogram development is started by tilting the glass strip (3) to the edge of the chromatographic plate. Then a planar capillary is formed between the glass strip and the wall of the solvent reservoir in which the solvent instantaneously rises, feeding the chromatographic plate. The maximum distance of development with this chamber is 10 cm. The chambers are offered for 10 × 10 and 20 × 10 mm plates.

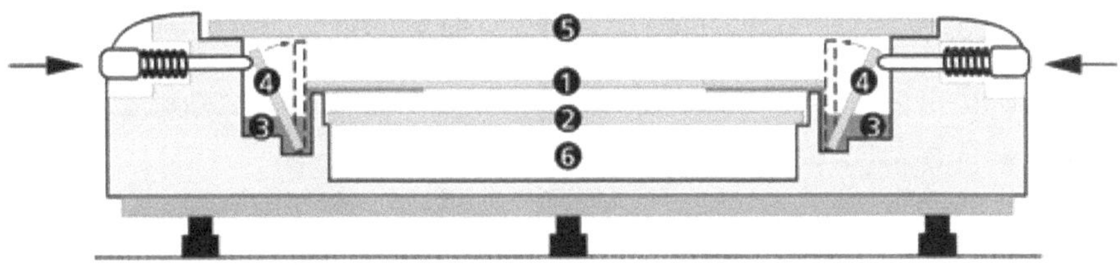

Fig. 14.4 Horizontal developing chamber (Camag): 1 – chromatographic plate with layer face down, 2 – counter plate (removable), 3 – troughs for solution of the mobile phase, 4 – glass strip for transfer of the mobile phase by capillary action to the chromatographic plate, 5 – cover glass plate. Adopted from http://www.camag.com

The two horizontal chambers types described above possess the following methodological possibilities (Dzido and Tuzimski 2008):

- Double number of samples in comparison to conventional chambers can be separated on one plate (due to two solvent reservoirs on both sides of the plate, which enable simultaneous development of two chromatograms from two opposite edges) (Fenimore and Davis 1981)
- Saturation of the adsorbent layer with vapors of the mobile phase or another solvent (Fenimore and Davis 1981; Dzido and Polak 1993)
- Two-dimensional separation of four samples on one plate simultaneously (De Brabander et al. 1988)
- Multiple development (Markowski and Matysik 1993; Matysik 1996)
- Stepwise gradient elution (Matysik and Soczewiński 1996; Matysik et al. 1992)
- Zonal sample application for preparative separation (in Horizontal DS Chamber) (Matysik et al. 1994; Dzido and Polak 2006)
- Continuous development (in SB/CD and Horizontal DS chambers)* (Dzido and Polak 2006)
- Short bed-continuous development (in SB/CD and Horizontal DS chambers)* (Dzido and Polak 2006; Soczewiński 1986)
- Development of six different chromatograms on one plate simultaneously (HPTLC Vario Chamber from Camag or Horizontal DS-M Chamber from Chromdes) (Dzido and Polak 2006).

More detailed description of these methodical possibilities can be found by the reader in the following references – Geiss (1987), Nyiredy (2001), Dzido and Polak (2006), Soczewiński (1986).

(*) SB/CD chamber is not manufactured at present. Another horizontal chamber (H-separating chamber) for TLC is manufactured by Desaga (Kraus et al. 1996). Its principle of action is based on the Brenner–Niedervieser (BN) chamber (Brenner and Niederwieser 1961). The chromatographic plate is fed with solvent from the reservoir using a wick made of porous glass. The chambers are manufactured for 5 × 5 cm and 10 × 10 cm plates. Other horizontal chambers such as Vario-KS-Chamber (Geiss et al. 1965; Geiss and Schlitt 1968), SB/CD-Chamber (Regis Technologies Inc.), Sequence-TLC Developing Chamber (Buncak 1982), ES-Chamber

(Soczewiński 1986), and ES-Chamber modified by Rumiński (1988) and by Wang et al. have been described in the literature and applied in some laboratories; however, these are not commercially offered at present (Su et al. 2001; Lan et al. 2003).

Horizontal Chambers for Radial (Circular and Anticircular) Development

The mode of radial development of planar chromatograms is rarely applied in laboratory practice. Radial development using circular mode can be easily performed with a Petri dish (Blome 1977). The so-called U-chamber was used for circular and anticircular developments. This mode, including both types of development, can be carried out using commercial U-chambers (Camag) (Kaiser 1977). However, at present, these chambers are not in the commercial offer of this firm probably due to low interest of the customers. In spite of some advantages regarding separation efficiency, the practitioners prefer to apply linear development rather than radial development. Two main reasons explain this status: at first, too sophisticated chamber construction and its maintenance to develop chromatograms, and the second reason is about shortage of equipment and software for chromatogram evaluation. Subsequently, the methods have been developed by a number of researchers to control the mobile phase movement (Tyihák and Mincsovics 1988).

Sample Application

Resolution of the chromatographic system is dependent on the size of the starting zone (spot) of the solute. Sample shape as streaks or bands is advantageous with regard to resolution and quantitative analysis. The sample can be applied to the stationary phase by spotting, dipping, spraying, or sampling through a syringe. Conventional application of sample mixture on the chromatographic plate can be performed with calibrated capillary or microsyringe. More advantageous modes of sample application can be performed with a semiautomatic applicator or fully automated device. All these modes can be applied for analytical and preparative separations as well.

Chromatogram Development

As it is was mentioned above, chromatogram development in TLC can be performed applying linear or radial modes. Both modes can be performed in a simple way using conventional chamber and applying very complicated procedures, including sophisticated devices. Involved operations and procedures depend on various variables about properties of the sample, adsorbent layer, solvent, mode of detection, and evaluation of chromatogram.

Mobile Phases Applied in TLC

Mobile phases used for TLC have to fulfill various requirements. It must not chemically affect and/or dissolve the stationary phase because this leads to modification of properties of the chromatographic system. It must not produce chemical transformations of the separated compounds. Multicomponent mobile phase applied in TLC must be used only once, not repeatedly, because the volatility of solvents produces a continuous modification of quantitative composition of the mobile phase, which negatively affects the chromatographic repeatability. The mobile phase must be easily eliminated from the adsorbent layer and must be compatible with detection methods. The reproducibility can be greatly affected by the conditions and the time of preservation of the mobile phase solution.

Chemical information about mobile phase properties is essential to the initial selection of chromatographic system and detection properties. Choice of the mobile phase (and also the stationary phase) is dependent on many factors about the properties of the compounds to be separated (Table 14.1) (Poole and Dias 2000; Snyder et al. 1997). When properties of the mobile phase and stationary phase of TLC systems are considerably different, then the separation selectivity is expected to be high. In general, if the stationary phase is polar, the mobile phase should be apolar or slightly polar, and then such system is named a normal-phase (NP) system. If the stationary phase is nonpolar, then the mobile phase should be polar, and such chromatographic system is named as reversed-phase one (RP). The choice of the mobile phase is dependent not only on the properties of the adsorbent and its activity but also on the structure and the type of separated analytes. Various solvents can be used as the components of the mobile phase in planar chromatography and their choice to the chromatographic process is based on eluotropic and isoelutropic series. The mobile phase applied in planar chromatography can be composed of one, two, or more solvents.

Classification of the Modes of Chromatogram Development

As it was mentioned above, chromatogram development can be performed applying linear or radial modes.

Linear development can be performed as:

- Isocratic linear development
- Continuous isocratic development (Nyiredy 2002b)

Table 14.1 Hints for stationary and mobile phase selection with respect to the type of sample components (Poole and Dias 2000; Snyder et al. 1997)

Sample information			
1. Polarity of compound			
Normal-phase chromatography (NP) with nonaqueous mobile phases			
Polarity of Compound	Stationary-phase	Mobile-phase	Comments
Low (hydrophobic)	Polar adsorbents (silica or, less often, alumina)	Nonpolar mobile phase (nonaqueous)	It is difficult to separate the compounds of low polarity on silica gel due to their weak retention. Selection of solvents to the mobile phase is limited because most solvents demonstrate too high elution strength (compounds show very high values of R_F or migrate with front of the mobile phase)
High (hydrophilic)			Compounds of high polarity are difficult to separate on silica gel because of strong retention. Selection of solvents to the mobile phase is limited because most solvents are of too lower elution strength for these solutes (compounds have very low values of R_F, stay on the start line of chromatogram)
Normal-phase chromatography (NP) with aqueous mobile phases ("pseudo-reversed phase system")			
Polarity of Compound	Stationary-phase	Mobile-phase	Comments
Very high (very hydrophilic)	Polar adsorbents (silica or, less often, alumina)	Polar mobile phase (aqueous)	Compounds of very high polarity (very high hydrophilic, e.g., alkaloids, which additionally show strong interactions with silanol groups of silica based stationary phases) are difficult to separate on silica gel with nonpolar mobile phases because of strong retention (stay on the start line of chromatogram even when 100% methanol as the mobile phase is applied) and should be chromatographed with more polar eluents containing water.
Normal-phase chromatography (NP) with nonaqueous mobile phases and reversed-phase chromatography (RP) with aqueous mobile phases			
Polarity of Compound	Stationary-phase	Mobile-phase	Comments
Intermediate polarity	Moderately polar bonded phases, chemically bonded on silica support: cyanopropyl – $(CH_2)_3$–CN, diol –$(CH_2)_3$–O–CH_2–$CHOHCH_2$-OH, or aminopropyl–$(CH_2)_3$–NH_2	Nonpolar mobile phase (nonaqueous) and polar mobile phase (aqueous)	Compounds of intermediate polarity are separated on polar chemically bonded stationary phases because their molecules can interact with silanol groups of silica gel. They demonstrate good separation selectivity in both normal-phase (nonaqueous) and reversed-phase (aqueous) systems. The

			moderately polar stationary phases are compatible with water mobile phase of whole concentration range. Many solvents can be selected to prepare the mobile phase. The only limitation is concerned with miscibility of the mobile phase components. In addition, migration velocity of front of the mobile phase varies less with solvent composition in comparison to typical RP systems.

Reversed-phase chromatography (RP) with aqueous mobile phases

Polarity of Compound	Stationary-phase	Mobile-phase	Comments
Low (hydrophobic)	Nonpolar adsorbents (chemical modification is based on reactions of the silanol (≡Si–OH) groups on the silica surface with organosilanes to obtain stationary phases of the type ≡ Si–R, where R is aliphatic chain of the type – C_1, C_2, C_8, C_{18})	Polar mobile phase (aqueous)	Compounds of low polarity are difficult to separate in systems with nonpolar adsorbents because of very strong retention. Mobile phase selection is limited because most solvents show too low elution strength for these separations
High (hydrophilic)			Compounds of high polarity are difficult to separate with nonpolar adsorbents because of weak retention. Appropriate mobile phase selection is restricted because most solvents show too strong elution strength for these solutes

Reversed-phase chromatography (RP) with nonaqueous mobile phases

Polarity of Compound	Stationary-phase	Mobile-phase	Comments
Low (very hydrophobic)	Nonpolar adsorbents The stationary phase of the type ≡ Si–R (R is aliphatic chain of various length e.g., – C_1, C_2, C_8, C_{18}) is formed after reaction of the silanol (≡Si–OH) groups of the silica surface with organosilanes	Polar mobile phase (nonaqueous)	The separation of compounds of low polarity (very hydrophobic samples) is difficult to separate with aqueous mobile phases on nonpolar adsorbents because of very strong retention stay on the start line of chromatogram even when acetonitrile–water (99:1, v/v) mobile phase is applied. The solutes (e.g., lipids) can be chromatographed using the mobile phase composed of more polar (acetonitrile, methanol) and less polar (tetrahydrofuran, chloroform, methylene chloride, acetone, methyl-*t*-butyl ether) organic solvents or various mixtures of these solvents. The retention decreases with increasing concentration of the less-polar solvent in the mobile phase (multicomponent eluent may contain even hexane or heptane)

(continued)

Table 14.1 (continued)

Sample information

2. Molecular Mass (MW)

Compound	MW <1,000	Organic soluble	NP systems with silica or with chemically bonded stationary phase (aminopropyl, cyanopropyl, diol)	
		Water soluble	Nonionic	RP systems with chemically bonded stationary phase (aminopropyl, cyanopropyl, diol) and alkylsiloxane-bonded stationary phases (C_2, C_6, C_8, C_{18})
				NP systems on silica and on chemically bonded stationary phases (aminopropyl, cyanopropyl, diol).
			Ionic	RP systems on chemically bonded stationary phases (aminopropyl, cyanopropyl, diol) and alkylsiloxane-bonded stationary phases with ligands of the type C_2, C_6, C_8, C_{18}
				IPC (Ion-pair chromatography) in RP systems with chemically bonded stationary phases (aminopropyl, cyanopropyl, diol) and alkylsiloxane-bonded stationary phases with ligands of the type C_2, C_6, C_8, C_{18}.
	MW >1,000	Organic soluble	Precipitation Chromatography	
		Water soluble	Cellulose	

3. pK_a value of compound

pK_a value of compound	Acid–base behavior	Stationary-phase	Mobile-phase	Comments
Low values of pK_a High values of pK_a	Strong acid or weak base Strong base or weak acid	Nonpolar adsorbents: alkylsiloxane-bonded stationary phases (with ligands of the type C_2, C_6, C_8, C_{18}, phenyl) or chemically bonded stationary phases of the type (aminopropyl, cyanopropyl, diol)	RP systems: buffered polar mobile phase with controlled Ph	When an acidic or basic molecule undergoes ionization (i.e., is converted from an uncharged species into charged one), it becomes much less hydrophobic (more hydrophilic). When pH value of the mobile phase is equal to pK_a of the compounds of interest then values of concentration of its ionized and unionized forms are identical (i.e., the values of concentration of B and BH^+ or HA and A^- in the mobile phases are equal). It means that retention changes of these solutes in principle take place in the pH range from the value pK_a – 1.5 to the value pK_a + 1.5. The relationship between retention of the solute and mobile phase pH in RP systems is more complicated for compounds with two or more acidic and/or basic groups

- Short bed-continuous development (Soczewiński 1986; Perry 1979; Matysik and Soczewiński 1988a)
- 2D mode of development
- Multiple development
- Gradient (stepwise and continuous) development (Soczewiński 1986; Markowski 1996, 2005; Gołkiewicz 2003; Matysik and Soczewiński 1988b, c; Matysik 1997; Soczewński and Matysik 1985
- Sequence development (Buncak 1982; Rumiński 1988; Su et al. 2001; Lan et al. 2003)
- With temperature control (Zarzycki 2001, 2002; Dzido 2001b; Zarzycki et al. 1997, 1999a, b).

Radial development of planar chromatograms can be performed as circular (Izmailov and Schraiber 1938; Ripphahn and Halpaap 1977; Botz et al. 1990; Nyiredy 2003a) and anticircular (Kaiser 1978; Studer and Traitler 1986; Issaq 1980).

There are some books that help to overcome these problems (Geiss 1987; Nyiredy 2001; Hahn-Deinstrop 2000; Fried and Sherma 1999; Dzido and Tuzimski 2008; Tuzimski 2010a). Some of them (multiple development techniques, 2D mode development, multidimensional mode development) are described in the next part of this chapter.

Sample Preparation

The analysis of biomolecules such as proteins requires their hydrolysis. Other biomolecules before their determination require specific operations. Planar chromatography belongs to methods for qualitative and quantitative analysis of free amino acids either before or after protein hydrolysis. Proteins are hydrolyzed to peptides and amino acids by treatment with acid, alkali, or enzymes. Each method has certain disadvantages (Bhushan 2008; Bhushan and Martens 2003a):

- *In acid hydrolysis by 8 M H_2SO_4 at 110°C for 18 h*: tryptophan is destroyed; serine and threonine are partially destroyed; presence of carbohydrates leads to formation of black material, humin
- *In acid hydrolysis by 6 M HCl at 110°C for 18 h*: tryptophan, asparagine are destroyed; serine, threonine, tyrosine are partially lost; cysteine and methionine are either partially destroyed or oxidized to cysteic acid and Met-S, S-dioxide, respectively
- *In alkali hydrolysis by $Ba(OH)_2$*: arginine, cysteine, serine, and threonine are partially or completely decomposed; racemization and some deamination occurs

- *In alkali hydrolysis by NaOH or LiOH*: LiOH is reported to be best for tryptophan determination
- *Enzymes pepsin, trypsin, papain, chymotrypsin*: each enzyme is generally specific for a particular peptide bond; hydrolysis of enzymes may occur, which would interfere with amino acid analysis

The most commonly used method for total hydrolysis are (Bhushan 2008; Bhushan and Martens 2003a):

Acid Hydrolysis

Some of the methods for clear hydrolysis of proteins for amino acid analysis were described by Light and Smith (1963), Moore and Stein and al. (Smyth et al. 1963), Savoy et al. (1975), or Perham (1978). A sample (50–100 mg) of air-dried or lyophilized protein is weighed into a tube, and 6 M HCl (1 ml for 5 mg of protein) is added (Bhushan 2008; Bhushan and Martens 2003a). The tube is evacuated using a vacuum desiccator (Phillips 1981), sealed, and placed in a circulating air oven at 110°C with good temperature control. After hydrolysis for the appropriate period of time (24, 48, or 71 h), it is centrifuged gently. Next, the tubes are cracked open and the HCl is removed as quickly as possible using a stream of N_2. The HCl can alternatively be neutralized by adding solid $Ba(OH)_2$ (up to pH 7) and removing sediment $BaSO_4$ by filtration or centrifugation. The clear hydrolysate may be frozen in an acetone–solid CO_2 bath, placed in a vacuum desiccator over NaOH or KOH, and lyophilized (Bhushan 2008; Bhushan and Martens 2003a).

Method for Sulfur-Containing Amino Acids

Cysteine and cystine as cysteic acid were determined by performic acid oxidation (Moore 1963). Methionine can also be determined as methionine SS-dioxide. Performic acid is prepared by adding H_2O_2 (1 ml, 30%) to formic acid (9 ml, 88%) and allowing the mixture to stand at room temperature for 1 h. It is then cooled to 0°C. Performic acid (2 ml) is added to the protein (containing about 0.1 mg cystine) in a Pyrex tube, which is then allowed to stand at 0°C for 4 h for soluble proteins or overnight for insoluble proteins. Next, HBr (0.30 ml, 48%) is added with swirling, the mixture is evaporated to dryness at 40°C using a rotary evaporator, and the protein is hydrolyzed in vacuums with HCl (3 ml, 6 M) at 110°C for 18 h. The hydrolysate is treated as mentioned earlier, before analysis (Bhushan 2008; Bhushan and Martens 2003a). Chiou and Wang (1990) described a method of protein hydrolysis by microwave irradiation, and the results were compared with those conventional heating methods.

Depending on the protein nature and source, various methods have been reported in the literature for their digestion before applying them to TLC experiments. Two of these methods are described below (Bhushan and Martens 2003b):

- Proteins are dissolved in ammonium bicarbonate (0.5%, pH 8.0) and digested with trypsin (1% w/w) for 4 h at 37°C. Chymotrypsin (1% w/w) may be added for trypsin–chymotrypsin digest and the digestion continued for a further 4 h. The peptides are recovered by freeze-drying (Bates et al. 1975).
- Burns and Turner (1967) subjected proteins to either alkylation with iodoacetic acid (Harris and Hindly 1965) or performic acid oxidation (Hirs 1956) to render them susceptible to enzymatic digestion. The treated proteins were then dissolved in ammonium bicarbonate buffer (0.05 M, pH 8.4) to a concentration of 2 mg ml^{-1}, and TPCK-treated trypsin [L-(1-tosylamide-2-phenylethyl chloromethyl ketone)] was added to give a final enzyme/substrate ratio of 1:75. The digest was incubated for 5 h at 30°C, freeze-dried, and redissolved in 10% isopropanol for application to the plates.
- For TLC of smaller peptides, the samples have been either synthesized (Hubert and Dellacherie 1973) or obtained commercially. The readers may find application details for TLC analysis of proteins in literature (Bhushan and Martens 2003b).

Detection and Identification of Biomolecules in TLC

Various chemical, physical, and biochemical methods have been used in the detection and identification of biomolecules. Some of them are very simple. For the detection of UV-absorbing substances simply visually by eye, TLC plates are prepared with fluorescence indicators (e.g., manganese-activated zinc-silicate). This dye absorbs light at 254 nm showing a green florescence at ~520 nm; the sample molecules inhibit light absorption on the plate with a fluorescence indicator. In the case of uncovered plates, dark spots or zones on a bright fluorescent background will indicate the positions of the components. The commonly used method is to expose the chromatographic plate to iodine vapor in a closed chamber that contains some iodine crystals. Sample spots show a dark brown color on a plate with yellow-brown background. The use of biological methods in planar chromatography is justified because they are highly specific and the detection limits are lower than those in other methods.

The possibility of using reagents (especially color reagents) is one of the essential advantages of planar chromatography. Stahl and Mangold (1975) reviewed the generally used common spraying "reagents." A basic source of reagents is given in the book that was edited by Jork et al. (1990).

Ninhydrin is the most commonly used reagent for the detection of amino acids, and a very large number of ninhydrin reagent compositions have been reported in the literature. Some of the ninhydrin compositions and their applications are listed (Bhushan 2008; Bhushan and Martens 2003a):

- A solution of ninhydrin (0.2% in acetone) is prepared with the addition of a few drops of collidine or glacial acetic acid. The chromatogram is dipped or sprayed with it and dried at 60°C for about 20 min or at 100°C for 5–10 min. Excessive heating causes a dark background. The sensitivity limit is 1 μg. Most amino acids give a violet color, whereas aspartic acid gives bluish-red, and proline and hydroxyproline give yellow.
- Ninhydrin (0.3 g) in *n*-butanol (100 ml) containing acetic acid (3 ml) is sprayed on a dried, solvent-free layer, which is then heated for 30 min at 60°C or for 10 min at 110°C (Dzido 2001a). Detection limits range from 0.001 μg for alanine to 0.1 μg for proline and aspartic acid (Srivastava et al. 1984).
- Ninhydrin (0.3 g), glacial acetic acid (20 ml), and collidine (5 ml) are made up to 100 ml with ethanol (Srivastava et al. 1985a) or ninhydrin (0.1% w/v) in acetone acetic acid–collidine (100:30:4) (Srivastava et al. 1985b).
- A solution of cadmium acetate (0.5 g) in water (50 ml) and glacial acetic acid (10 ml) is made up to 500 ml with acetone. Portions of this solution are taken, and solid ninhydrin is added to give a final concentration of 0.2% g. The chromatogram is sprayed and heated at 60°C for 15 min. It is interesting to note the results immediately and again after 24 h, at room temperature (Bhushan et al. 1985). Alternatively, the layer is impregnated thoroughly with the reagent, and the colors are allowed to develop in the dark at room temperature for 24 h (Bhushan and Ali 1986). This reagent gives permanent colors, mainly red but yellow for proline. Sensitivity is 0.5 nmol.
- Ninhydrin (1.0 g) in absolute ethanol (700 ml), 2,4,6-collidine (29 ml), and acetic acid (210 ml) has been used for spraying on solvent-free cellulose layers. The chromatogram is then dried for 20 min at 90°C.
- Development of ion-exchange resin layers in ninhydrin (1%) in acetone containing collidine (10%) at room temperature for 24 h or 70°C for 10 min has also been recommended.
- A spray of ninhydrin (0.1% or 0.2%) in acetone on chromatograms followed by heating at 60°C or 90°C for 10–20 min has also been used (Bhushan and Reddy 1987a, b; Bhushan and Ali 1987a, b, c, d).
- *Polychromatic reagents*: Mosffat and Lytle developed a polychromatic ninhydrin reagent. It consisted of (a) ninhydrin (0.2%) in ethanol (50 ml) + acetic acid (10 ml) + 2,4,6-collidine (2 ml) and (b) a solution of copper nitrate (1.0%) in absolute ethanol (Bhushan et al. 1987). The two solutions are mixed in a ratio of 50:3 before use. Krauss and Reinbothe replaced ethanol by methanol and also achieved polychromatic amino acid detection by joint application of ninhydrin and primary, secondary, or tertiary amines (Bhushan and Reddy 1987c). The layers were first sprayed with diethylamine, dried for 3 min at 110°C, when the spots of amino acids appeared on a pale blue background. Use of ninhydrin (0.27 g), isatin (0.13 g), and triethylamine (2 ml) in methanol (100 ml) gave spots of amino acids on a yellow background.

In the analysis of biomolecules such as amino acids, the developed chromatograms are dried in an oven between 60 and 100°C and the cooled plates are usually sprayed with ninhydrin reagent. Heating at 90–100°C for 5–10 min produces blue to purple zones of all amino acids except proline, which gives yellow spot.

Chromatograms sprayed with ninhydrin (0.03 g ninhydrin in 100 ml of *n*-butanol plus 3 ml of glacial acetic acid) show air-sensitivity, stability, and color differentiation in comparison with different recipes of ninhydrin and fluorescamine sprays (Norfolk et al. 1994). Another possibility for detection of amino acids was also described (Devani et al. 1991; Laskar et al. 1991; Laskar and Basak 1990a, b; Sinhababu et al. 1994). TLC analysis of amino acids extracted from various plants were described (Gaspar and Males 2005; Males et al. 2001, 2004, 2005; Kopyt'ko 2003). Separation and estimation of amino acids in crude plant extracts was carried out by thin-layer electrophoresis (TLE) and chromatography (Bieleski and Turner 1966). A method for detecting *N*-acyl amino acid (substrate of aminoacylase) in vivo was described (Umebayashi 1968). The readers may be find a large number of successful application details for TLC analysis of amino acids and their derivatives (and also the enantiomeric resolution) in literature (Bhushan 2008; Bhushan and Martens 2000a, b, 2003a).

Detection of peptides and proteins on cellulose or silica plate can be performed with (Bhushan and Martens 2003b):

- *Morin Reaction*: The dried plates are sprayed with a 0.05% solution of morin (3,5,7,2′,4-pentahydroxyflavone) in methanol and heated for 2 min at 100°C. The N-protected amino acids and peptide derivatives give yellowish-green fluorescence on a green fluorescent background or dark absorption spots under UV. The detection limits is just about 2 μg/spot (Shellenberg 1962).
- *Iodine–Starch Reaction*: The chromatogram is placed in a strong iodine vapor atmosphere for 5 min. The excess iodine is removed by leaving the plate in the open air, and then layer is sprayed with 1% aqueous starch solution. The peptides (and amino acids as well) give blue spots (Barrett 1962).

Other details for determination of peptides and proteins were described (Bhushan and Martens 2003b). In another chapter (Steinberg 2003) the method of detection and their applications for TLC analysis of nucleic acids and their derivatives were described. Sample pretreatment and detection methods for biomolecules in plants were described in a book (Waksmundzka-Hajnos et al. 2008).

Quantitative Determination of Biomolecules

Detection of separated components on a TLC plate and generating analog curves of the chromatogram tracks for qualitative and quantitative evaluation are generally called densitometry. Densitometry can be performed in absorbance or fluorescence mode. Unlike scanning densitometry, which is based on sequential evaluation of the individual chromatogram tracks, video densitometry is based on grouping the pixels of the image according to the tracks and evaluating them on a gray scale (Reich and Schibli 2006). All pixels of the track, which have the same R_F value, are averaged and can be plotted as a function of distance in the direction of development. Because monochromatic light in the range of 190–800 nm can be used and tuned to the absorption or fluorescence maximum of the individual compounds, the measurement is highly sensitive. Typical detection limits are in the low nanogram range (absorbance) or medium picogram range (fluorescence). Densitometry is usually performed before derivatization. Only substances without chromophoric groups must be chemically altered to render them detectable (Reich and Schibli 2006).

Hyphenated Techniques

Rapid detection of biomolecules plays a strategical role in their investigation. Hyphenated techniques such as TLC coupled to UV diode array detection and to mass spectrometry provide on-line extensive structural information on the metabolites prior to isolation. In this chapter, the combination of planar chromatography (TLC) with biomolecules specific detection by diode array scanning (DAD), mass spectrometry (MS), nuclear magnetic resonance (NMR), and Fourier transform infrared spectroscopy (TLC-FTIR) is discussed. In the last part of the chapter, the reader will gain useful information about a new method of planar chromatography – multidimensional planar chromatography (MD-PC) and information on application of different modes of multidimensional planar chromatography and combination of this technique with diode array detection (MD-PC–DAD) and HPLC–DAD for separation, detection, and qualitative and quantitative determination of biomolecules.

Combined Thin Layer Chromatography/Fourier Transform Infrared Spectroscopy (TLC-FTIR) and Nuclear Magnetic Resonance (TLC-NMR)

TLC was first coupled with FTIR in 1989 (Glauninger et al. 1989), thereby providing a power tool for in situ identification of separated compounds, with applications in biological, pharmaceutical, environmental, and related sciences. Although densitometric evaluation enables rapid quantification with excellent precision and low limits of detection, the information potential of UV/Vis spectra is rather poor and rarely enables unambiguous identification of a substance. Furthermore, only chromophores can be detected by UV/Vis absorption and, because of the nature of dispersion instruments, spectra cannot be recorded on-line. Even the compounds that do not absorb in UV have IR spectrum, so that the TLC-FTIR technique is universal. The quality of TLC-FTIR spectra is sufficient for identification of unknown substances and discrimination between closely related substances

(Kovar et al. 1991). Because IR absorption is observed for nearly all chemical compounds, it is possible to detect and quantify non-UV-absorbing substances on TLC plates, which makes this hyphenated technique universally applicable (Rager and Kovar 2001). In comparison with the KBr technique, fewer absorbance bands are observed and – on commercially available plates – only the region between 3,550 cm^{-1} and 1,370 cm^{-1} can be evaluated (Rager and Kovar 2001). Diffuse reflectance Fourier transform (DRIFT) infrared IR spectra can be recorded from HPTLC/TLC plates (Kovar et al. 1991). Development of an optimized sorbent for direct HPTLC-FTIR on-line coupling was described (Bauer et al. 1998).

Short introduction to the theory, instrumentation, possibilities, and limitations for the direct on-line coupling of planar chromatography and infrared spectroscopy was described (Rager and Kovar 2001). Ten-year report on HPTLC-FTIR online coupling was reviewed (Stahlmann 1999). Qualitative and quantitative analysis by FTIR coupled with HPTLC was reported (Cimpoiu 2005). Figure 14.5 shows schematic diagram of the DRIFT unit constructed for on-line measurement. The special mirror arrangement was constructed in collaboration with the Bruker Company from Germany. Spectrometer is connected to the external unit by means of the interface mirror, which diverts the infrared beam through a one-sided mirror with an aperture to the parabolic mirror, which in turns directs the infrared beam on the plate at an exact angle of 90°. The parabolic mirror reflects the direct surface reflectance through the aperture back into the spectrometer, so that most of Fresnel reflectance is not detected (because it is eliminated in the

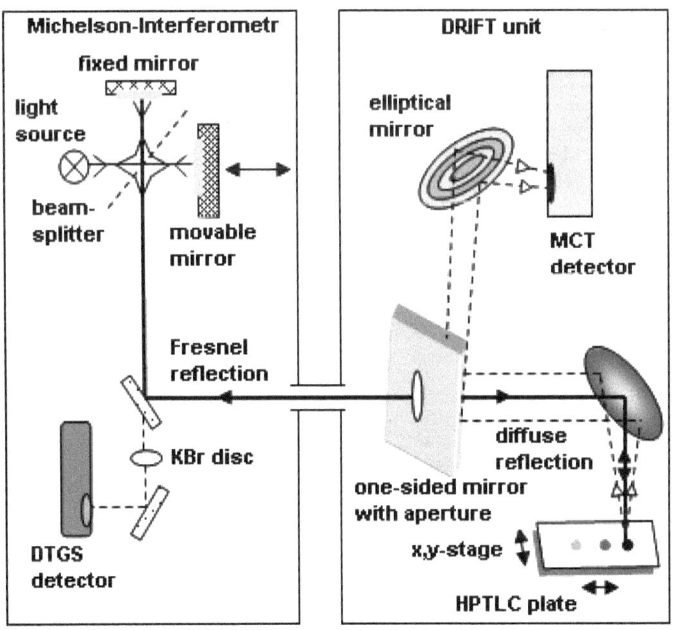

Fig. 14.5 Schematic diagram of DRIFT unit constructed for on-line measurement [Adapted from Rager and Kovar (2001)]

3,600–1,350 cm^{-1} region). The diffuse reflectance is gathered at the edges of the parabolic mirror and reflected by the one-sided mirror to the elliptical mirror, which again collects the diffuse reflectance and directs the converged beam to the MCT (mercury, cadmium, telluride) detector (Rager and Kovar 2001; Stahlmann 1999). By the use of commercially available GC-IR software, it is possible to obtain chromatogram from HPTLC plate. There are several different ways of presenting the huge amount of data acquired: as the *Gram-Schmidt* trace or as frequency-dependent chromatograms at any wavelength of interest. In addition to the three-dimensional (3D) plot of spectra, a contour plot can be derived from 3D stacked plot by acquiring a series of horizontal sections at different altitudes. This method enables rapid checking of peak purity. The DRIFT spectra of the substances can be calculated for reflectance and for quasi absorbance, by spectra division of the single beam spectra of reference and analyte or in *Kubelka–Munk* units (Rager and Kovar 2001; Stahlmann 1999).

On-line coupling of high-performance TLC with Fourier transform infrared spectroscopy (HPTLC-FTIR) was used successfully for identification of adenosine and related substances such as adenine, cAMP, AMP, IMP, xanthine, hypoxanthine, and inosine in biological material (Pfeifer et al. 1996). Adenosine and cAMP were detected in homogenates of rat insulinoma cells by in situ FTIR spectroscopy after TLC chromatographic development with water–methanol–ammonia–chloroform (Pfeifer et al. 1996).

The characterization of substances separated by TLC can also be obtained by NMR spectroscopy (TLC-NMR). TLC combined with high-resolution solid-state NMR for compound identification without substance solution was described (Wilson et al. 1997). The result presented show that it is possible to obtain NMR spectra, capable of interpretation, for analytes separated on RP TLC plate without recourse to exhaustive isolation procedures (Wilson et al. 1997).

More efficient identification of substances separated by planar chromatography is possible by modern scanning densitometry and the best characterization of analytes is possible by TLC coupled with modern mass spectrometry.

Thin Layer Chromatography with Diode Array Scanning (TLC–DAD)

In planar chromatography, light is used to detect separated sample spots by illuminating the layer plate from the top with light of known intensity (I_0). If the illuminating light shows higher intensity than the reflected light (J), a fraction of light must have been absorbed by the analyte and/or layer of TLC/HPTLC plate. The definition of the total absorption coefficient a, which describes the plate and sample light absorptions, is (Spangenberg 2006; Hiegel and Spangenberg 2009):

$$I_{\text{abs}} = I_0 - J = aI_0. \quad (14.1)$$

Theoretical considerations lead to the following equation for transformation purposes, showing linearity between the transformed measurement data (TDM) and the absorption coefficient (Spangenberg 2006; Hiegel and Spangenberg 2009).

$$\mathrm{TDM}(\lambda) = k(I_0/J - J/I_0) + (J/I_0 - 1) = a/1 - a, \tag{14.2}$$

where k is the backscattering factor, which has values in the range from 0 to $1, I_0$ the illuminating light intensity at different wavelengths, J the intensity of reflected light at different wavelengths, and a the total absorption coefficient.

- For $k = 0$, no incident light is reflected on the plate top (Spangenberg 2006; Hiegel and Spangenberg 2009). All light is scattered "forward," in the direction on the incident light, and in this situation, the light leaving the TLC/HPTLC plate at top must therefore be fluorescent light.

$$\mathrm{TDM}(\lambda, k = 0) = (R - 1). \tag{14.3}$$

- For $k = ½$, the Kubelka–Munk expression results from (Nyiredy 2001; Hahn-Deinstrop 2000) (14.2):

$$\mathrm{TDM}(\lambda, k = 1/2) = (1 - R)^2/2R = a/s \tag{14.4}$$

- For $k = 1$, (14.5) can be derived from (14.2):

$$\mathrm{TDM}(\lambda, k = 1) = (1/R - 1) = a/s \tag{14.5}$$

where R is the reflectance ($R = J/I_0$), and s the scattering coefficient ($s = 1 - a$).

With $k = 1$, (14.2) describes situation where all the light is reflected from the plate surface (Hiegel and Spangenberg 2009).

In general, fluorescence measurements are more specific than absorption measurements, due to the restricted number of analytes showing fluorescence (Hiegel and Spangenberg 2009).

TLC combined with modern scanning densitometry provides the possibility of quantitative analysis (Spangenberg 2006; Hiegel and Spangenberg 2009; Spangenberg and Klein 2000, 2001; Spangenberg et al. 2003; Ahrens et al. 2002). The method offers a simple and economical alternative to other chromatographic techniques, especially column high-performance liquid chromatography (HPLC). Application of modern fiber optic TLC scanner with a diode array detector (DAD) has several advantages (Tuzimski 2010a; Spangenberg 2006; Hiegel and Spangenberg 2009; Spangenberg and Klein 2000, 2001; Spangenberg et al. 2003; Ahrens et al. 2002; Tuzimski 2010b), for example, the scanner can measure TLC plates simultaneously at different wavelengths without destroying the plate surface and permits parallel recording of chromatograms and in situ UV spectra in the range 191–1,033 nm; therefore, it is possible to obtain doubly credible correct identification of the compounds on a chromatogram. The TLC scanner permits analysis of each compound at its optimum wavelength, thus offering optimum sensitivity for detection of each component. The TLC–DAD scanner permits measurement of a three-dimensional chromatogram, $A = f(\lambda, t)$, with absorbance as a function of wavelength and distance. The TLC–DAD scanner can compare parallel UV spectra of an unknown compound and a standard from library of spectra.

Software is available that allows the user access to all common parameters used in HPLC–DAD: peak purity, resolution, identification via spectral library match, etc. The TLC–DAD scanner is especially useful for correct identification of components of difficult, complicated mixtures, such as in plant extract and toxicological analysis.

At present, only a limited number of papers describe fiber optical scanning in TLC. An analytical procedure using HPTLC in combination with fiber optic scanning densitometry for identification of drugs in biological samples was described (Spangenberg and Klein 2001). An HPTLC–DAD method for dequalinium chloride (antibacterial and antifungal agent, active against many Gram-positive and Gram-negative bacteria and *Borrelia vincenti*, *Candida albicans*, and some species of Trichophyton) was described (Hiegel and Spangenberg 2009). Application of SPE and TLC with DAD for qualitative and quantitative analysis of dyes in beverages was also published (Tuzimski and Woźniak 2008). In another work, fiber optic scanning densitometry was used for identification and quantitative analysis of fenitrothion in fresh apple juice (Tuzimski 2005a). Figure 14.6 shows an example of the three-dimensional plot (scanning range × trace distance × absorbency) obtained from an apple extract (Tuzimski 2005a). Identification was achieved by comparing the UV spectrum obtained from the extract and a fenitrothion standard. Figure 14.7 shows UV spectra obtained from fenitrothion standards at eight concentrations (100–1,000 μg ml^{-1}) and the UV spectrum obtained

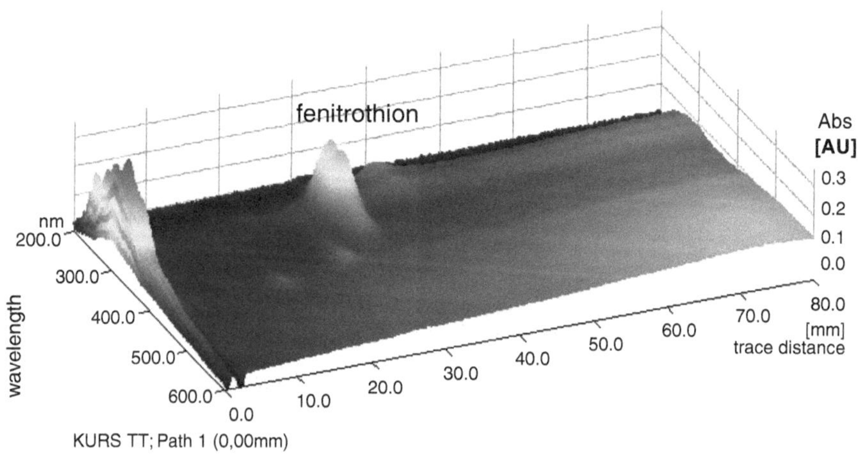

Fig. 14.6 Three-dimensional plot obtained from an apple extract containing fenitrothion. [Adopted from Tuzimski (2005a)]

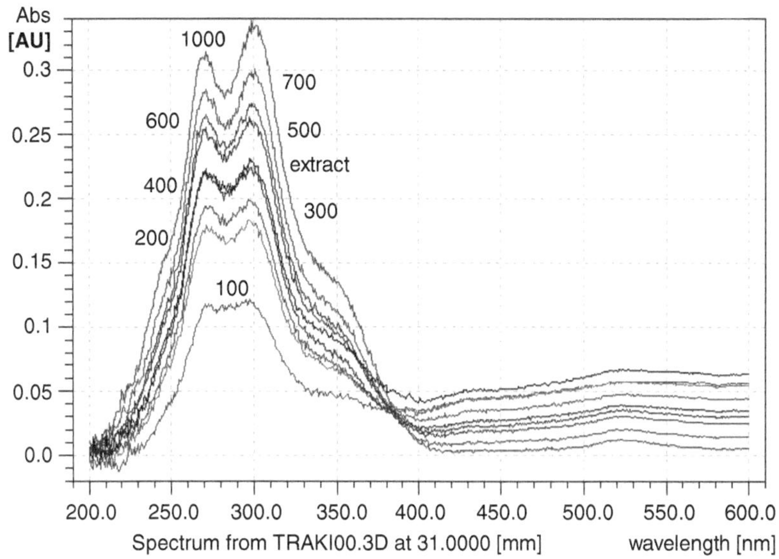

Fig. 14.7 UV spectra of obtained from fenitrothion standards at eight concentrations (100–1,000 μg/ml) and from an extract of freshly squeezed apple juice containing fenitrothion. [Adopted from Tuzimski (2005a)]

from fenitrothion in an extract from freshly squeezed apple juice (Tuzimski 2005a). The instrument detection limit (IDL) for fenitrothion was also determined. The IDL for fenitrothion was 10 μg ml^{-1}. The concentration of fenitrothion in the extract from fresh squeeze apple juice 45 days after spraying apples was below the method detection limit (MDL) for fenitrothion (Tuzimski 2005a).

The peak purity index is a numerical index for the quality of the coincidence between two datasets. It is given by the least-squares-fit coefficient calculated for all intensity pairs in the two datasets under consideration. The following equation is applied:

$$P = \frac{\sum_i (s_i - \bar{s})(r_i - \bar{r})}{\sqrt{\sum_i (s_i - \bar{s})^2 \sum (r_i - \bar{r})^2}}, \tag{14.6}$$

where s_i and r_i are the respective intensities for the same abscissa value, i is the number of data points, and $\bar{s}$ and $\bar{r}$ are the average intensities of the first and second dataset.

A peak purity index has values in the range from 0 to 1. A peak-purity index of 1 indicates that the compared spectra are identical. The components of two mixtures of pesticides, which were separated by 2D-TLC with adsorbent gradients of the type silica-wettable with water octadecyl silica or silica-cyanopropyl, were identified by R_F in both chromatographic systems and by comparison of UV spectra (Tuzimski 2005b). In other papers, an application of fiber optical scanning densitometry

(TLC–DAD) and HPLC–DAD for identification and quantitative analysis of pesticides in water samples from nine lakes and from Wieprz-Krzna Canal from Łęczyńsko-Włodawskie Lake District (South-East Poland) were demonstrated (Tuzimski 2008a). Atrazine, clofentezine, chlorfenvinphos, hexaflumuron, terbuthylazine, lenacyl, neburon, bitertanol, and metamitron were enriched from canal water samples by solid-phase extraction (SPE) on C18/SDB-1, C18, C18 Polar Plus, and CN cartridges. The recovery rates were high for all extraction materials except for all pesticides on cyanopropyl (CN) plates, for which the values were lower. SPE was used not only for the preconcentration of analytes but also for their fractionation. The analytes were eluted with methanol and next with dichloromethane. Methanol eluates were analyzed by HPLC–DAD (Fig. 14.8), the dichloromethane eluates with TLC–DAD. The method was validated for precision, repeatability, and accuracy. The calibration plots were linear between 0.1 and 50.0 $\mu g\ ml^{-1}$ for all pesticides, and the correlation coefficients, r, were between 0.9994 and 1.000 as determined by HPLC–DAD. Calibration plots were linear between 0.1 and 1.5 μg/spot for all pesticides, and the correlation coefficients, r, were between 0.9899 and 0.9987 determined by TLC–DAD. The limit of detection (LOD) was between 0.04 and 0.23 μg/spot (TLC–DAD) and between 0.02 and 0.45 $\mu g\ ml^{-1}$ (HPLC–DAD) (Tuzimski 2008a). Application of SPE and HPLC–DAD and/or TLC–DAD to the determination of pesticides in water samples from lakes from Łęczyńsko-Włodawskie Lake District (South-Eastern Poland) was also described (Tuzimski 2008b, 2009a; Tuzimski and Sobczyński 2009). Dichloromethane eluates were analyzed by TLC–DAD (Fig. 14.9). The identities of the bands of analytes in the water samples were confirmed by overlaying their UV

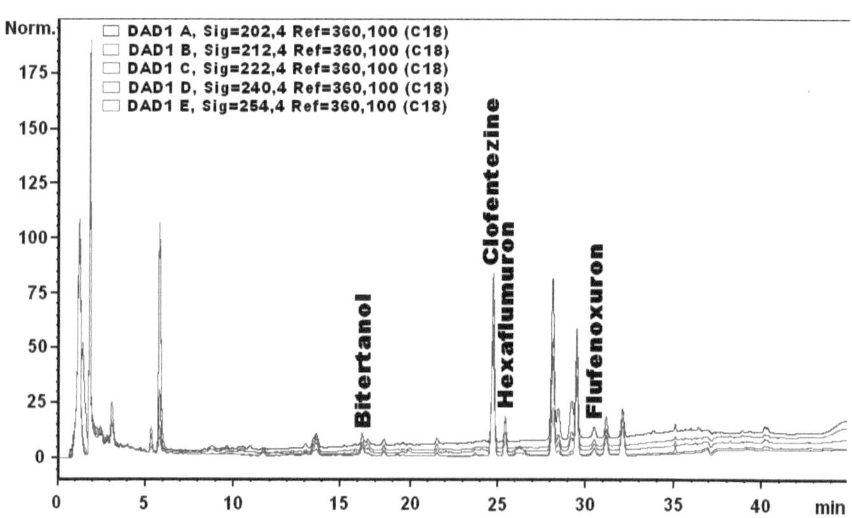

Fig. 14.8 Chromatogram of water obtained from Wieprz-Krzna Canal (July 2007) showing four detected and quantified pesticides. [Adopted from Tuzimski (2008a)]

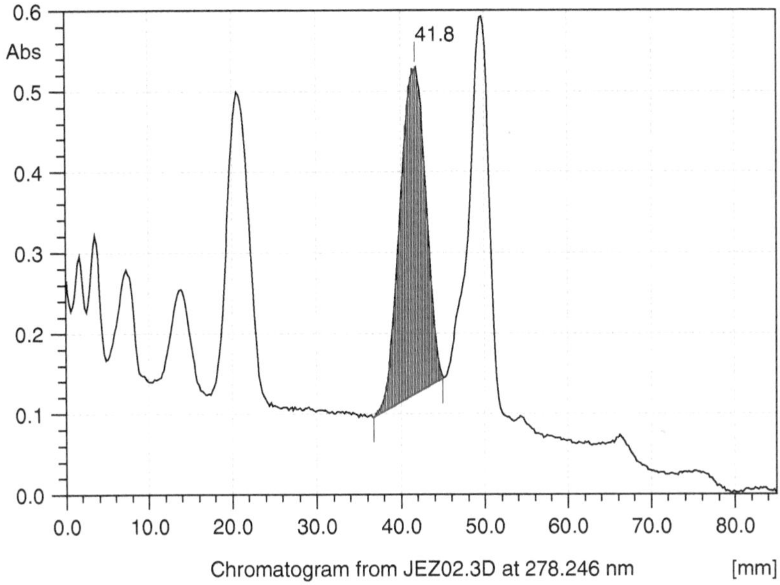

Fig. 14.9 Chromatogram obtained of fortified water sample after SPE and TLC–DAD for optimal wavelength for clofentezine ($\lambda = 278.246$ nm). [Adopted from Tuzimski (2008a)]

absorption spectra with those of the standards of these compounds (Fig. 14.10). A peak-purity index of 1 indicates that the compared spectra are identical. Least squares fit value of spectrum from fortified sample of water and spectrum from clofentezine standard was presented also (Fig. 14.11).

TLC in combination with fiber optical (diode array) scanning densitometry for identification of flavonol aglycones (quercetin and kaempferol) in extracts from rose leaves was described (Nowak and Tuzimski 2005). For quantitative determination, silica gel plates were used with 1,4-dioxane–toluene–85% acetic acid (6 + 24 + 1, v/v), as mobile phase. Figure 14.12 shows densitogram recorded at $\lambda = 364.8$ (maximum absorption of kaempferol). Chromatograms were developed on the distance of 75 or 80 mm. The tracks were scanned in the range 200 to 550 nm. Calibration plots were constructed by linear regression of peak area against concentration in the calibration range from 10 to 0.1 μg per spot, and quantification of quercetin and kaempferol was performed by measurement of the areas of their peaks from extracts of nine *Rosa L. taxa* (Nowak and Tuzimski 2005). Separation of the ginsenosides fraction obtained from the roots of *Panax quinquefolium L.* cultivated in Poland was also described (Ludwiczuk et al. 2005). Figure 14.13 shows 3D chromatogram of ginsenoside fraction from roots of *P. quinquefolium.* The identification of selected coumarins fractions from fruits of *Archangelica officinalis* Hoffm and *Heracleum species* separated by 2D-TLC was also confirmed by the comparison of the UV spectra of the mentioned extracts and the UV spectra of corresponding standards. Figure 14.14 shows UV spectra of bergapten (A),

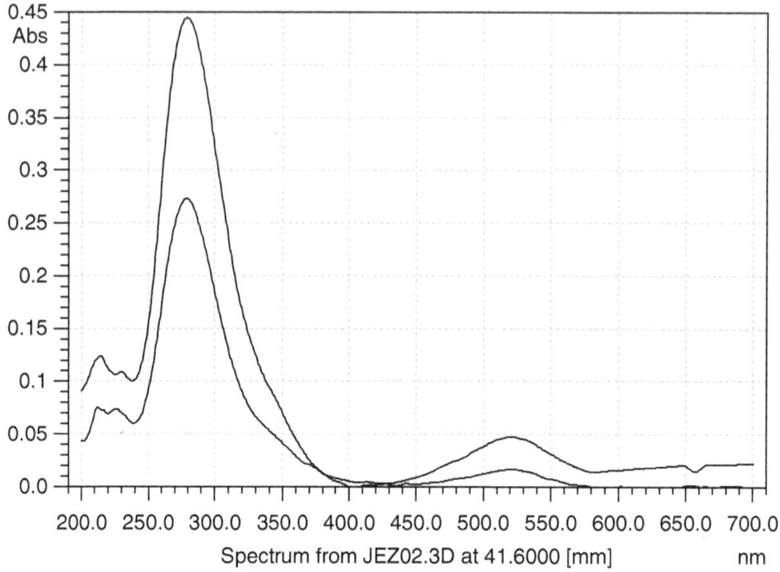

Fig. 14.10 Comparison of UV spectrum of clofentezine standard with in-situ spectrum of fortified water sample after SPE and TLC–DAD (Purity index (Pearson's r) $P = 0.9959$). [Adopted from Tuzimski (2008a)]

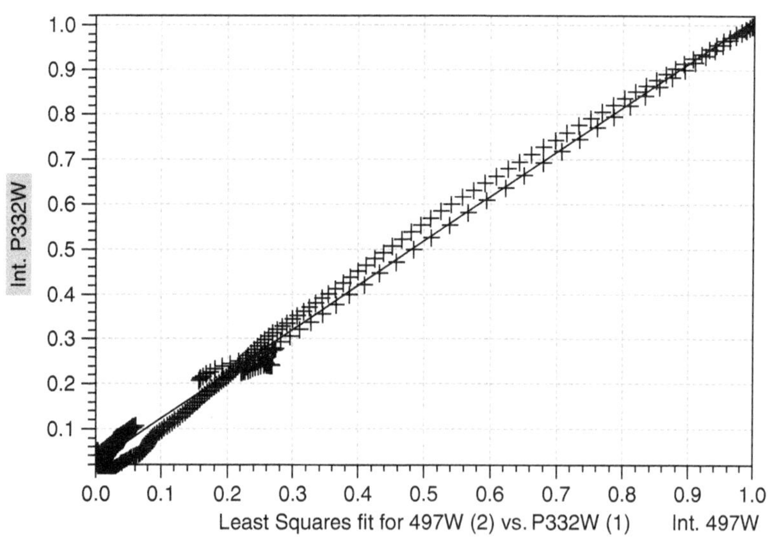

Fig. 14.11 Least squares fit value (obtained by cross-correlation) of spectrum from fortified sample of water and spectrum from clofentezine standard. [Adopted from Ref. Tuzimski (2008a)]

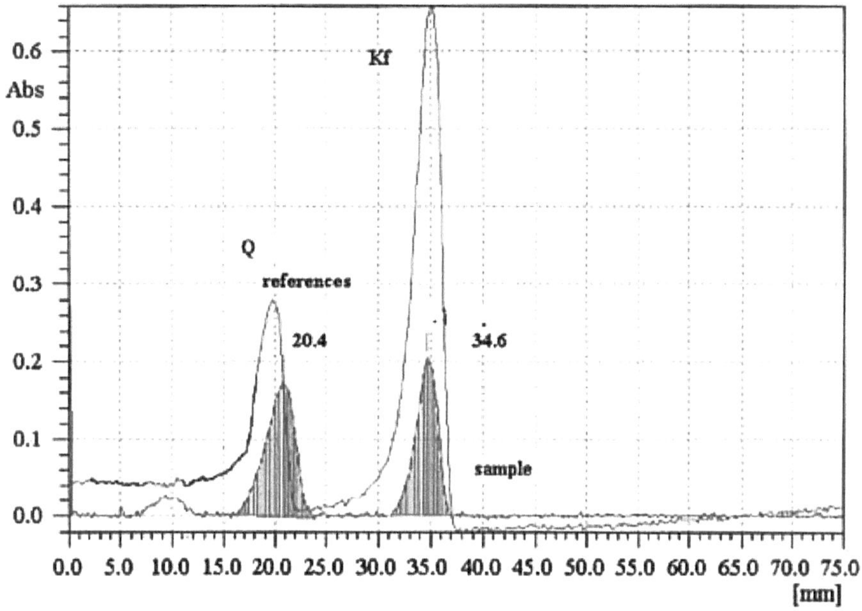

Fig. 14.12 Densitometric thin-layer chromatograms obtained from a mixture of kaempferol and quercetin standards and illustrating resolution of kaempferol and quercetin in an extract from *R. villosa*; scanning was performed at $\lambda = 364.8$ nm. [From Ref. Nowak and Tuzimski (2005) with permission]

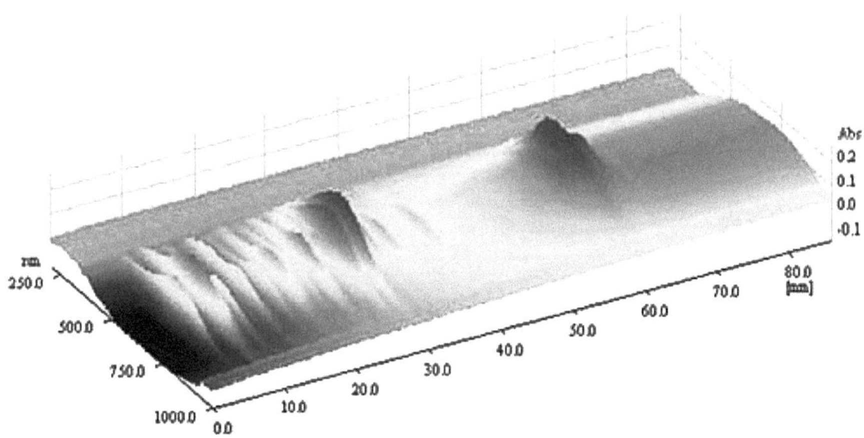

Fig. 14.13 3D chromatogram of ginsenoside fraction from roots of *P. quinquefolium*. Mobile phase chloroform-methanol-ethyl acetate-water-hexane, 20 + 22 + 60 + 8 + 4 (v/v). [From Ludwiczuk et al. (2005) with permission]

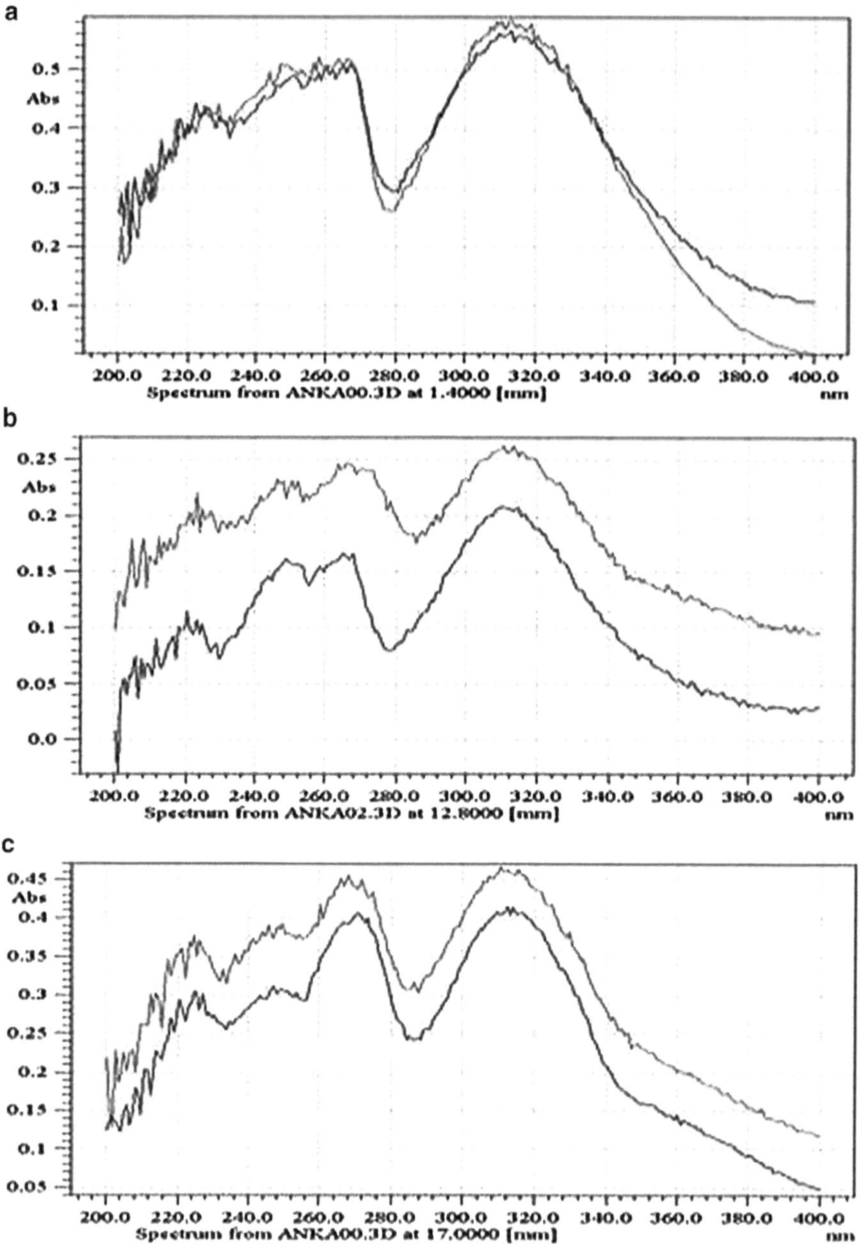

Fig. 14.14 UV spectra of some coumarin standards and *H. sibiricum* extract components obtained by DAD densitometer for bergapten (**a**); xanthotoxol (**b**); and isopimpinellin (**c**). [From Ref. Waksmundzka-Hajnos et al. (2006) with permission]

xanthotoxol (B), and isompimpinellin (C) (Waksmundzka-Hajnos et al. 2006). Isoquinoline alkaloids present in extracts from *Chelidonium majus*, *Fumaria officinalis*, and *Glaucium flavum* herbs separated by 2D-TLC with DAD-densitometry were also described (Petruczynik et al. 2007, 2008a).

Separation of the enantiomers of some amino acids on commercial chiral TLC plates in reversed-phase mode was described (Polak et al. 2006). This study shows that in chiral ligand-exchange chromatography (CLETLC), both retention and selectivity depend on the pH of the aqueous–organic mobile phase. Retention and selectivity differences are quite large, so chromatography should be conducted at pH 3–4 or 6–7, because at these pH values, the highest enantioselectivity can be obtained and the chiral layer is least damaged by the mobile phase (Polak et al. 2006).

Thin Layer Chromatography/Mass Spectrometry (TLC-MS)

Not all samples may be processed by HPLC-MS or HPLC–DAD systems due to zero or low detectability of the compounds or impurities in the UV range, a heavy matrix load or lack of MS compatible solvents. On the other hand, HPTLC is another very fast and convenient method to separate samples. The on-line coupling between planar chromatography and mass spectrometry provides a powerful combination for the detection and identification of compounds separated by TLC. TLC combined with mass spectrometry (TLC-MS) has been reviewed by Bush (2003, 2004), Somsen et al. (1995), and Wilson (1999). Different methods are reported in literature, but the analytical principle is the same: the sample is ionized from the layer surface by means of a laser beam, under vacuum, and in the presence of an energy-buffering matrix. In thin-layer chromatography, the sample is fixed on the layer and can be found in three-dimensional space. The interface between the layer and MS must separate the compound molecules from the layer and must present a way to transport them into the MS, either via a liquid or via a gaseous phase. Surveys have shown that the samples may be removed from the layer-by-layer desorption or matrix-assisted laser desorption ionization (MALDI), which can be coupled with time-of-flight (TOF), ion trap, or Fourier-transform mass spectrometer. The electron ionization (EI) or chemical ionization (CI) methods are most commonly used to break the sample molecules into fragments (ions) in the gaseous phase. The obtained spectra can be used for sample identification by comparison with a spectral library. The laser desorption ionization assisted by a matrix (MALDI and SALDI techniques) uses the ionization of the sample at the surface of the thin layer, and the matrix plays a role of energy buffer. Surface-assisted laser desorption ionization (SALDI) is differentiated from MALDI, in that the added matrix is thought to mediate the high energy of desorbing laser through a different process. Electrospray ionization (ESI) is another method of transferring the sample to MS by spraying organic solvents containing the molecules of the sample (Cimpan 2004).

Fast-atom bombardment (FAB) ionization and liquid secondary-ion mass spectrometry (LSIMS) are methods of ionization in which organic molecules are sputtered from surfaces by the impact of an energetic particle beam. Laser desorption (LD) is a mode of ionization in which sputtering of organic molecules from a surface occurs as a result of the high thermal energy imparted by the laser beam to the surface (Bush 2003). The sample ions formed by the FAB, LSIMS, and LD are usually the same even-electron ions such as $(M + H)^{+}$ formed in chemical ionization, and spectral interpretation proceeds along the same lines. A difference between the FAB, LSIMS, and LD on the one hand, and EI and CI on the other, is that the sample is not evaporated in separate step, and both volatile and nonvolatile "elements" can be sampled (Bush 2003).

A key difference between EI and CI on the one hand and fast-atom bombardment (FAB), LSIMS, and laser desorption (LD) on the other is the fact that the sampling in FAB and LSIMS is from a specified location that corresponds to the impact footprint of primary particle beam. If the sample is a solution, as it often is for FAB and LSIMS mass spectra of discrete samples, then diffusion within the solution blurs the spatial resolution of the ionization method. If the sample is held in a solid state, in a diffusion-controlled liquid state, or within a substrate such as a thin-layer chromatogram, the spatial resolution inherent to the sampling method is preserved. The natural compatibility of FAB, LSIMS, and LD with the direct mass spectrometric analysis of TLC plates is readily apparent (Bush 2003).

The need constant for reliable and sensitive analytical methods for nonvolatile and thermally labile compounds, including those of high molecular weight such as biomolecules, has encouraged development of new ionization methods in mass spectrometry. The key to successful adoption of TLC and mass spectrometry is produce a viable and useful interface, which will be a simple interface device that transforms the distribution of samples on an *xy* plane into a sequence of sample molecules in gas or liquid stream, mimicking LC/MS or GC/MS analysis. For hyphenation with ESI, different online approaches were developed. Anderson and Busch (Anderson and Busch 1998) reported the use of ESI coupled with TLC, and their publication includes a discussion of one-dimensional and two-dimensional interface designs. Wachs and Henion (2001) and Van Berkel et al. (2002) successfully demonstrated the employment of the surface sampling/elestrospray emitter system in the field of planar chromatography as an online method. Hsu et al. (2003) described an interface based on a channeled plate with inserted optical fibers at channel end or TLC stripe with a sharpened tip and the supply of a makeup solution conducting high voltage. Prosek et al. (2004) constructed a plunger-based online interface. Luftmann (2004) developed as well a plunger-based online extraction device. Van Berkel et al. (2005) reported HPTLC with MS with desorption electrospray ionization (DESI) based on the work Takats et al. (2004). Van Berkel and Kertesz (2006) illustrated some of the practical applications of TLC-DESI-MS. Ford et al. (2005) succeeded in using a stable isotope internal standard method to cope with instrumental drawbacks of HPTLC/ESI-MS. Morlock et al. (Jautz and Morlock 2006) described HPTLC/MS by a plunger-based extraction device to quantitative planar chromatography. Morlock et al. (Morlock and Ueda 2007) described also the coupling planar chromatography with direct analysis in real time time-of-flight mass spectrometry (DART-TOF-MS). Figure 14.15 shows the plate positioning, which led to the best response, which was a distance of approximately 1 mm aside orifice 1 and an angle of about 160° vertical to the gas flow. The analytical response regarding DART depended on the distance and angle of the plate relative to the DART gas stream. Desorption and ionization of substances from different stationary phases such as silica gel, amino phases, or RP18 phases were possible. Spatial resolution of DART was proven to be better than 3 mm (Fig. 14.16), then the plate was positioned in an in-house built plate holder system (Fig. 14.17) and direct into DART gas stream (Morlock and Ueda 2007; Morlock and Jautz 2008). Comparison of two different plunger geometries for HPTLC-MS coupling via an extractor-based interface was also described (Aranda and Morlock 2006).

Camag has released a new TLC-MS interface, an instrument to aid analysis of unknown compounds from a TLC/HPTLC plate by introducing the sample into a mass spectrometer for substance identification or structure elucidation. In the past, unknown substances were scraped off from the TLC/HPTLC plate, eluted into a

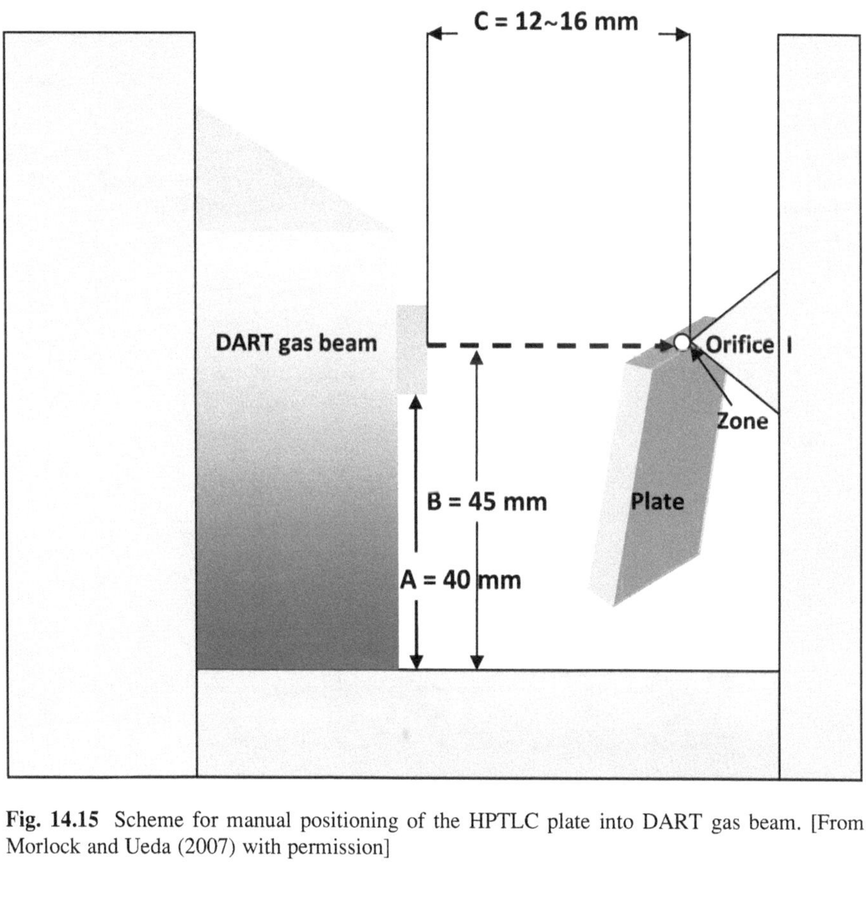

Fig. 14.15 Scheme for manual positioning of the HPTLC plate into DART gas beam. [From Morlock and Ueda (2007) with permission]

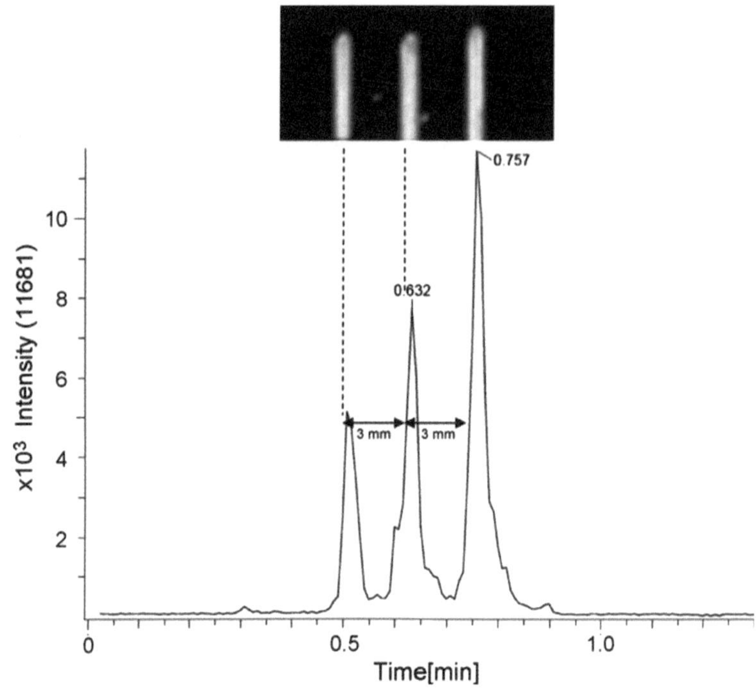

Fig. 14.16 Mass chromatogram at *m/z* 255.05936 to 255.10936 of three zones of 100 ng ITX spaced 3 mm apart; the plate was adjusted and fixed in a plate holder system and manually shifted at a speed of about 2 mm/10 s. [From Morlock and Ueda (2007) with permission]

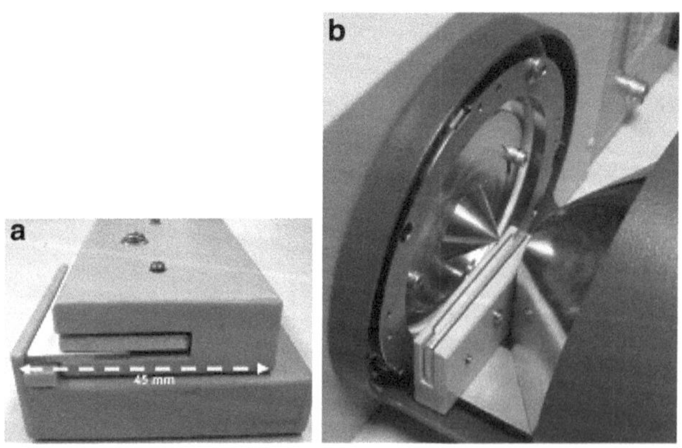

Fig. 14.17 In-house built plate holder system: (**a**) plate alignment tool to obtain the required positioning height of 45 mm from plate holder bottom to plate edge, (**b**). [From Morlock and Ueda (2007) with permission]

tube, and transferred into the MS. Scraping the spot of the analyte of interest from TLC/HPTLC plate and extracting the compounds into a solvent is a very useful method and provides the desired information, although this approach is quite tedious and time-consuming. Additionally, compounds with very similar R_F values can be hardly differentiated by scrapping. Now a very convenient and universal TLC-MS interface is available, which can semiautomatically extract zones of interest and direct them online into any brand of HPLC-MS system. Application of modern TLC-MS interface has several advantages, for example (Jautz and Morlock 2006, 2007; Morlock and Ueda 2007; Morlock and Jautz 2008; Aranda and Morlock 2006; Morlock and Schwack 2006; Alpmann and Morlock 2006, 2008; Morlock and Dytkiewitz 2008):

- The TLC-MS interface is compatible with all common HPLC-MS system
- The TLC-MS is quickly and easily connected (by two fittings) to any LC-coupled mass spectrometer without adjustments or mass spectrometer modifications
- The interface permits rapid and convenient extraction directly into most types of MS system (APCI-MS, APPI-MS, or ESI-MS)
- The interface is especially useful for identification of unknown substances
- The TLC-MS interface gives the possibility of confirmation of target compounds
- The TLC-MS enables to solve the problem by semiautomatic extraction of zones of interest directed on-line into most types of MS systems
- The TLC-MS is compatible with the most common TLC/HPTLC layers
- The TLC-MS interface gives the possibility of extraction into vials for NMR or (ATR-)FTIR, static nanospray, direct inlet EI-MS, and MALDI
- The TLC-MS interface is the solution to an ongoing analysis issue, where traditional TLC/HPLC methods are unable to identify unknown compounds
- The TLC-MS interface eliminates scraping off the plate.
- The TLC-MS interface gives the possibility of automatic cleaning of the piston between the extractions.

According to the producer, the TLC-MS interface enables substances to be directly extracted from a TLC/HPTLC plate, and sensitive mass spectrometric signals are obtained within a minute per substance zone. The interface extracts the complete substance zone with its depth profile and thus allows detections comparable to HPLC down to the pg/zone range. The interface has been proven to be one of the most reliable and versatile interfaces for TLC/HPTLC-MS coupling (Jautz and Morlock 2006, 2007; Morlock and Ueda 2007; Morlock and Jautz 2008; Aranda and Morlock 2006; Morlock and Schwack 2006; Alpmann and Morlock 2006, 2008; Morlock and Dytkiewitz 2008).

Examples of TLC-MS analysis of biomolecules (e.g., proteins, peptides, DNA, nucleic acids, and lipids) are presented below. The paper (Reinders et al. 2004) described application of TLC combined with matrix-assisted laser desorption (TLC-MALDI-MS) for analysis of peptides, which were historically used as first components for evaluating the capabilities and limitations of TLC-MALDI. The combination of TLC with MALDI mass spectrometry was described (Gusev et al.

1995a). Various MALDI matrices were evaluated for use, including silica gel and cellulose TLC plates, for the detection of smaller detection proteins (bradykinin, angiotensin, and enkephalin that possess molecular masses between about 500 and 1,500 Da). The best detection limits were obtained with ferulic acid (ferulic acid/fucose) and sinapinic acid matrices. These matrices also showed high sodium and potassium salt tolerance. Use of salts in the developing solvents caused an increase in the cationized peaks without impacting sensitivity. The best crystal homogeneity on silica gel and cellulose as well as surface coverage was obtained using 2-(4-hydroxyphenylazo) benzoic acid (HABA) matrix. An absolute detection limit of 2–4 ng was demonstrated for bradykinin, angiotensin, and enkephalin derivatives. The peptides are particularly important because they are used as calibrants for MALDI, which can be coupled with the time-of-flight technique (MALDI-TOF) (Hillekamp and Peter-Katalinić 2007). The method also worked for larger peptides and small proteins like bovine insulin chain B, insulin, horse heart cytochrome *c*, and myoglobin, but TLC separation was poor (Gusev et al. 1995a). Spatial resolutions of about 250–500 μm and absolute detection limits in the analytes range were described (Gusev et al. 1995b). TLC separation of larger peptides with detection by MALDI-TOF-MS was described as well (Bakry et al. 2007). The detection of smaller oligonucleotides is possible by using 2,4,6-trihydroxyacetophone as matrix and ammonium hydroxide/methanol as solvent system, and then the detection limits in the range 10 pg were reported (Isbell et al. 1999). In the next example, capabilities and disadvantages of combined MALDI-TOF MS and HPTLC for analysis of egg yolk lipids was described (Fuchs et al. 2009). TLC coupled with MALDI-MS was used in the analysis of lipids (Schiller et al. 2003). HPTLC with MALDI-MS by use of liquid matrix (glycerol) for the analysis of biomolecules (GM3 gangliosides) was described (Dreisewerd et al. 2005).

An important problem particularly related to MS imaging of biological tissues is stemming from the interface of matrix peaks with biological molecules of interest. In order to overcome this problem, it was suggested to use graphite-assisted laser desorption/ionization mass spectrometry (GALDI-MS), and this technique was applied for the analysis of cerebrosides in a complex total brain lipid extract (Cha and Yeung 2007).

Tandem mass spectrometry (MS-MS) offers further advantages to the chromatographer, in that it can provide more detailed information on the mass spectrum, in the absence of interferences from ions resulting from co-chromatographing contaminations and the background. HPTLC or TLC in combination with tandem mass spectrometry (MS-MS) was reviewed (Wilson and Morden 1996).

Automated Multiple Development of Biomolecules

A special device for an automated multiple developments (AMD) of a chromatogram was described by Perry et al. (1973) and a programmable setup constructed by Burger (1984) and produced by Camag is known as AMD or AMD 2 system. A full separation process comprising 20–25 steps takes a long time. However, this is compensated by simultaneous separation of many samples on one chromatographic plate and using the system outside working hours without inspection. Therefore, the final analysis is characterized by a relatively high throughput. This throughput can be increased by reduction of the number of steps of the AMD procedure. Application of special software for the simulation of the planar chromatography process can additionally enhance this procedure (Markowski 1996, 2005).

Advantages of this device are the following (http://www camag com):

- The mobile phase for each development is prepared automatically by mixing appropriate portions of solvents from up to five different reservoirs.
- Gradient development can be accomplished with a similar number of the mobile-phase components.
- Chromatography is monitored, and the run stops when the selected developing distance is reached.
- Chromatographic plate (usually HPTLC plate) is developed repeatedly in the same direction.
- Each step of the chromatogram development follows complete evaporation of the mobile phase from the chromatographic plate and is performed over longer migration distance of the solvent front than the one before.
- Each step of the chromatogram development uses a solvent of lower elution strength than the one used in the preceding run; it means that a complete separation process proceeds under conditions of gradient elution.
- The focusing effect of the solute bands takes place during the separation process, which leads to very narrow component zones and high efficiency of the chromatographic system comparable to HPLC.
- AMD is highly reproducible.
- AMD has major strength for separation of components spanning a wide polarity range or that are similar. In the first case, a steep gradient provides best results. In the second case, the focusing effect of multiple developments combined with a shallow gradient, small increases of developing distance, and a large number of steps gives the best separation.
- If the experimentation is realized with the regulation of the temperature, all the conditions are got together to increase the reproducibility in planar chromatography analysis.
- The fully automated development of the plates (preconditioning time, automated eluent gradient, drying time) determines a good reproducibility of the analysis and the accurate mixture eluent for each step permit a sharper separation in well defined experimental conditions with no spot diffusion in the adsorbent and also reproducible R_F values.
- AMD permits the analysis on very small quantities and obtains sharper separations because the absence of diffusion in the adsorbent at the upper R_F; this makes it a very interesting method for densitometry.

Poole and Belay (1991) reviewed the essential methods and parameters of multiple development techniques in planar chromatography (including also

AMD). Evaluation of parameters such as change in the zone width vs. number of developments, zone separation vs. number of developments through AMD, and several typical applications of AMD are described.

The chromatogram is developed several times on the same plate and each step of the development follows the complete evaporation of the mobile phase from the chromatographic plate of the previous development. On the basis of the development distance and the composition of the mobile phase used for consecutive development steps, multiple development techniques are classified into four categories (Szabady and Nyiredy 1996a):

- Unidimensional multiple developments (UMD), in which each step of chromatogram development is performed with the same mobile phase and the same migration distance of eluent front.
- Incremental multiple development (IMD), in which the same mobile phase but an increasing development distance in each subsequent step is applied.
- Gradient multiple development (GMD), in which the same development distance but a different composition of the mobile phase in each step is applied.
- Bivariant multiple development (BMD), in which the composition and development distance is varied in each step of the chromatogram development.

These modes of chromatogram development are mainly applied for analytical separations due to the good efficiency, which is comparable to HPLC.

AMD mode enables both isocratic and GMD. In a typical isocratic AMD mode, the development distance is increased during consecutive development steps, whereas the mobile-phase strength is constant. In the initial stage of the AMD gradient procedure, the solvent of the highest strength is used (e.g., methanol, acetonitrile, or acetone); in the next stages – an intermediate or base solvent of medium strength (e.g., chlorinated hydrocarbons, ethers, esters, or ketones); and in the final stage – a nonpolar solvent (e.g., heptane, hexane) (Ebel and Volkl 1990).

Several parameters must be considered to obtain the best separation in AMD mode: choice of solvents, gradient profile of solvents, and a number of steps. All modes of multiple developments can be easily performed using chambers for automatic development, which are manufactured by some firms. However, these devices are relatively expensive. Typical horizontal chambers for planar chromatography should be considered for application in multiple developments in spite of more manual operations in comparison with the automatic chromatogram development. Especially horizontal DS chamber could be considered for separations with multiple developments. This chamber can be easily operated due to its convenient maintenance, including cleaning the eluent reservoir. For the separation of a more complicated sample mixture, a computer simulation could be used to enhance the efficiency of the optimization procedure (Markowski 1993, 1996, 2005; Markowski and Soczewiński 1992, 1993; Markowski et al. 1994).

The automated multiple development (AMD) technique is suitable for the analysis of very complex mixtures. AMD has proved to be an efficient planar chromatographic technique that provides increased separation for compounds with neighboring structures. Analysis of three types of opiate alkaloids (the poppy alkaloids: morphine, codeine, noscapine, and papaverine; the semisynthetic and synthetic derivates such as pholcodine, ethylmorphine, and dextromethorphan; the narcotic compounds, diacetylmorphine (heroin)) and opiates employed as substitutes in treatment of addiction (buprenorphine and methadone) by AMD was described and compared with results obtained by TLC (Pothier and Galand 2005). The AMD system enabled a clean separation of each of three opiate groups (antitussives and substitutes) studied, and the best results have been obtained with universal gradient. Two reagents were used for the detection of alkaloids by spraying: Dragendorff and iodoplatine reagents. In the first experimentation, the universal gradient: methanol 100, methanol, dichloromethane 50/50, dichloromethane 100, dichloromethane 100, and hexane 100 during 20 steps were used; the term universal characterizes a gradient that performs a separation of a mixture with a large polarity scale. This gradient permitted a very clean separation of the major alkaloids from opium extract: morphine, codeine, thebaine, noscapine, and papaverine. This one shows zones corresponding to these of the standard solutions, orange–red or red with dragendorff or other colors with iodoplatinate reagent's (Fig. 14.18). For antitussives and substitutes, the gradient used was: 5% of 28% ammonia in methanol 100, acetone 100, acetone 100, ethyl acetate 100, ethyl acetate–dichloromethane 50/50, dichloromethane 100, and with this eluent, a clean separation was obtained with antitussives (Fig. 14.19) and opiate derivatives and substitutes (Fig. 14.20). AMD can have applications to control the quality of opium according to the norms required by the *Pharmacopeia*. The best resolution and the lack of diffusion makes AMD a method that is the most efficient for densitometry and can be an alternative to isocratic chromatography described in many pharmacopeias (Pothier and Galand 2005).

Planar chromatography combined with AMD is a suitable technique to separate constituents in crude plant extracts. HPTLC method for the separation of boswellic acids (β-boswellic acid, acetyl- β-boswellic acid, 11-keto- β-boswellic acid, and

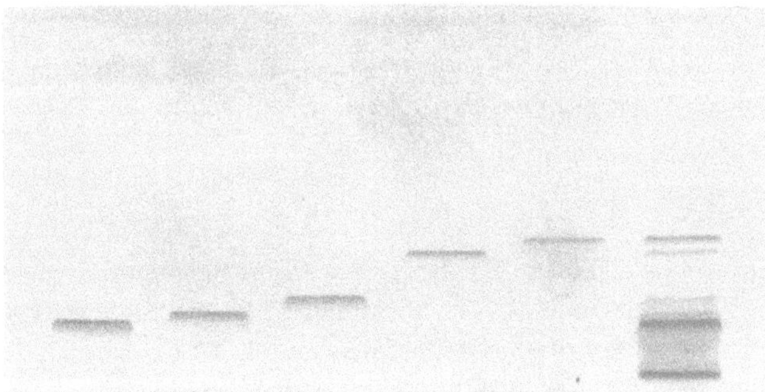

Fig. 14.18 HPTLC chromatogram by AMD of opium extract and standard alkaloids of opium: (1) morphine; (2) codeine; (3) thebaine; (4) papaverine; (5) noscapine; (6) opium extract; eluent used was universal gradient: methanol 100, methanol–dichloromethane 50/50, dichloromethane 100, dichloromethane 100, hexane; 100; derivatization by Dragendorff reagent. [From Pothier and Galand (2005) with permission]

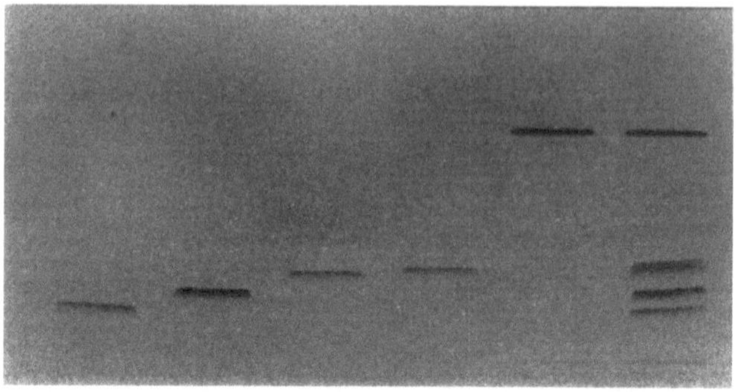

Fig. 14.19 HPTLC chromatogram by AMD of antitussives: (1) pholcodine; (2) dextromethorphan; (3) ethylmorphine; (4) codeine; (5) noscapine; (6) mixture of the five standards; gradient used was gradient A: 5% of 28% ammonia in methanol 100, acetone 100, acetone 100, ethyl acetate 100, ethyl acetate–dichloromethane 50/50, dichloromethane 100; derivatization by iodoplatinate of potassium. [From Pothier and Galand (2005) with permission]

Fig. 14.20 HPTLC chromatogram by AMD of opiates, derivatives, and substitutes: (1) morphine; (2) codeine; (3) diacetylmorphine; (4) methadone; (5) buprenorphine; (6) mixture of the five standards; gradient used was gradient A: 5% of 28% ammonia in methanol 100, acetone 100, acetone 100, ethyl acetate 100, ethyl acetate–dichloromethane 50/50, dichloromethane 100; derivatization by iodoplatinate of potassium. [From Pothier and Galand (2005) with permission]

acetyl-11-keto β-boswellic acid), the active constituents in *Boswellia serrata* extract, was developed, and TLC of these compounds on silica by AMD using solvent gradients was described (Pozharitskaya et al. 2006).

A technique reciprocal to AMD was introduced by Markowski and Matysik (1993), Matysik (1996) and Matysik and Soczewiński (1988b) called multiple gradient development (MGD): the plate is developed several times (with intermittent evaporation of solvents), starting with a weak solvent on full distance, then with eluents of increasing eluent strengths on decreasing distances, retaining separation of spots of higher R_F values achieved in the preceding developments.

Multidimensional Planar Chromatography (MD-PC) Methods in Biomolecules Analysis

The application of multidimensional planar chromatography combined with different separation systems and modes of chromatogram development is often necessary for performing the separation of more complicated multicomponent mixtures. High separation efficiency can be obtained using modern planar chromatographic techniques, which comprise two-dimensional development, chromatographic plates with different properties, a variety of solvent combinations for mobile phase preparation, various forced-flow techniques, and multiple development modes. By combination of these possibilities, multidimensional planar chromatography (MDPC) can be performed in various ways. Giddings defined multidimensional chromatography as a technique that includes two criteria (Giddings 1990):

- The components of the mixture are subjected to two or more separation steps in which their migration depends on different factors.
- When two components are separated in any single step, they always remain separated until completion of the separation.
- Divided multidimensional planar chromatography techniques as follows (Tuzimski 2010a; Tuzimski and Nyiredy 2005; Nyiredy 2002a, 2003b).
- Comprehensive two-dimensional planar chromatography (PC × PC) – multidimensional development on the same monolayer stationary phase and two developments with different mobile phases or using a bilayer stationary phase and two developments with the same or different mobile phases.
- Targeted or selective two-dimensional planar chromatography (PC + PC) – technique, in which following the first development from the stationary phase, a heart-cut spot is applied to a second stationary phase for subsequent analysis to separate the compounds of interest.
- Targeted or selective two-dimensional planar chromatography (PC + PC) – second mode – technique, in which following the first development, which is finished and the plate dried, two lines must be scraped into the layer perpendicular to the first development and the plate developed with another mobile phase, to separate the compounds that are between the two lines. For the analysis of multicomponent mixtures containing more than one fraction, separation of components of the next fractions should be performed with suitable mobile phases.
- Modulated two-dimensional planar chromatography (^{n}PC) – technique, in which on the same stationary phase, the mobile phases of decreasing solvent strengths and different selectivities are used.
- Coupled-layer planar chromatography (PC–PC) is a technique in which two plates with different stationary phases are turned face to face (one stationary phase to second stationary phase) and pressed together so that when a narrow zone of the layers overlaps, the compounds from the first stationary phase are transferred to the second plate and separated with a different mobile phase.

- Combination of multidimensional planar chromatography methods – technique, in which the best separation of multicomponent mixture is realized by parallel combination of stationary and mobile phases, which are changed simultaneously. By use of this technique, for example, after separation of compounds in the first dimension with changed mobile phases, the plate is dried and separation process is continued in perpendicular direction by use of the grafted technique with changed mobile phase (based on the idea of coupled TLC plates, denoted as graft TLC in 1979 (Pandey et al. 1979)).

Comprehensive Two-Dimensional (2D) Chromatography on One Adsorbent

One of the most attractive features of planar chromatography is the ability to operate in the two-dimensional (2D) mode. Two-dimensional TLC (2D-TLC) is performed by spotting the sample in the corner of a square chromatographic plate and by development in the first direction with the first eluent. After the development is completed, the chromatographic plate is then removed from the developing chamber and the solvent is allowed to evaporate from the layer. The plate is rotated through 90° and then developed with the second solvent in the second direction, which is perpendicular to the direction of the first development. In 2D-TLC, the layer is usually of continuous composition, but two different mobile phases must be applied to obtain a better separation of the components. If these two solvent systems are of approximately the same strength but of optimally different selectivity, then the spots will be distributed over the entire plate area, and in the ideal case, the spot capacity of the two-dimensional system will be the product of the spot capacity of the two constituent one-dimensional systems. If the two constituent solvent systems are of the same selectivity but of different strengths, the spots will lie along a straight line; if both strength and selectivity are identical, the spots will lie along the diagonal.

Computer-aided techniques enable identification and selection of the optimum mobile phases for separation of different groups of compounds. The first report on this approach was by Guiochon and co-workers, who evaluated ten solvents of fixed composition in two-dimensional separation of 19 dinitrophenyl amino acids chromatographed on polyamide layers (Gonnord et al. 1983). The authors introduced two equations for calculation of the separation quality – the sum of the squared distances between all the spots, D_A, and the inverse of the sum of the squared distances between all the spots, D_B. Steinbrunner et al. (1986) proposed other functions for identification of the most appropriate mobile phases – the distance function *DF* and the inverse distance function *IDF*, which are the same form as D_A and D_B, respectively, but which use distances rather than the squares of distances. The planar response function *PRF* has been used as optimization criterion by Nurok et al. (1987). Strategies for optimizing the mobile phase in planar chromatography (including two-dimensional separation) (Nurok 1989) and overpressured layer chromatography (including two-dimensional overpressured layer chromatography) (Nurok et al. 1997) has also been described. Another powerful tool is the use of graphical correlation plots of retention data for two chromatographic systems, which differ with regard to modifiers and/or adsorbents (De Spiegeleer et al. 1987). The interpretation of plots is illustrated in Fig. 14.21. The plots on Fig. 14.21

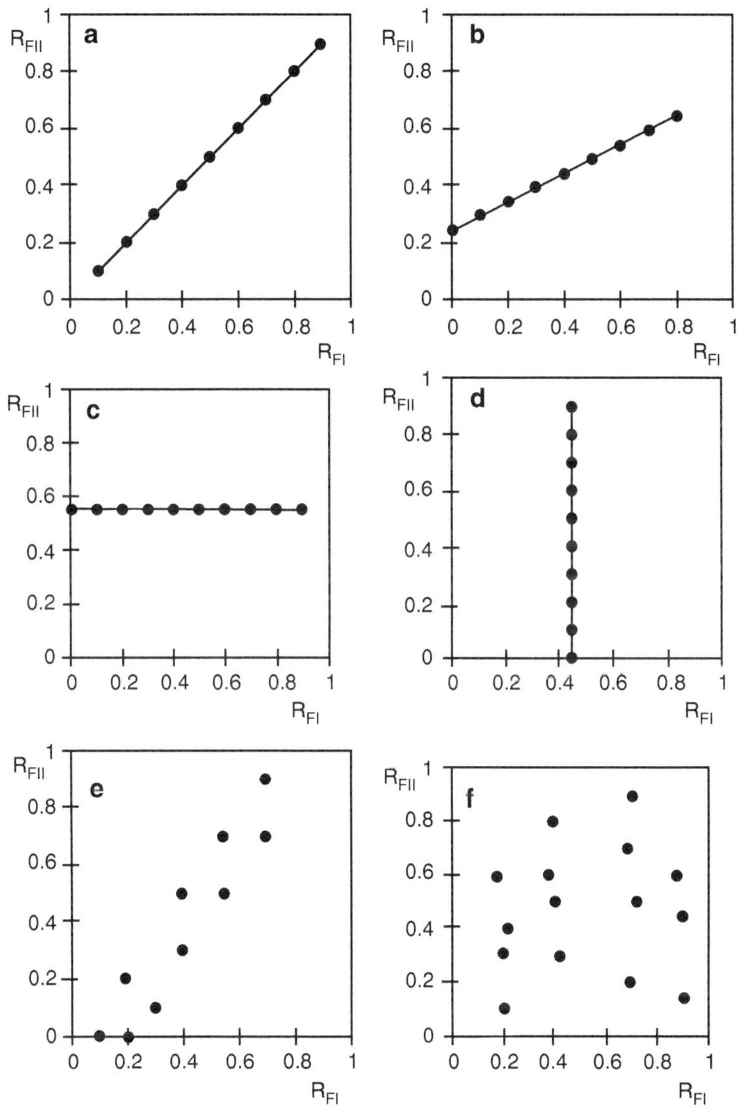

Fig. 14.21 Characteristic correlations $R_{F\,II}$ vs. $R_{F\,I}$. [From Tuzimski and Soczewiński (2002d) with permission]

indicate then directly the positions of spots on two-dimensional chromatograms (2D-TLC). As shown in Fig. 14.21f, if the best separation of complex mixtures by 2D-TLC is possible with differentiated R_F values in both systems, then the correlation plots of retention parameters for two chromatographic systems are poor (Tuzimski and Soczewiński 2002d). Good separation can be achieved when the spots are spread over the whole of the chromatographic plate (Tuzimski 2004; Tuzimski and Soczewiński 2002a, b, c; Tuzimski and Bartosiewicz 2003). The largest differences were obtained by combination of normal-phase (NP) and reversed-phase (RP) systems with the same chromatographic layer, e.g., cyanopropyl (Tuzimski 2004; Tuzimski and Soczewiński 2004). An example of this type of 2D development is illustrated in Fig. 14.22d.The first comprehensive 2D separations were achieved with paper chromatography in 1944 (Consden et al. 1944). This discovery led to a number of other multidimensional techniques, with many of these involving electrodriven separations. Hugaard and Kroner separated amino acids by coupling paper chromatography with paper electrophoresis, thus performing the original multidimensional electrodriven separation (Hugaard and Kroner 1948). This study, published in 1948, involved the use a 100 V electric field applied across one dimension of the paper, while a phosphate buffer used as a chromatographic eluent to move substances in the orthogonal direction (Hugaard and Kroner 1948). 2D planar electrophoresis was first used in 1951 (Duruum 1951), while electrophoresis was coupled with TLC in 1964 to separate mixtures of nucleotides and nucleosides (Keck and Hagen 1964). O'Farrell determined that high-resolution separation of protein mixtures could be achieved with polyacrylamide gel isoelectric focusing (IEF) in one dimension and polyacrylamide slab gel electrophoresis (PAGE) in the other (O'Farrell 1975). This technique was used to separate 1,100 different components from *Escherichia coli*, some of which differed by as little as single charge, while being similar in molecular weight. Two-dimensional thin-layer electrophoresis/TLC (2D TLE/TLC) remains one of the most popular methods for analysis of peptides. 2D TLE/TLC coupled to MS detection involves sequencing, identification, and quantification of phosphorylated peptides at femtomolar levels based upon extraction of individual spots from a cellulose TLC plate and Edman degradation (Kochin et al. 2006). Fractionation of complex peptide mixtures by 2D TLE/TLC with MALDI-TOF-MS or MALDI-TOF-MS/MS was described also (Panchagnula et al. 2007).

Two-Dimensional Thin-Layer Chromatography (2D-TLC)

In 2D development, the mixtures can be simultaneously spotted at each corner of the chromatographic plate so that the number of separated samples can be higher in comparison to the "classical 2D development" (Dzido 2001a). An example of this type of 2D development is illustrated in Fig. 14.22a–d. Figure 14.22d shows a videoscan of the plate, which shows separation of three fractions of the mixture of nine pesticides by 2D planar chromatography with NP/RP systems on a chemically bonded cyanopropyl stationary phase.

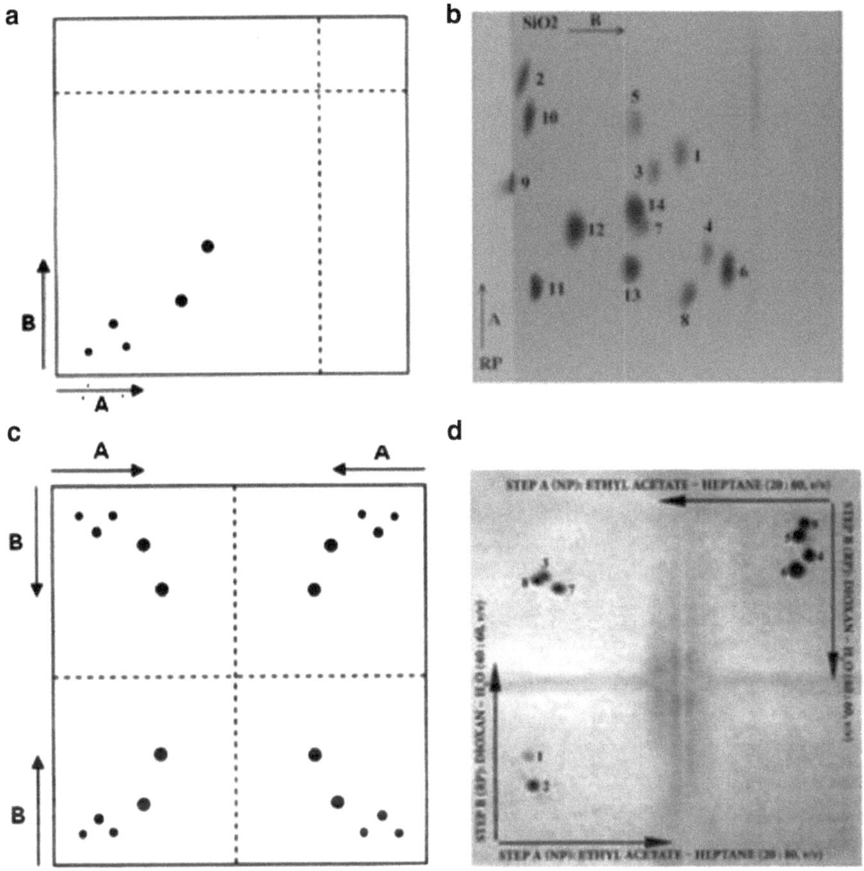

Fig. 14.22 Two-dimensional development, (**a**) schematic presentation of 2D-chromatogram [Adapted from Dzido (2001a)], (**b**) 2D-chromatogram of the 14-component mixture of pesticides presented as videoscan of dual-phase Multi-K CS5 plate in systems: A (first direction): methanol–water (60:40, v/v) on octadecyl silica adsorbent, B (second direction): tetrahydrofuran-*n*-heptane (20:80, v/v) on silica gel [From Tuzimski and Soczewiński (2002b) with permission], (**c**) schematic presentation of 2D-chromatogram of four samples simultaneously separated on the plate [Adapted from Dzido (2001a)], (**d**) 2D-chromatograms of three fractions of the mixture of nine pesticides presented as videoscan of the HPTLC plate (cyanopropyl) in systems with A (first direction): ethyl acetate-*n*-heptane (20:80, v/v), B (second direction): dioxane–water (40:60, v/v). [From Tuzimski and Soczewiński (2004) with permission]

Nyiredy (2001) and Szabady and Nyiredy (1996b) described the technique of joining two different adsorbent layers to form a single plate. Also, the largest differences were obtained by the combination of normal-phase systems of the type silica/nonaqueous eluent and reversed-phase systems of the type octadecyl silica/water + organic modifier (methanol, acetonitrile, dioxane) on multiphase plates with a narrow zone of SiO_2 and a wide zone of RP-18 (or vice versa), which are commercially available from Whatman (Multi K SC5 or CS5 plates) (Tuzimski and Soczewiński 2002a, b, c; Tuzimski and Bartosiewicz 2003). Tuzimski and Soczewiński as first used bilayer Multi K plates for separation of complex mixtures (Fig. 14.22b) (Tuzimski and Soczewiński 2002a, b, Tuzimski and Bartosiewicz 2003).

Method development for 2D-TLC of complex mixtures can be formulated as follows (Tuzimski and Soczewiński 2002d):

- Determine R_F vs. % modifier plots for polar adsorbent and nonaqueous eluents composed of heptanes (or hexane) and 2–3 polar modifiers; choose compositions of eluents for optimal differentiated retention of the components (in the range 0.05–0.70)
- Determine R_F vs. % modifier concentration plots for aqueous RP systems (octadecyl, cyanopropyl silica, or other polar adsorbents) for methanol and acetonitrile modifiers and choose optimal concentration of modifier
- Correlate the R_F values for NP/RP combinations and choose that corresponding to optimal spacing of spots on the plate area
- Use the optimal combination of NP/RP eluents for a bilayer or monolayer plate (silica, cyanopropyl silica, etc.)

Horizontal chambers can be easily used for two-dimensional separations. The only problem seems to be the sample size. In a conventional two-dimensional separation used for analytical purposes, the sample size is small. The quantity of the sample can be considerably increased when using a spray-on technique with an automatic applicator. Soczewiński and Wawrzynowicz have proposed a simple mode to enhance the size of the sample mixture with the ES horizontal chamber (Soczewiński and Wawrzynowicz 1981).

Graft Thin-Layer Chromatography

The multidimensional separation can be performed using different mobile phases in systems with single-layer or bilayer plates. Graft thin-layer chromatography is a multiple system in which chromatographic plates with similar or different stationary phases are used. Compounds from the first chromatographic plate after chromatogram development can be transferred to the second plate, without scraping, extraction, or re-spotting the bands by use of a strong mobile phase (Nyiredy 2001). Graft-thin layer chromatography, a novel multiplate system with layers of the same or different adsorbents for isolation of the components of natural and synthetic mixtures on preparative scale, was first described by Pandey et al. (1979). The procedure of performing reproducible graft-TLC analysis was described in detail by Tuzimski (2007a). An example of this technique is demonstrated in Fig. 14.23 (Tuzimski 2007a). In graft-TLC experiments with connected adsorbent layers several mixtures can be applied as spots 1 cm from the edge of the first adsorbent, e.g., silica gel plate. Several samples can be developed at the same time in the first

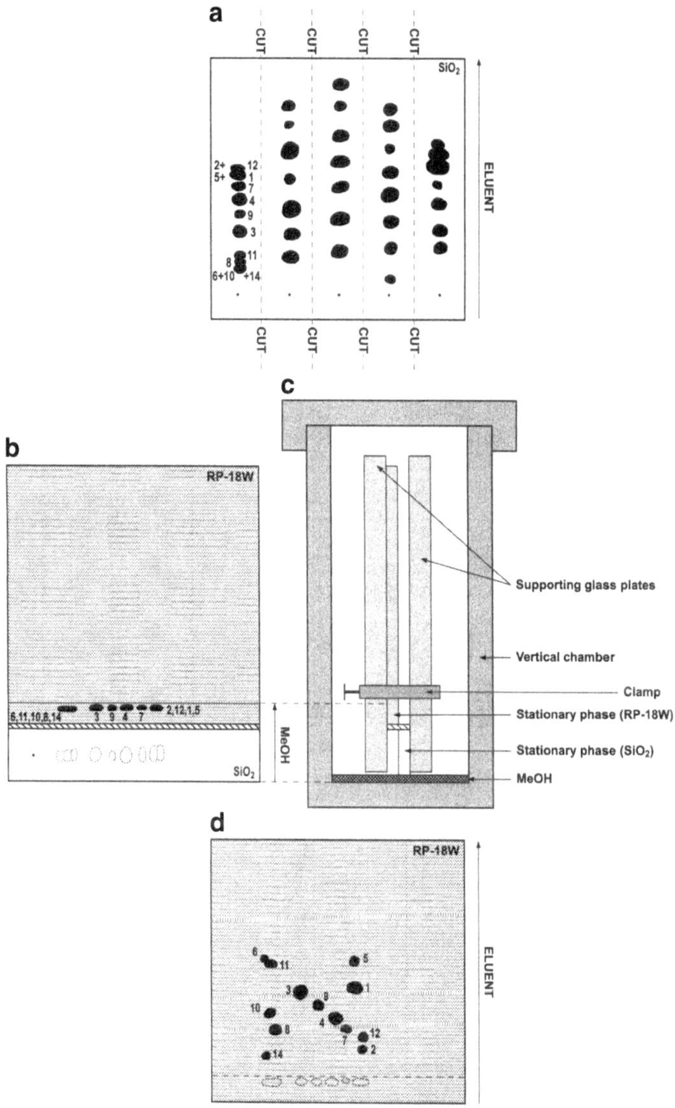

Fig. 14.23 Transfer of the mixture of pesticides from the first plate to the second one. (**a**) First development with partly separated mixtures of pesticides on silica plate. After development, the silica plate was dried and cut along the dashed lines into 2 cm × 10 cm strips. (**b**) A narrow strip (2 cm × 10 cm) was connected (2 mm overlap – hatched area) to 10 cm × 10 cm HPTLC RP-18W plate along the longer (10 cm) side of the strip. The partly separated mixture of pesticides was transferred in a vertical chamber to the second plate using methanol as strong eluent to the distance of about 1 cm. (**c**) Schematic diagram of cross section of connected two adsorbents layers. (**d**) The HPTLC RP-18W plate was developed in the second dimension with organic – water eluent in the DS chamber. [From Tuzimski (2007a) with permission]

direction, on the first adsorbent (up to ten in case of 20 cm × 10 cm plates). After drying, the plate used in the first run is cut into 2 cm × 10 cm strips. The cut strips should have smooth edges, without irregularities resulting from partial loss of adsorbent, because such irregularities may lead to deformation of the zones during the transfer to the second adsorbent layer. If the adsorbent edge is uneven, it should be smoothed before attachment to the second adsorbent (Tuzimski 2007a). Then individual strips are clamped to other plates and compounds are transferred. Individual strips should be connected (2 mm overlap) to 10 cm × 10 cm HPTLC plates along the longer (10-cm) side of the strip. It is essential that the two plates are in close contact but without disintegration of the overlapping layers. To achieve the HPTLC, plates were placed between thicker glass plates pressed together with screw clamps. The transfer of analyzed compounds is performed in vertical glass chamber as the joined plates are difficult to be developed in horizontal chambers. The most important issue in graft-TLC is to choose appropriate solvent to transfer compounds from the first adsorbent to another. The choice of this solvent depends on the choice of the first and second adsorbents and character of the transferred substances (whether polar or nonpolar). MeOH is usually applied for transferring compounds from the first adsorbent to another layer (Tuzimski 2007a). If the analyzed compounds are strongly adsorbed on the first adsorbent, the addition of organic acids, and also water, to transferring solvent, is advised. If after transfer from, e.g., the silica layer, the spots are spread along the 1-cm transfer distance, and the second HPTLC plate can be developed to a distance of 1 cm with a strong solvent to improve their shape. The application of narrow strip of the first adsorbent may also play the same role as the preconcentrating zone in case of multiphase (bilayer) plate.

The sample components are not only separated on the first step of graft-TLC experiment but also concentrated, and as such, developed in the second direction. The concentration is also performed during the transfer as the strong mobile phase used in this procedure introduce the analyzed substances to another adsorbent as very thin bands.

Graft thin-layer chromatographic separation (2D planar chromatography on connected layers) of mixture of phenolic acids (Glensk et al. 2002), coumarins (Cieśla et al. 2008a, b, c), alkaloids (Petruczynik et al. 2008b; Luczkiewicz et al. 2004), saponins (Glensk and Cisowski 2000), and three mixtures of pesticides were described also (Tuzimski 2005a, 2007a).

Examples of graft-TLC for separation of complex mixtures of compounds are listed in Table 14.2:

Combination of Multidimensional Planar Chromatography (MDPC) Methods – Hyphenated Techniques

Very difficult separations of multicomponent mixtures of compounds require the application of multidimensional planar chromatography combining different separation

Table 14.2 Examples of graft-TLC for separation of complex mixtures of various compounds

Substance	Stationary and mobile phases used in graft-TLC				Reference
	First direction		Second direction		
	Stationary phase	Mobile phase (v/v)	Stationary phase	Mobile phase (v/v)	
Phenolic acids	RP-18W	$MeOH/H_2O$ (40:60)	Silica	Multiple gradient development: diisopropyl ether/HCOOH (80:20) – 2 cm, diisopropyl ether/HCOOH/ cyclohexane (78:2:20) – 9 cm (2×)	Glensk et al. (2002)
Coumarins	CN-silica	$MeCN/H_2O$ (30 + 70) (triple development)	Silica	Ethyl acetate/*n*-heptane (35 + 65) (triple or double development)	Cieśla et al. (2008a, b)
Coumarins	Silica	ethyl acetate/*n*-heptane (35 + 65) (triple development)	RP-18W	$MeOH/H_2O$ (55 + 45)	Cieśla et al. (2008a, b, c)
Alkaloids	CN-silica	10% MeOH + diisopropyl ether + 2% ammonia	RP-18W	80% MeOH + acetate buffer at pH 3.5 + 0.05 ML^{-1} diethylamine	Petruczynik et al. (2008b)
Alkaloids	CN-silica	10% MeOH + diisopropyl ether + 2% ammonia	Silica:	1:5% acetone + diisopropyl ether + 0.1 ML^{-1} diethylamine; 2:5% acetone + 5% MeOH + 90% diisopropyl ether + 0.1 ML^{-1} diethylamine	Petruczynik et al. (2008b)
Saponins	RP-18W	1% aq. HCOOH–MeOH (30 + 70)	Silica	$CHCl_3$–MeOH–HCOOH–H_2O (10 + 4 + 1 + 1)	Glensk and Cisowski (2000)
Alkaloids	Diol-silica	CH_3Cl–MeOH–25% ammonia (85 + 15 + 1)	RP-18W	30% MeCN–H_2O–36% hydrochloric acid (30 + 100 +7)	Luczkiewicz et al. (2004)
Pesticides	Silica	THF–*n*-heptane (20 + 80)	RP-18W	$MeOH/H_2O$ (60 + 40)	Tuzimski (2005b)
Pesticides	Silica	Ethyl acetate–*n*-heptane (20 + 80 or 30 + 70)	RP-18W	$MeOH/H_2O$ (60 + 40 or 75 + 25)	Tuzimski (2007a)

systems. A new procedure for separation of complex mixtures by combination of different modes of multidimensional planar chromatography was described (Tuzimski 2007b, 2008c). By the help of this new procedure, 14 or 22 compounds from a complex mixture were separated on 10 cm × 10 cm TLC and HPTLC plates (Tuzimski 2007b, 2008c). In Fig. 14.24, an example of this procedure is presented step by step for separation of 22 compounds from complex mixtures on TLC plate (Tuzimski 2008c).

Silica is the most popular and least expensive adsorbent used in planar chromatography and has an excellent separation power. In the first part of experiments, solvent systems on silica stationary phase were selected to assign the investigated pesticides to groups of compounds based on the solvent classification by Snyder (1978) and the Prisma Method' described by Nyiredy and coworkers (Dallenbach-Tölke et al. 1986; Nyiredy et al. 1998; Nyiredy 1989; Nyiredy and Fater 1995). For the selection of suitable mobile phases, the first experiments were carried out on TLC silica plates in unsaturated chromatographic chambers with ten solvents (diethyl ether, 2-propanol ethanol, tetrahydrofuran, acetic acid, dichloromethane, ethyl acetate, dioxane toluene, and chloroform) from the eight groups of Snyder's solvent-selectivity triangle for normal phase (NP) chromatography according to their properties as proton acceptors, proton donors, and their dipole interactions. The literature data (Stahl 1967) show that the most widely applied solvent classes for NP planar chromatography are from the corners of Snyder's solvent-selectivity triangle (groups I, VII, and VIII) and from group VI (ethyl acetate, methyl ethyl ketone, dioxane, acetone, acetonitrile) where all three effects (proton acceptor, proton donor and dipole interaction) are practically equalized; the solvents from group VI have a special function in the optimization of the mobile phase (Nyiredy 1997).

Next, the solvent strength has to be either reduced or increased, so that the R_F values of investigated compounds lie between 0.2 and 0.8. If the solvents afford good separation, other solvents from the same group were also tested. Good results (Fig. 14.25) were obtained with ethyl acetate–*n*-heptane, 40:60 (v/v) as mobile phase in first direction; this separated the pesticides into five groups, the first containing compounds 1–4, the second containing 5–7, the third containing 8–12, the fourth containing 13–17, and the fifth containing 18–22.

The objective of the next series of experiments of the study reported in this paper was to investigate the separation of components of the five groups of pesticides. For optimization of the mobile phases for separation of pesticides in all five groups, the suitable solvents are selected by the procedure described above (Dallenbach-Tölke et al. 1986; Nyiredy et al. 1998; Nyiredy 1989; Nyiredy and Fater 1995). If the spots of pesticides on the plate migrate too far (R_F values of investigated compounds are above 0.7), the solvent strength must be reduced by addition of heptane (or hexane) to the mobile phase. Conversely, if the spots of compounds on the plate do not migrate far enough (R_F values of pesticides are between 0 and 0.3), the solvent strength must be increased by addition of a stronger solvent (e.g., methanol, acetic acid, water).

After optimization of the mobile phases for separation of the components of all groups of pesticides, these mobile phases were used for multidimensional planar chromatography (Fig.14.24a–f).

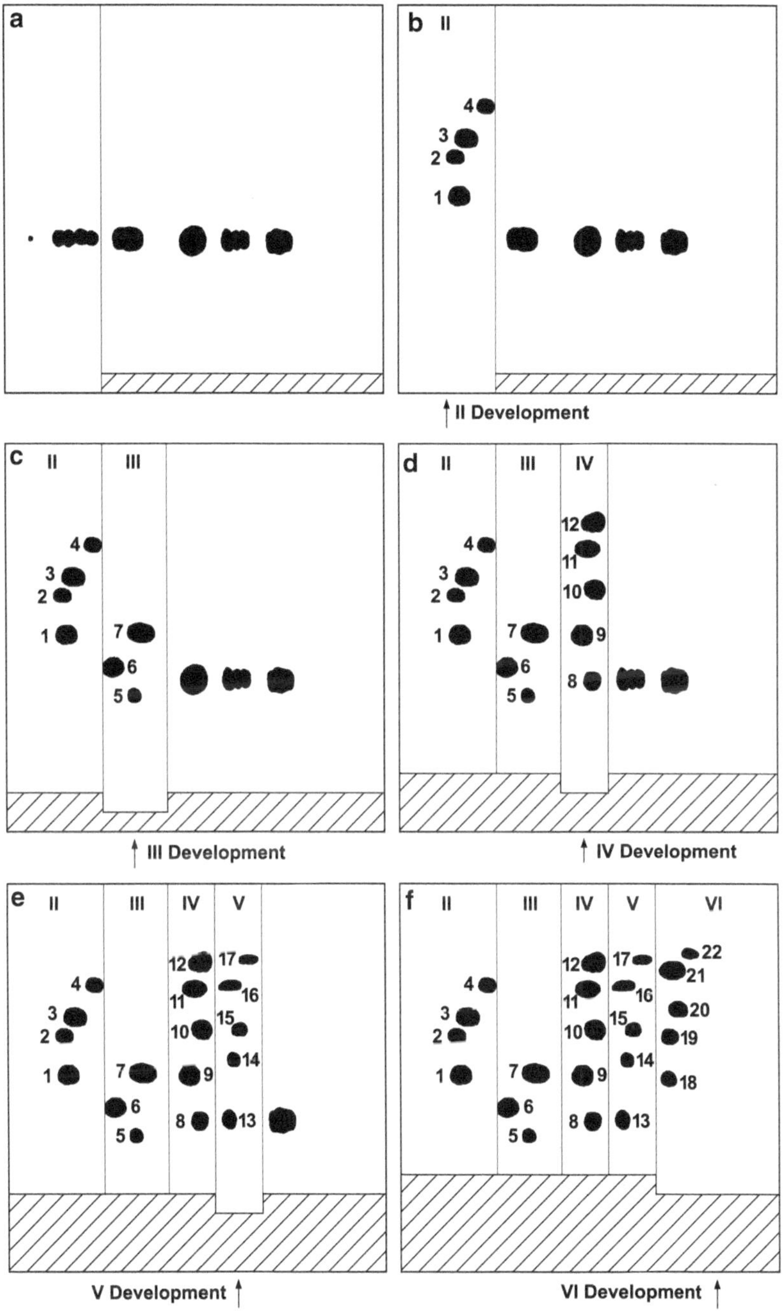

Fig. 14.24 Illustration of step by step selective multidimensional planar chromatographic separation. (**a**) The dried plate after the first separation (1st development) prepared for separation of the first group of compounds. One line (approx. 1 mm thick) is scraped in the stationary phase

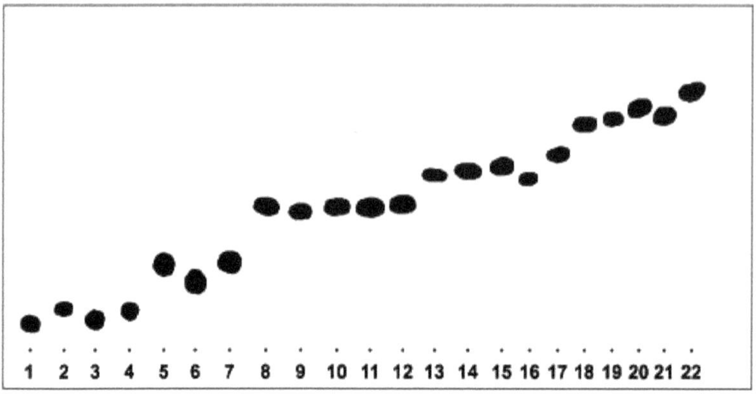

Fig. 14.25 Separation of 22 pesticides into five groups – compounds 1–4, 5–7, 8–12, 13–17, and 18–22 on a silica TLC plate with ethyl acetate–*n*-heptane, 40:60 (v/v), as mobile phase. [From Tuzimski (2008c) with permission]

Mixtures of pesticides were applied as spots, 1 cm from the bottom and 3.5 cm from the left edge of the plate. TLC plates were developed in the first dimension (step I) with ethyl acetate–*n*-heptane, 40:60 (v/v), as normal-phase eluent. HPTLC plates were developed in the first dimension (step I) with ethyl acetate–*n*-heptane, 50:50 (v/v), as normal-phase eluent. After drying in air for 20 min, the plates were turned by 90° (so that the partly separated components of the complex mixture of compounds were on the start line of the next step). Next, one or two lines (approx. 1 mm wide) were scraped in the adsorbent layer perpendicular to the direction of the first development, so the spot(s) of the target compounds were between the lines. Some of the adsorbent layer (approx. 5 mm wide) must be removed to ensure that the mobile phase of the second development (step II) develops only the spot(s) of the target compounds between the two lines. Next, all the plate and stationary phase except the part to be developed was covered by glass plates, which were fixed

Fig. 14.24 (continued) perpendicular to the first development, in such a way that the spot(s) of the target compounds are between the line and the edge of plate. For separation of the first group of compounds, another 5-mm wide region of the silica gel layer was removed from the bottom of the plate (*the hatched lines* indicate the stationary phase removed) so that in the next step, mobile phase runs only up a narrow strip of adsorbent. (**b**) The dried plate after separation of the first group of pesticides (1–4) by use of acetonitrile–chloroform, 15:85 (v/v), as mobile phase in the second development. (**c**) The prepared and dried plate after separation of the components of the second group of pesticides (5–7) by development with 100% chloroform twice over the same distance (UMD). (**d**) The prepared and dried plate after separation of the five components of the third group of pesticides (8–12) with nitromethane–dichloromethane, 5:95 (v/v), as mobile phase in the fourth development. (**e**) The prepared and dried plate after separation of the five components of next group of pesticides (13–17) with nitromethane–chloroform, 5:95 (v/v), as mobile phase in the fifth development. (**f**) The prepared and dried plate after separation of the five components of last group of pesticides (18–22) with toluene–*n*-heptane, 70:30 (v/v), as mobile phase in the sixth development. [From Tuzimski (2008c) with permission] with clamps. This procedure was repeated in subsequent steps (steps III–VI). Before each of steps III to VI, a region of the adsorbent layer must again be removed from the plate to ensure that the mobile phase only develops the zone of group of compounds of interest. (The regions removed before each development is shown by the hatched lines in Fig. 14.24). The plates were developed in an unsaturated vertical chamber in MDPC experiment (Fig. 14.24a–f). To prevent the eluent from migration on adsorbent with constituents that are not aimed to be chromatographed during the particular step, the part of the plate, from which the adsorbent was removed, can be covered with lipophilic substance (wax). In this case, the plates can be developed in horizontal chamber in MDPC experiments.

The compounds from the first group (1–4) were chromatographed with acetonitrile–chloroform, 15:85 (v/v), as mobile phase (Fig. 14.24b). The pesticides in the second group (5–7) were chromatographed twice with chloroform as mobile phase over the same distance (Fig. 14.24c). The plate was dried for approximately 5 min between the two steps. The plate was then dried after the second separation of compounds of this group. Another portion of the stationary phase (the next 5 mm – hatched lines on Fig. 14.24d) was then removed to ensure that the mobile phase used for development IV (Fig. 14.24d) affected only the spots of the next group of compounds (8–12) between the two lines. Separation of components 13–17 with nitromethane–chloroform, 5:95 (v/v), as mobile phase in the next step on the TLC plate is depicted in Fig. 14.24e. Separation of pesticides of the last group (18–22) with toluene–*n*-heptane, 70:30 (v/v), as mobile phase is depicted in Fig. 14.26f. Separation of 22 components from a complex mixture by developments I–VI by multidimensional planar chromatography was also achieved on a silica HPTLC plate. The best results were obtained with ethyl acetate–*n*-heptane, 50:50 (v/v), as mobile phase in first direction, which separated the pesticides into five groups (1–4, 5–7, 8–12, 13–17, and 18–22). A videoscan of the multidimensional separation of the 22 components on a silica HPTLC plate is shown in Fig. 14.26 (Tuzimski 2008c). The separation can be characterized as "PC × (PC + *n*PC + PC + PC + PC)."

Also, videoscan (Fig. 14.27) was performed showing a complete separation of the next 14-component mixture of pesticides in the Ist–Vth developments

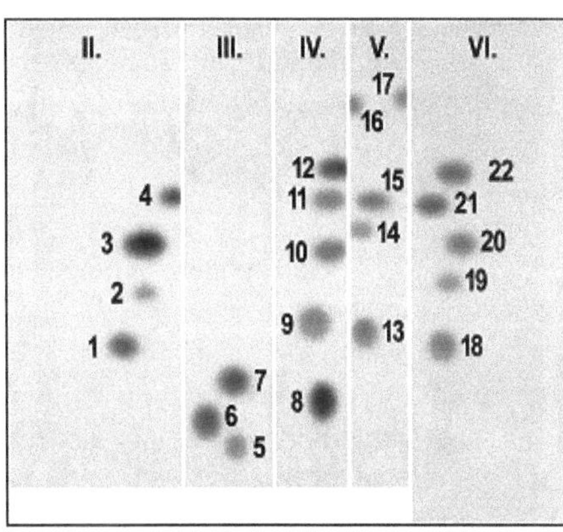

Fig. 14.26 Videoscan at 254 nm of the silica HPTLC plate showing separation of the 22 components of a complex mixture by developments I–VI in multidimensional planar chromatography. [From Tuzimski (2008c) with permission]

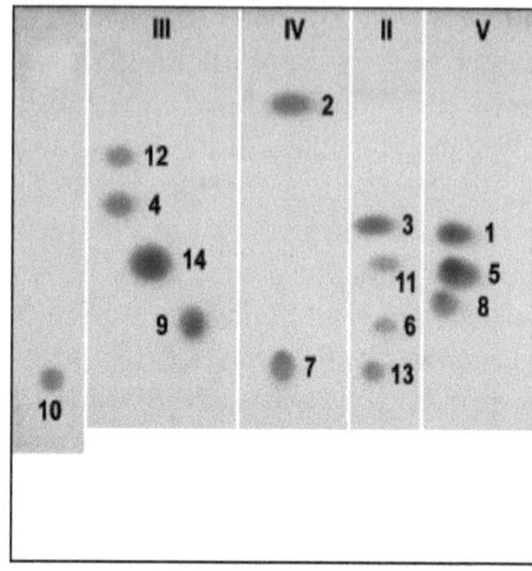

Fig. 14.27 Videoscan at 254 nm showing complete separation of the mixture of 14 pesticides by multidimensional planar chromatography (MDPC). [From Tuzimski (2007b) with permission]

by multidimensional planar chromatography [PC × (*n*PC + PC + PC + PC)] (Tuzimski 2007b).

Future Aspects of Multidimensional Planar Chromatography (MD-PC): Combined MD-PC–DAD with HPLC–DAD

The best combination for multidimensional planar chromatography is the parallel combination of stationary and mobile phases. In the next mode of multidimensional planar chromatography, the separation of multicomponent mixtures was realized on multiphase plates. Also, the largest differences were obtained by combination of normal-phase systems of the type silica/nonaqueous eluent in the first step of MDPC and reversed-phase systems of the type octadecyl silica/water + organic modifier (methanol, acetonitrile, dioxane) in the next steps of MDPC on multiphase plates, e.g., with a narrow zone of SiO_2 and a wide zone of RP-18 (or vice versa), which are commercially available from Whatman (Multi K SC5 or CS5 plates) (Tuzimski 2008d, 2009c). Multidimensional planar chromatography (MDPC) is performed by spotting the multicomponent mixture in the corner of a square chromatographic plate (20 cm × 20 cm) and by development in the first direction with the first eluent on the narrow zone of SiO_2 (Multi K SC5 plate) or octadecyl silica (Multi K CS5 plate). After the development is completed, the chromatographic plate is then removed from the developing chamber and the solvent is allowed to evaporate from the layer. The plate is rotated through 90° and then compounds are transferred on a wide zone of layer: octadecyl silica (Multi K SC5 plate) or silica (Multi K CS5 plate) with a strong solvent. After transferring, the narrow zone of adsorbent should be separated from the wide zone of layer. Different solvents are used in the next steps, which are perpendicular to the direction of the first development. The procedure for these steps is the same as the procedure described above and is shown in Fig. 14.24a–f (Tuzimski 2009a, b).

Planar chromatography combined with different modes of scanning enables quantitative analysis. Application of multidimensional planar chromatography and modern fiber optical TLC densitometric scanners with DAD is especially useful for correct identification of components of difficult, complicated mixtures, e.g., plant extracts. The procedure described for the separation of complex mixtures of compounds is inexpensive and can be applied to routine analysis of analytes in samples of natural origin, e.g., in water or plant extracts, after preliminary clean-up and concentration, e.g., by solid-phase extraction (SPE). The identification of analytes was confirmed by the comparison of the UV spectra of the components of plant extracts and standards of analytes by DAD densitometer. The analytes can be confirmed by TLC-MS also (Tuzimski 2009a, b).

Heart-cut spots of analytes from the stationary phase were also injected on, e.g., a C18 column and analyzed by HPLC–DAD. Analyte identification was accomplished on the basis of the retention times of the analytes and by comparison of the UV spectrum of the reference compounds in the library and the UV spectrum of the detected peak the sample. A match equal or higher than 990 was fixed to confirm identification between both spectra for all the analytes determined (Tuzimski 2008d; Tuzimski 2009a, b). Determination of analytes in medical herbs extracts by SPE coupled with multidimensional planar chromatography in combination with DAD densitometry (MDPC–DAD) and HPLC–DAD were also described (Tuzimski 2010c, d).

Notes

Summing up, planar chromatography is one of the principal analytical techniques. It can be used for identification of known and unknown compounds, and at least equally important for correct identification of biomolecules in various samples. TLC has many advantages, such as wide optimization possibilities with the chromatographic systems, special development modes, diverse detection methods, and low-cost analysis of samples, requiring minimal sample cleanup. TLC is a chromatographic technique widely used for separation, detection, and qualitative and quantitative determination of biomolecules belonging to different chemical classes.

References

Ahrens B, Blankenhorn D, Spangenberg B. Advanced fiber optical scanning in thin-layer chromatography for drug identification. J Chromatogr B 2002; 772, 11–18.

Alpmann A, Morlock G. Improved online coupling of planar chromatography with electrospray mass spectrometry: extraction of zones from glass plates. Anal Bioanal Chem 2006; 386 (5) 1543–1551.

Alpmann A, Morlock G. Rapid and sensitive determination of acrylamide in drinking water by planar chromatography and fluorescence detection after derivatization with dansulfinic acid. J Sep Sci 2008; 31 (1) 71–77.

Anderson RM, Busch KL. Thin-layer chromatography coupled with mass spectrometry: interfaces to electrospray ionization. J Planar Chromatogr – Mod TLC 1998; 11 (5) 336–341.

Aranda M, Morlock G. Simultaneous determination of riboflavin, pyridoxine, nicotinamide, caffeine and taurine in energy drinks by planar chromatography-multiple detection with confirmation by electrospray ionization mass spectrometry. J Chromatogr A 2006; 1131 (1–2) 253–260.

Bakry R, Bonn GK, Mair D, Svec F. Monolithic porous polymer layer for the separation of peptides and proteins using thin-layer chromatography coupled with MALDI-TOF-MS. Anal Chem 2007; 79 (2) 486–493.

Barrett GC. Iodine as a 'non-destructive' colour reagent in paper- and thin-layer chromatography. Nature 1962; 194, 1171–1172.

Bates DL, Perham RN, Coggins JR. Methods for obtaining peptide maps of proteins on a subnanomole scale. Anal Biochem 1975; 68, 175–184.

Bauer GK, Pfeifer AM, Hauck HE, Kovar K-A. Development of an optimized sorbent for direct HPTLC-FTIR on-line coupling. J Planar Chromatogr – Mod TLC 1998; 11 (2) 84–89.

Bhushan R (2008). Amino Acids. Chapter 13 in Thin-Layer Chromatography in Phytochemistry. Waksmundzka-Hajnos M, Sherma J, Kowalska T (eds) published by CRC Press, Taylor and Francis Group, Boca Raton, pp. 299–329.

Bhushan R, Ali I. TLC resolution of PTH-amino acid on Zn^{2+}, Cd^{2+}, Hg^{2+}, impregnated silica plates. J Liq Chromatogr 1986; 9, 3479–3487.

Bhushan R, Ali I. TLC resolution of enantiomeric mixtures of amino acids. Chromatographia, 1987; 23, 141–142.

Bhushan R, Ali I. TLC resolution of racemic mixtures of PTH-amino acids on (+)-tartaric acid and impregnated silica gel plates. J Chromatogr 1987; 392, 460–463.

Bhushan R, Ali I. Effects of halides on TLC resolution of amino acids, below their isoelectric points. Chromatographia 1987; 23, 207–208.

Bhushan R, Ali I. TLC resolution of amino acids in a new solvent and effect of alkaline earth metals. J Liq Chromatogr 1987; 10, 3647–3652.

Bhushan R, Martens, J (2000). Peptides and proteins, thin layer (planar) chromatography. In: Wilson, ID, Adlard ER, Cook M, Poole CF (eds), Encyclopedia of Separation Science, Vol. 8, pp. 3626–3636.

Bhushan R, Martens J (2000). Amino acids: thin layer (planar) chromatography. In: Wilson, ID, Adlard ER, Cook M, Poole CF (eds), Encyclopedia of Separation Science, Vol. 5, pp. 2013–2033.

Bhushan R, Martens J (2003a). Amino acids and their derivatives. Chapter 14 in: Handbook of Thin-Layer Chromatography, third edition, Sherma J, Fried B (eds) published by Marcel Dekker, Inc., New York, Basel, pp. 373–415.

Bhushan R, Martens J (2003b). Peptides and proteins. Chapter 26 in Handbook of Thin-Layer Chromatography, third edition, Sherma J, Fried B (eds) published by Marcel Dekker, Inc., New York, Basel, pp. 749–766.

Bhushan R, Reddy GP. Resolution of PTH-amino acids on impregnated silica plates. Fresenius Z Anal Chem 1987; 327, 798–799.

Bhushan R, Reddy GP. Rapid resolution of PTH-amino acids by TLC on silica gel plates impregnated with transition metal ions. Anal Biochem 1987; 162, 427–430.

Bhushan R, Reddy GP. TLC of PTH-amino acids on silica gel thin layers impregnated with zinc salts and sulphates of some metals. Analyst (London) 1987; 112, 1467–1469.

Bhushan R, Srivastava SP, Chauhan RS. Thin layer chromatographic separation of thorium, titanium, zirconium and some rare earths on cellulose impregnated layers. Anal Lett 1985; 18A, 1549–1553.

Bhushan R, Chauhan RS, Reena A, Ali I. A Comparison of amino acid separation of Zn^{2+}, Cd^{2+} and Hg^{2+} impregnated silica layers. J Liq Chromatogr 1987; 10, 3653–3657.

Bieleski RL, Turner NA. Separation and estimation of amino acids in crude plant extracts by thin-layer electrophoresis and chromatography. Anal Biochem 1966; 17 (2), 278–293.

Blome J (1977). High Performance Thin-Layer Chromatography, Zlatkis A, Kaiser RE (eds) published by Elsevier, Institute of Chromatography, Bad Dürkheim.

Botz L, Nyiredy S, Sticher O. A new device for circular preparative planar chromatography. J Planar Chromatogr – Mod TLC 1990; 3, 401–406.

Brenner M, Niederwieser A. Durchlaufende Dünnschicht-Chromatographie. Experientia 1961; 17, 237–238.

Buncak P. Chromatographie, Suppl.GIT, Darmstadt, 1982, 3–8.

Burger K. Fresenius Z Anal Chem 1984; 318, 228–233.

Burns DJW, Turner NA. J Chromatogr 1967; 30, 469.

Bush KM (2003). Thin-Layer Chromatography coupled with mass spectrometry. In: Handbook of thin-layer chromatography, third edition, Sherma J, Fried B (eds) published by Marcel Dekker, Inc., New York, Basel, Chapter 9, pp. 239–275.

Bush KM. Planar separation and mass spectrometric detection. J Planar Chromatogr – Mod TLC 2004; 17 (6) 398–403.

Cha S, Yeung ES. Colloidal graphite-assisted laser desorption/ionization mass spectrometry and MSn of small molecules. 1. Imaging of cerebrosides directly from rat brain tissue. Anal Chem 2007; 79, 2373–2385.

Chiou SH, Wang KT (1990). in Villafranca J (ed) Current Research in Protein Chemistry: Techniques, Structure & Function, published by Academic Press, San Diego, pp. 3–10.

Cieśla Ł, Petruczynik A, Hajnos M, Bogucka-Kocka A, Waksmundzka-Hajnos M. Two-dimensional thin-layer chromatography of structural analogs. Part I: Graft TLC of selected coumarins. J Planar Chromatogr – Mod TLC 2008a; 21 (4) 237–241.

Cieśla Ł, Petruczynik A, Hajnos M, Bogucka-Kocka A, Waksmundzka-Hajnos M. Two-dimensional thin-layer chromatography of structural analogs. Part II: Method for quantitative analysis of selected coumarins in plant materials. J Planar Chromatogr – Mod TLC 2008b; 21 (6) 447–452.

Cieśla Ł, Bogucka-Kocka A, Hajnos M, Petruczynik A, Waksmundzka-Hajnos M. Two-dimensional thin-layer chromatography with adsorbent gradient as a method of chromatographic fingerprinting of furanocoumarins for distinguishing selected varieties and forms of *Heracleum* spp. J Chromatogr A 2008c; 1207 (1–2) 160–168.

Cimpan G (2004). Solute identification in TLC. In: Encyclopedia of Chromatography, published by Marcel Dekker, Inc., New York.

Cimpoiu C. Qualitative and quantitative analysis by hyphenated (HP)TLC FTIR Technique. J Liq chromatogr Relat Technol 2005; 28 (7–8) 1203–1213.

Consden R, Gordon AH, Martin AJP. Qualitative analysis of proteins: partition chromatographic method using paper. J Biochem 1944; 38, 224–232.

Dallenbach-Tölke K, Nyiredy Sz, Meier B, Sticher O. Optimization of overpressured layer chromatography of polar, naturally occurring compounds by the "Prisma" model. J Chromatogr 1986; 365, 63–72.

De Brabander HF; Smets F; Pottie G. Faster and cheaper two-dimensional HPTLC using the '4 × 4' mode. J Planar Chromatogr – Mod TLC 1988; 1, 369–371.

De Spiegeleer B, Van den Bossche W, De Moerlose P, Massart D. A strategy for two-dimensional, high-performance thin-layer chromatography, applied to local anesthetics. Chromatographia 1987; 23, 407–411.

Devani MB, Shishoo CJ, Shah SA, Soni KP, Shah RS. Localization of amino acids on thin-layer chromatograms with acetylacetone–formaldehyde reagent. J Chromatogr 1991; 537 (1–2), 494–496.

Dreisewerd K, Müthing J, Rohlfing A, Meisen I, Vukelić Ž, Peter-Katalinić J, Hillenkamp F, Berkenkamp S. Analysis of gangliosides directly from thin-layer chromatography plates by infrared matrix-assisted laser desorption/ionization orthogonal time-of-flight mass spectrometry with a glycerol matrix. Anal Chem 2005; 77 (13) 4098–4107.

Duruum EL. Two-dimensional electrophoresis and ionophoresis. J Colloid Sci 1951; 6, 274–290.

Dzido TH (2001a). Chapter 4 in Planar Chromatography, a Retrospective View for the Third Millennium, Nyiredy Sz (ed), published by Springer Scientific Publisher, Budapest.

Dzido TH. Effect of temperature on the retention of aromatic hydrocarbons with polar groups in binary reversed-phase TLC. J Planar Chromatogr – Mod TLC 2001b; 14, 237–245.

Dzido TH, Polak B. A coadsorption effect in a liquid–solid TLC system of the type silica – heptane + dioxane in horizontal DS chamber. J Planar Chromatogr – Mod TLC 1993; 6, 378–381.

Dzido TH, Polak B (2006) On methodical possibilities of the horizontal chambers in PLC, in: Preparative Layer Chromatography, Kowalska T, Sherma J (ed), published by CRC Press, Taylor and Francis Group, Boca Raton.

Dzido TH, Tuzimski T (2008) Chapter 7 in Thin-Layer Chromatography in Phytochemistry, Waksmundzka-Hajnos M, Sherma J, Kowalska T (eds), published by CRC Press, Taylor and Francis Group, Boca Raton.

Ebel S, Volkl S. Dtsch Apoth Ztg 1990; 130, 2162–2169.

Fenimore DC, Davis CM. High performance thin-layer chromatography. Anal Chem 1981; 53, 252A–266A.

Ford MJ, Deibel MA, Tomkins BA, Van Berkel GJ. mass spectrometry analysis of caffeine using a surface sampling probe electrospray ionization tandem mass spectrometry system/mass spectrometry analysis of caffeine using a surface sampling probe electrospray ionization tandem mass spectrometry system. Anal Chem 2005; 77 (77) 4385–4389.

Fried B, Sherma J (1999). Thin-Layer Chromatography, fourth edition, revised and expanded, published by Marcel Dekker Inc., New York, Basel.

Fuchs B, Schiller J, Süβ R, Nimptsch A, Schürenberg M, Suckau D. Capabilities and disadvantages of combined matrix-assisted laser desorption/ionization time-of-flight mass spectrometry (MALDI-TOF-MS) and high-performance thin-layer chromatography (HPTLC): analysis of egg yolk lipids. J Planar Chromatogr – Mod TLC 2009; 22 (1) 35–42.

Gaspar RZ, Males, Z. Determination of free amino acids in valerian (*Valeriana officinalis* L.) by thin-layer chromatography. Farmaceutski Glasnik 2005; 61 (12), 691–697.

Geiss F (1987). Fundamentals of Thin-Layer Chromatography, published by Huethig, Heidelberg.

Geiss F, Schlitt H. A new and versatile thin-layer chromatography separation system ('KS-Vario-Chamber'). Chromatographia 1968; 1, 392–402.

Geiss F, Schlitt H, Klose A. Fresenius Z Anal Chem 1965; 213, 331–346.

Giddings, JC (1990). Use of Multiple Dimensions in Analytical Separations, In: Multidimensional Chromatography, Cortes, H.J., (ed) published by Marcel Dekker, New York, pp. 251–299.

Glauninger G, Kovar K-A, Hoffmann V (1989). Untersuchungen zum online DC-IR Transfer. In Gauglitz G (Ed) Software Entwicklung in der Chemie 3, Proc 3rd Workshops Comput Chem, published by Springer, Berlin, pp. 171–180.

Glensk M, Cisowski W. Two-dimensional TLC with an adsorbent gradient for analysis of saponins in *Silene vulgaris* Garcke. J Planar Chromatogr – Mod TLC 2000; 13, 9–11.

Glensk M, Sawicka U, Mażol I, Cisowski W. 2D TLC – graft planar chromatography in the analysis of a mixture of phenolic acids. J Planar Chromatogr – Mod TLC 2002; 15 (6) 463–465.

Gołkiewicz W (2003). Gradient development in thin layer chromatography. Chapter 6 in Handbook of Thin-Layer Chromatography, third edition, Sherma J, Fried B (eds), published by Marcel Dekker, Inc., New York, Basel, pp. 153–173.

Gonnord MF, Levi F, Guiochon G. Computer assistance in the selection of the optimum combination of systems for two-dimensional chromatography. J Chromatogr 1983; 264, 1–6.

Gusev AI, Proctor A, Rabinovich YI, Hercules DM. Thin-layer chromatography combined with matrix-assisted laser desorption/ionization mass spectrometry. Anal Chem 1995; 67 (11) 1805–1814.

Gusev AI, Vasseur OJ, Proctor A, Sharkey AG, Hercules DM. Anal Chem 1995; 67, 4565–4570.

Hahn-Deinstrop E (2000). Applied Thin-Layer Chromatography, published by Wiley-VCH, Weinheim.

Harris JI, Hindly J. J Mol Biol 1965; 13, 864.

Hiegel K, Spangenberg B. New method for the quantification of dequalinium cations in pharmaceutical samples by absorption and fluorescence diode array thin-layer chromatography. J Chromatogr A 2009; 1216, 5052–5056.

Hillekamp F, Peter-Katalinić J (2007). MALDI MS – a practical guide to instrumentation, methods and application, published by Wiley-VCH, Weinheim.

Hirs CHW. Oxidation of ribonuclease. J Biol Chem 1956; 219, 611–621.

Hsu FL, Chen CH, Yuan CH, Shiea J. Interface to connect thin-layer chromatography with electrospray ionization mass spectrometry. Anal Chem 2003; 75 (10) 2493–2498.

http://www.camag.com.

http://www.chromdes.com.

Hubert P, Dellacherie E. J Chromatogr 1973; 80, 144.

Hugaard G, Kroner TD. Partition chromatography of amino acids with applied voltage. J Am Chem Soc 1948; 70, 2135–2137.

Isbell DT, Gusev AI, Taranenko NI, Chen CH, Hercules DM. Fresenius J Anal Chem 1999; 365, 625–630.

Issaq HJ. Triangular thin-layer chromatography. J Liq Chromatogr 1980; 3, 789–796.

Izmailov NA, Schraiber MS. Farmatzija 1938; 3, 1–7.

Jaenchen DE (1996). Handbook of Thin Layer Chromatography, second edition, Sherma J, and Fried B (eds), published by Marcel Dekker Inc., New York.

Jautz U, Morlock G. Efficacy of planar chromatography coupled to (tandem) mass spectrometry for employment in trace analysis. J Chromatogr A 2006; 1128 (1–2) 244–250.

Jautz U, Morlock G. Validation of a new planar chromatographic method for quantification of the heterocyclic aromatic amines most frequently found in meat. Anal Bioanal Chem 2007; 387 (3) 1083–1093.

Jork H, Funk W, Fischer W, Wimmer H (eds) (1990). Thin-Layer Chromatography, published by VCH, Verlagsgesellschaft, Weinheim.

Kaiser RE (1977). High Performance Thin-Layer Chromatography, Zlatkis A, Kaiser RE (eds) published by Elsevier, Institute of Chromatography, Bad Dürkheim.

Kaiser RE. J High Resol Chromatogr Chromatogr Commun 1978; 3, 164–168.

Keck K, Hagen U. Separation of the DNA [deoxyribonucleic acid] units on cellulose layers. Biochem Biophys Acta 1964; 87, 685–687.

Kochin V, Imanishi SY, Eriksson JE. Fast track to a phosphoprotein sketch – MALDI-TOF characterization of TLC-based tryptic phosphopeptide maps at femtomolar detection sensitivity. Proteomics 2006; 6, 5676–5682.

Kopyt'ko YaF. Amino acids and fatty acids in homeopathic matrix tinctures of grass of Parnassus (*Parnassia palustris*). Pharmceut Chem J 2003; 37 (7), 347–349.

Kovar K-A, Enβlin HK, Frey OR, Rienas S, Wolff SC. Applications of on-line coupling of thin-layer chromatography and FTIR spectroscopy. J Planar Chromatogr – Mod TLC 1991; 4 (3) 246–250.

Kraus L, Koch A, Hoffstetter-Kuhn S (1996). Dünnschichtchromatographie, published by Springer-Verlag, Berlin, Heidelberg.

Lan M, Wang D, Han J. A new distributor for half-way development. J Planar Chromatogr – Mod TLC 2003; 16, 402–404.

Laskar S, Basak BK. Identification of amino acids with ninhydrin in acid media. J Planar Chromatog – Mod TLC 1990; 3, 535–536.

Laskar S, Basak BK. Identification of amino acids on thin-layer chromatography plates. J Planar Chromatogr – Mod TLC 1990; 3, 275–276.

Laskar S, Bhattacharya U, Basak B. Modified ninhydrin spray reagent for the identification of amino acids on thin-layer chromatography plates. Analyst 1991; 116 (6), 625–626.

Light A, Smith EL (1963). Amino acid analysis of peptides and proteins. In: Hans Neurath (ed) Proteins, Composition, Structure, and Function, Vol. 1, published by Academic Press: New York, pp. 1–44.

Łuczkiewicz M, Migas P, Kokotkiewicz A, Walijewska M, Cisowski W. Two-dimensional TLC with adsorbent gradient for separation of quinolizidine alkaloids in the herb and in-vitro cultures of several *Genista* species. J Planar Chromatogr – Mod TLC 2004; 17, 89–94.

Ludwiczuk A, Nyiredy Sz, Wolski T. Separation of the ginsenosides fraction obtained from the roots of *Panax quinquefolium* L. cultivated in Poland. J Planar Chromatogr – Mod TLC 2005; 18 (105) 104–107.

Luftmann H. A simple device for the extraction of TLC spots: direct coupling with an electrospray mass spectrometer. Anal Bioanal Chem 2004; 378 (4) 964–968.

Males Z, Plazibat M, Petlevski R. Thin-layer chromatographic analysis of amino acids of *Crithmum maritimum* L. Farm Glas 2001; 57 (5), 175–180.

Males Z, Plazibat M, Vundac VB, Zuntar I, Pilepic KH. Thin-layer chromatographic analysis of flavonoids, phenolic acids and amino acids in some croatian *Hypericum taxa*. J Planar Chromatogr – Mod TLC 2004; 17 (4), 280–285.

Males Z, Plazibat M, Bilusic VV, Kremer D, Alar J. Study of flavonoids and amino acids of Spanish broom (*Spartium junceum* L., *Fabaceae*). Farm Glas 2005; 61 (9), 499–509.

Markowski W. Computer-aided optimization of gradient multiple-development thin-layer chromatography. I. Multi-stage development. J Chromatogr 1993; 635 283–289.

Markowski W. Computer-aided optimization of gradient multiple development thin-layer chromatography. III. Multi-stage development over a constant distance, J Chromatogr A 1996; 726, 185–192.

Markowski W (2005). Gradient development in TLC. Encyclopedia of Chromatography, J Cazes (ed), published by Marcel Dekker, Inc., New York, Basel, pp. 699–713.

Markowski W, Matysik G. Analysis of plant extracts by multiple development thin-layer chromatography. J Chromatogr 1993; 646, 434–438.

Markowski W, Soczewiński E. Computer-aided optimization of gradient multiple-development thin layer Chromatography. I. Two-stage development. J Chromatogr 1992; 623, 139–147.

Markowski W, Soczewiński E. Computer-aided optimization of stepwise gradient and multiple-development thin-layer chromatography. Chromatographia 1993; 36, 330–336.

Markowski W, Czapińska KL, Błaszczak M. Determination of the constants of the Snyder-Soczewiński equation by means of gradient multiple development. J Liq Chromatogr 1994; 17, 999–1009.

Matysik G. Modified programmed multiple gradient development (MGD) in the analysis of complex plant extracts. Chromatographia 1996; 43, 39–43.

Matysik G (1997). Problemy optymalizacji chromatografii cienkowarstwowej (in Polish). Medical University of Lublin, Lublin 1997, published by BACCARAT, Lublin.

Matysik G, Soczewiński E. Stepwise gradient development in thin-layer chromatography. IV. Miniaturized generators for continuous and stepwise gradients. J Chromatogr 1988; 446, 275–282.

Matysik G, Soczewiński E. A miniaturized for the generation of eluent composition gradients for sandwich thin-layer chromatography. Anal Chem (Warsaw) 1988; 33, 363–369.

Matysik G, Soczewiński E. Gradient thin-layer chromatography of plant extracts. Chromatographia 1988; 26, 178–180.

Matysik G. Soczewiński E. Computer-aided optimization of stepwise gradient TLC of plant extracts. J Planar Chromatogr – Mod TLC 1996; 9, 404–411.

Matysik G, Markowski W, Soczewiński E, Polak B. Computer-aided optimization of stepwise gradient profiles in thin-layer chromatography. Chromatographia 1992; 34, 303–307.

Matysik G, Soczewiński E, Polak B. Improvement of separation in zonal preparative thin-layer chromatography by gradient elution. Chromatographia 1994; 39, 497–504.

Moore S. On determination of cystine as cysteic acid. J Biol Chem 1963; 238, 235–237.

Morlock G, Dytkiewitz E. analytical strategy for rapid identification and quantification of lubricant additives in mineral oil by high-performance thin-layer chromatography with UV absorption and fluorescence detection combined with mass spectrometry and infrared spectroscopy. J AOAC Int 2008; 91 (5) 1237–1244.

Morlock G, Jautz U. Comparison of two different plunger geometries for HPTLC-MS coupling via an extractor-based interface. J Planar Chromatogr – Mod TLC 2008; 21 (5) 367–371.

Morlock G, Schwack W. Determination of isopropylthioxanthone (ITX) in milk, yoghurt and fat by HPTLC-FLD, HPTLC-ESI/MS and HPTLC-DART/MS. Anal Bioanal Chem 2006; 385 (3) 586–595.

Morlock G, Ueda Y. New coupling of planar chromatography with direct analysis in real time mass spectrometry. J Chromatogr A 2007; 1143 (1–2) 243–251.

Norfolk E, Khan SH, Fried B, Sherma J. Comparison of amino acid separations on high performance silica gel, cellulose, and C-18 reversed phase layers and application of HPTLC to the determination of amino acids in *Biomphalaria glabrata* snails. J Liq Chromatogr 1994; 17 (6), 1317–1326.

Nowak R, Tuzimski T. A solid-phase extraction-thin-layer chromatographic-fiber optical scanning densitometric method for determination of flavonol aglycones in extracts of rose leaves. J Planar Chromatogr – Mod TLC 2005; 18 (105) 437–442.

Nurok D. Strategies for optimizing the mobile phase in planar chromatography. Chem Rev 1989; 89, 363–375.

Nurok D, Habibi-Goudarzi S, Kleyle R. Statistical approach to solvent selection as applied to two-dimensional thin-layer chromatography. Anal Chem 1987; 59, 2424–2428.

Nurok D, Kleyle R, McCain CL, Risley DS, Ruterbories KJ. Statistical method for quantifying mobile phase selectivity in one- and two-dimensional overpressured layer chromatography. Anal Chem 1997; 69, 1398–1405.

Nyiredy Sz, Dallenbach-Tölke K, Sticher O. Correlation and prediction of the k′ values for mobile phase optimization in HPLC. J Liq Chromatogr 1989; 12, 95–116.

Nyiredy Sz (1997). In: Kaiser O, Kaiser RE, Gunz H, Günther W, (eds), Chromatography – Celebrating Michael Tswett's 125th Birthay, InCom, Duesseldorf, pp. 231–239.

Nyiredy Sz (2001). Planar Chromatography, a Retrospective View for the Third Millennium, published by Springer Scientific Publisher, Budapest.

Nyiredy Sz (2002a) Multidimensional Planar Chromatography. In: Multidimensional Chromatography. Mondello L, Lewis AC, Bartle KD (eds) published by Chichester, Chapter 8, pp. 171–196.

Nyiredy Sz, Fully on-line TLC/HPTLC with diode-array detection and continuous development. Part 1: description of the method and basic possibilities. J Planar Chromatogr – Mod TLC 2002b; 15, 454–457.

Nyiredy S (2003). Handbook of Thin-Layer Chromatography, third edition, Sherma J, Fried B (eds) published by Marcel Dekker, Inc., New York, Basel, pp. 307–337.

Nyiredy Sz. Multidimensional planar chromatography. LC-GC Eur 2003; 16, 52–59.

Nyiredy Sz, Fatér Zs. Automatic mobile phase optimization, using the 'PRISMA' model, for the TLC separation of apolar compounds. J Planar Chromatogr 1995; 8, 341–345.

Nyiredy Sz, Dallenbach-Tölke K, Sticher O. The "PRISMA" optimization system in planar chromatography. J Planar Chromatogr 1998; 1, 336–342.

O'Farrell PH. High resolution two-dimensional electrophoresis of proteins. J Biol Chem 1975; 250, 4007–4021.

Panchagnula V, Mikulskis A, Songa L, Wang Y, Wang M, Knubovets T, Scrivener E, Golenko E, Krull IS, Schulz M, Hauck HE, Patton WF. Phosphopeptide analysis by directly coupling two-dimensional planar electrochromatography/thin-layer chromatography with matrix-assisted laser desorption/ionization time-of-flight mass spectrometry. J Chromatogr A 2007; 1155 (1) 112–123.

Pandey RC, Misra R, Rinehart Jr. KL. Graft thin-layer chromatography. J Chromatogr 1979; 169, 129–139.

Perham RN (1978). Techniques for determining the amino-acid composition and sequence of proteins. In Techniques in the Life Sciences, B110, pp. 1–39.
Perry JA. New look at solvent strength, selectivity, and continuous development. J Chromatogr, 1979; 165, 117–140.
Perry JA, Haag KW, Glunz LJ. Programmed multiple development in thin-layer chromatography. J Chromatogr Sci 1973; 11, 447–453.
Petruczynik A, Waksmundzka-Hajnos M, Michnikowski T, Plech T, Tuzimski T, Hajnos MŁ, Gadzikowska M, Jóźwiak G. Thin-layer chromatography of alkaloids on cyanopropyl bonded stationary phases. Part I. J Chromatogr Sci 2007; 45 (7) 447–454.
Petruczynik A, Waksmundzka-Hajnos M, Plech T, Tuzimski T, Hajnos MŁ, Jóźwiak G, Gadzikowska M, Rompała A. Thin-layer chromatography of alkaloids on cyanopropyl bonded stationary phases. Part II. J Chromatogr Sci 2008; 46 (4) 291–297.
Petruczynik A, Waksmundzka-Hajnos M, Plech T, Tuzimski T, Hajnos MŁ, Jóźwiak G, Gadzikowska M, Rompała A. TLC of alkaloids on cyanopropyl bonded stationary phases. Part II. Connection with RP18 and silica plates. J Chrom Sci 2008; 46 (4) 291–297.
Pfeifer AM, Tolimann G, Ammon HPT, Kovar K-A. Identification of adenosine in biological samples by HPTLC-FTIR on-line coupling. J Planar Chromatogr – Mod TLC 1996; 9 (1) 31–34.
Phillips RD. An apparatus for rapidly preparing protein samples for hydrolysis. Anal Biochem 1981; 113 (1) 102–107.
Polak B, Gołkiewicz W, Tuzimski T. Effect of mobile phase pH* on chromatographic behaviour in chiral ligand-exchange thin-layer chromatography (CLETLC) of amino acid enantiomers. Chromatographia 2006; 63 (3, 4) 197–201.
Poole CF, Belay MT. Progress in automated multiple development. J Planar Chromatogr – Mod TLC 1991; 4, 345–359.
Poole CF, Dias NC. Practitioner's guide to method development in thin-layer chromatography. J Chromatogr A 2000; 892, 123–142.
Pothier J, Galand N. Automated multiple development thin-layer chromatography for separation of opiate alkaloids and derivatives. J Chromatogr A 2005; 1080 (2) 186–191.
Pozharitskaya ON, Ivanova SA, Shikov AN, Makarov VG. Separation and quantification of terpenoids of *Boswellia serrata* Roxb. extract by planar chromatography techniques (TLC and AMD). J Sep Sci 2006; 29, 2245–2250.
Prosek M, Milivojevic L, Krizman M, Fir M. On-line TLC-MS. J Planar Chromatogr – Mod TLC 2004; 17 (6) 420–423.
Rager IOC, Kovar K-A (2001). Planar Chromatography and IR. In: Planar Chromatography, a Retrospective View for the Third Millennium. Nyiredy Sz (ed), published by Springer Scientific Publisher, Budapest, pp. 247–260.
Regis Technologies Inc., Morton Grove, IL, USA.
Reich E, Schibli A (2006). High-Performance Thin-Layer Chromatography for the Analysis of Medicinal Plants, published by Thieme Medicinal Publisher, Inc., New York.
Reinders J, Lewandrowski U, Moebius J, Wagner Y, Sickmann A. Proteomics 2004; 4, 3686–3703.
Ripphahn J, Halpaap H (1977). HPTLC High Performance Thin Layer Chromatography. Zlatkis A, Kaiser RE (eds) published by Elsevier, Amsterdam, The Netherlands, pp. 189–221.
Rumiński JK. Sandwich chamber for preparative layer chromatography. Chem Anal (Warsaw) 1988; 33, 479–481.
Savoy CF, Heinis JL, Seals RG. Improved methodology for rapid and reproducible acid hydrolysis of food and purified proteins. Anal Biochem 1975; 68 (2) 562–571.
Schiller J, Süß R, Fuchs B, Müller M, Zschörnig O, Arnold K. Combined application of TLC and matrix-assisted laser desorption and ionisation time-of-flight mass spectrometry (MALDI-TOF MS) to phospholipid analysis of brain. Chromatographia 2003; 57 (1) S297–S302.
Shellenberg P, Angew Chem 1962; 74, 118.

Sinhababu A, Basak B, Laskar S. Novel spray reagent for the identification of amino acids on thin-layer chromatography plates. Anal Proc 1994; 31 (2), 65–66.
Smyth DG, Stein WH, Moore S. The sequence of amino acid residues in bovine pancreatic ribonuclease: revisions and confirmation. J Biol Chem 1963; 238, 227–234.
Snyder LR. J Chromatogr Sci 1978; 16, 223–231.
Snyder LR, Kirkland JJ, Glajch JL (1997). Practical HPLC method development, second edition, published by John Wiley & Sons, New York, Chapter 6, pp. 233–291.
Soczewiński E (1986). Equilibrium sandwich TLC chamber for continuous development with a glass distributor. In: Chromatographic Methods. Planar Chromatography, Vol. 1, Kaiser RE (ed) published by Dr. Alfred Huetig Verlag, Heidelberg, Basel, New York, 79–117.
Soczewiński E, Wawrzynowicz T. Thin-layer chromatography as a pilot technique for optimization of preparative column chromatography. J Chromatogr 1981; 218, 729–732.
Soczewński E, Matysik G. A simple device for gradient elution in equilibrium sandwich chambers for continuous thin-layer chromatography. J Liq Chromatogr 1985; 8, 1225–1238.
Somsen GW, Morden W, Wilson ID. Planar chromatography coupled with spectroscopic techniques. J Chromatogr A 1995; 703 (1–2) 613–665.
Spangenberg B. Does the *Kubelka–Munk* theory describe TLC evaluations correctly? J Planar Chromatogr – Mod TLC 2006; 19 (111) 332–341.
Spangenberg B, Klein K-F. Fibre optical scanning with high resolution in thin-layer chromatography. J Chromatogr A 2000; 898, 265–269.
Spangenberg B, Klein K-F. New evaluation algorithm in diode-array thin-layer chromatography. J Planar Chromatogr – Mod TLC 2001; 14 (4) 260–265.
Spangenberg B, Lorenz K, Nasterlack S. Fluorescence enhancement of pyrene measured by thin-layer chromatography with diode-array detection. J Planar Chromatogr – Mod TLC 2003; 16 (5) 331–337.
Srivastava SP, Bhushan R, Chauhan RS. TLC separation of some metal ions on impregnated layers. J Liq Chromatogr 1984; 7, 1341–1344.
Srivastava SP, Bhushan R, Chauhan RS. Use of ligands for improved separation of metal ions on thin silica. J Liq Chromatogr 1985; 8, 571–574.
Srivastava SP, Bhushan R, Chauhan RS. TLC separation of some closely related synthetic dyes on impregnated silica gel layers. J Liq Chromatogr 1985; 8, 1255–1261.
Stahl E (1967) (ed) Dünnschicht-Chromatographie, Ein Laboratoriumschandbuch, published by Springer, Berlin.
Stahl E, Mangold HK (1975). In Chromatography, Van Nostrand-Reinhold, New York, p. 164.
Stahlmann SA. Ten-years report on HPTLC-FTIR online coupling. J Planar Chromatogr – Mod TLC 1999; 12 (1) 5–12.
Steinberg JJ (2003). Nucleic acids and their derivatives. Chapter 25 in Handbook of Thin-Layer Chromatography, third edition, Sherma J, Fried B (eds) published by Marcel Dekker, Inc., New York, Basel, pp. 733–747.
Steinbrunner JE, Johnson EK, Habibi-Goudarzi S, Nurok D (1986). In: Planar Chromatography Vol. 1, Kaiser RE (ed), published by Hüthig Verlag, Heidelberg.
Studer A, Traitler H. J High Resol Chromatogr Chromatogr Commun 1986; 9, 218–223.
Su P, Wang D, Lan M. Half-way development – the technique and the device. J Planar Chromatogr – Mod TLC 2001; 14, 203–207.
Szabady B, Nyiredy Sz (1996a). Dünnschicht-Chromatographie in Memorian Professor Dr. Hellmut Jork, Kaiser RE, Gunther W, Gunz H, Wulff G, (eds), published by InCom Sonderband, Dusseldorf, pp. 212–224.
Szabady B, Nyiredy Sz. (1996b) The versatility of multiple development. In: Dünnschicht-Chromatographie in Memorian Professor Dr. Hellmut Jork, Kaiser RE, Günther W, Gunz H, Wulff G (Eds.), InCom Sonderband, Düsseldorf, pp. 212–224.
Takats Z, Wiseman JM, Gologan B, Cooks RG. Mass spectrometry sampling under ambient conditions with desorption electrospray ionization. Science 2004; 306, 471.

Tuzimski T. Separation of a mixture of eighteen pesticides by two-dimensional thin-layer chromatography on a cyanopropyl-bonded polar stationary phase. J Planar Chromatogr – Mod TLC 2004; 17, 328–334.
Tuzimski T. Use of thin-layer chromatography in combination with diodearray scanning densitometry for identification of fenitrothion in apples. J Planar Chromatogr – Mod TLC 2005a; 18 (106) 419–422.
Tuzimski T. Two-dimensional TLC with adsorbent gradients of the type silica-octadecyl silica and silica-cyanopropyl for separation of mixtures of pesticides. J Planar Chromatogr – Mod TLC 2005b; 18 (105) 349–357.
Tuzimski T. Separation of multicomponent mixtures of pesticides by graft thin-layer chromatography on connected silica and octadecyl silica layers. J Planar Chromatogr – Mod TLC 2007a; 20, 13–18.
Tuzimski T. A new procedure for separation of complex mixtures of pesticides by multidimensional planar chromatography. J Sep Sci 2007b; 30, 964–970.
Tuzimski T. Determination of pesticides in water samples from the Wieprz-Krzna Canal in the Łęczyńsko-Włodawskie Lake District of Southeastern Poland by thin-layer chromatography with diode array scanning and high performance column liquid chromatography with diode array detection. J AOAC Int 2008a; 91 (5) 1203–1209.
Tuzimski T. Application of SPE–HPLC–DAD and SPE–TLC–DAD to the determination of pesticides in real water samples. J Sep Sci 2008b; 31 (20) 3537–3542.
Tuzimski T. Strategy for separation of complex mixtures by multidimensional planar chromatography. J Planar Chromatogr – Mod TLC 2008c; 21, 49–54.
Tuzimski T (2008d). A new strategy for separation of complex mixtures by multidimensional planar chromatography in combination with diode-array scanning densitometry and HPLC–DAD. In: HPLC 2008 32nd International Symposium on High Performance Liquid Phase Separations and Related Techniques, May 10–16, Baltimore MD, USA, p. 48.
Tuzimski T. Application of SPE–HPLC–DAD and SPE–HPTLC–DAD to the analysis of pesticides in lake water. J Planar Chromatogr – Mod TLC 2009a; 22 (4) 235–240.
Tuzimski T (2009b) A New Procedure for Separation of Complex Mixtures by Multidimensional Planar Chromatography in Combination with Diode Array Scanning Densitometry (MDPC–DAD) and High Performance Liquid Chromatography Coupled with DAD Detector (HPLC–DAD). In: 8th Balaton Symposium on High Performance Separation Methods and 15th International Symposium on Separation Sciences, September 2–4, Siófok, Hungary, p. 21.
Tuzimski T (2009c). Progress in Separation of Multicomponent Mixtures by Combination of Different Modes of Multidimensional Planar Chromatography (MDPC). in: 8th Balaton Symposium on High Performance Separation Methods and 15th International Symposium on Separation Sciences, September 2–4, Siófok, Hungary, p. 20.
Tuzimski T (2010a). Use of Planar Chromatography in Pesticide Residue Analysis, Chapter 9 in Handbook of Pesticides: Methods of Pesticides Residues Analysis, Nollet LML, Rathore H (eds) published by CRC Press, Taylor and Francis Group, Boca Raton, pp. 187–264.
Tuzimski T (2010b). Pesticides Residues in the Environment. In: Rathore HS, Nollet LML (eds) Pesticides: Evaluation of Environmental Pollution, published by CRC Press, Taylor & Francis Group, Chapter 5, in press.
Tuzimski T (2010c). New procedure for determination of analytes in complex mixtures by multidimensional planar chromatography in combination with diode array scanning densitometry (MDPC–DAD) and high performance liquid chromatography coupled with DAD detector (HPLC–DAD). J Planar Chromatogr – Mod TLC; 23 (3), 184–189.
Tuzimski T. Determination of clofentezine in medical herbs extracts by chromatographic methods combined with diode array scanning densitometry. J Sep Sci 2010d; 33, 1954–1958.
Tuzimski T, Bartosiewicz A, Correlation of retention parameters of pesticides in normal and RP systems and their utilization for the separation of a mixture of ten urea herbicides and fungicides by two-dimensional TLC on cyanopropyl-bonded polar stationary phase and two-adsorbent-layer Multi-K plate. Chromatographia 2003; 58, 781–788.

Tuzimski T, Nyiredy Sz. (2005) A new procedure for the separation of complex mixtures by multidimensional planar chromatography. in: 6th Balaton Symposium on High Performance Separation Methods, September 7–9, Siófok, Hungary, P.64.

Tuzimski T, Sobczyński J. Application of HPLC–DAD and TLC–DAD after SPE to the quantitative analysis of pesticides in water samples. J Liq Chromatogr Relat Technol 2009; 32 (9) 1241–1258

Tuzimski T, Soczewiński E. Chemometric characterization of the R_F values of pesticides in thin-layer chromatography on silica with mobile phases comprising a weakly polar diluent and a polar modifier. Part V. J Planar Chromatogr – Mod TLC 2002a; 15, 164–168.

Tuzimski T, Soczewiński E. Correlation of retention parameters of pesticides in normal- and reversed-phase systems and their utilization for the separation of a mixture of 14 triazines and urea herbicides by means of two-dimensional thin-layer chromatography. J Chromatogr A 2002b; 961, 277–283.

Tuzimski T, Soczewiński E. Use of database of plots of pesticide retention (R_F) against mobile-phase composition. Part I. Correlation of pesticide retention data in normal- and reversed-phase systems and their use to separate a mixture of ten pesticides by 2D-TLC. Chromatographia 2002c; 56, 219–223.

Tuzimski T, Soczewiński E (2002d). Retention and selectivity of liquid-solid chromatographic systems for the analysis of pesticides (Retention database of ca. 100 pesticides). In: Problems of Science, Teaching and Therapy. Published by Medical University of Lublin, Poland, No 12, Lublin, October 2002, 219 pp. (limited number available on request: tomasz.tuzimski@umlub.pl).

Tuzimski T, Soczewiński E. Use of database of plots of pesticide retention (R_F) against mobile-phase compositions for fractionation of a mixture of pesticides by micropreparative thin-layer chromatography. Chromatographia 2004; 59, 121–128.

Tuzimski T, Woźniak A. Application of solid-phase extraction and planar chromatography with diode-array detection to the qualitative and quantitative analysis of dyes in beverages. J Planar Chromatogr – Mod TLC 2008; 21 (2) 89–96.

Tyihák E, Mincsovics E. Forced-flow planar liquid chromatographic techniques. J Planar Chromatogr – Mod TLC 1988; 1, 6–19.

Umebayashi M. Aminoacylase in higher plants. III. New method for detection of N-acyl amino acid by thin-layer chromatography. Nippon Dojo Hiryogaku Zasshi 1968; 39 (2), 137–142.

Van Berkel GJ, Kertesz V. Automated sampling and imaging of analytes separated on thin-layer chromatography plates using desorption electrospray ionization mass spectrometry. Anal Chem 2006; 78 (14) 4938–4944.

Van Berkel GJ, Sanchez AD, Quirke JM. Thin-layer chromatography and electrospray mass spectrometry coupled using a surface sampling probe. Anal Chem 2002; 74 (24) 6216–6223.

Van Berkel GJ, Ford MJ, Deibel MA. Thin-layer chromatography and mass spectrometry coupled using desorption electrospray ionization. Anal Chem 2005; 77 (5) 1207–1215.

Wachs T, Henion J. Electrospray device for coupling microscale separations and other miniaturized devices with electrospray mass spectrometry. Anal Chem 2001; 73 (3) 632–638.

Waksmundzka-Hajnos M, Petruczynik A, Hajnos MŁ, Tuzimski T, Hawrył A, Bogucka-Kocka A. Two-dimensional thin-layer chromatography of selected coumarins. J Chromatogr Sci 2006; 44 (8) 510–517.

Waksmundzka-Hajnos M, Sherma J, Kowalska T (2008), Thin-Layer Chromatography in Phytochemistry. (ed) published by CRC Press, Taylor and Francis Group, Boca Raton.

Wilson ID. The state of art in thin-layer chromatography – mass spectrometry: a critical appraisal. J Chromatogr A 1999; 856 (1–2) 429–442.

Wilson ID, Morden W. Advances and application in the use of HPTLC-MS-MS. J Planar Chromatogr 1996; 9 (2) 84–91.

Wilson ID, Spraul M, Humpfer E. Thin-layer chromatography combined with high-resolution solid-state NMR for compound identification without substance elution: preliminary results. J Planar Chromatogr – Mod TLC 1997; 10 (3) 217–219.

Zarzycki PK. Some technical problems associated with temperature-controlled thin-layer chromatography. J Planar Chromatogr – Mod TLC 2001; 14, 63–65.

Zarzycki PK. Simple chamber for temperature-controlled planar chromatography, J Chromatogr A 2002; 971, 193–197.

Zarzycki PK, Nowakowska J, Chmielewska A, Wierzbowska M, Lamparczyk H. Thermodynamic study of the retention behaviour of selected macrocycles using reversed-phase high-performance thin-layer chromatography plates and methanol–water mobile phases. J Chromatogr A 1997; 787, 227–233.

Zarzycki PK, Wierzbowska M, Nowakowska J, Chmielewska A, Lamparczyk H. Interactions between native cyclodextrins and *n*-alcohols studied using thermostated thin-layer chromatography. J Chromatogr A 1999; 839, 149–156.

Zarzycki PK, Wierzbowska M, Lamparczyk H. Retention and separation studies of cholesterol and bile acids using thermostated thin-layer chromatography. J Chromatogr A 1999; 857, 255–262.

Chapter 15
HPTLC–MS Coupling: New Dimension of HPTLC

Ajai Prakash Gupta and Suphla Gupta

Abstract The direct coupling of TLC/HPTLC with mass spectrometry (MS) is of particular interest because of the later's high sensitivity, rapid analysis, and ability to aid structural characterization. The conventional ionization method requires heating of the analyte and was suitable only for the analyses of volatile and thermally stable compounds. In contrast, desorption/ionization (DI) techniques, such as fast atom bombardment (FAB), plasma desorption, laser desorption, and matrix-assisted laser desorption/ionization (MALDI) methods, appear to be better for directly characterizing nonvolatile and thermally labile chemical or biochemical compounds separated by HPTLC. HPTLC–MS is cost-effective because the chromatographic run is decoupled with the detection step. Van Berkel and co-workers introduced a novel TLC–ESI–MS, a surface sampling probe-to-TLC/HPTLC plate liquid microjunction, and a self-aspirating ES emitter for the direct detection of small organic compounds. Luftmann developed a plunger-based extraction device for TLC/HPTLC ESI–MS and used a direct electrospray probe to generate analyte ions directly from the end of a TLC plate. Recently, Van Berkel's group applied desorption electrospray ionization (DESI). Cooks et al. introduced the analysis of analytes on solids for coupling TLC with MS. Due to the great understanding of the subject, thin layer chromatography/high-performance thin-layer chromatography can be used interchangeably for methods developed in the twenty-first century.

A.P. Gupta (✉)
Plant Cell Division, Indian Institute of Integrative Medicine, Canal Road, Jammu and Kashmir, India
e-mail: ajai747@747@yahoo.co.in

S. Gupta
Plant Biotechnology Division, Indian Institute of Integrative Medicine, Canal Road, Jammu and Kashmir, India

MM. Srivastava (ed.), *High-Performance Thin-Layer Chromatography (HPTLC)*,
DOI 10.1007/978-3-642-14025-9_15,

Introduction

Mikhail Semenovich Tswett (1899–1901) contemplated the possibility of "chromatography" while carrying out his first research work on the physico-chemical structure of plant chlorophylls, and he reported "on a new category of adsorption analysis" in 1903. The evolution of chromatography followed the advances of this century: each decade brought new innovations based logically on the previous ones. By the end of the twentieth century, chromatography has become the most widely used separation technique in chemistry and biochemistry. Thus, it is no exaggeration to call it the separation technique of the twentieth century.

An essential condition for all fruitful research is to have at one's disposal a satisfactory technique.

M.S. Tswett

In his lifetime, Tswett's achievements were not appreciated by most of his peers who criticized both his results on pigment research and his claims that chromatography is a superior separation technique which can provide compounds as pure as crystallization. In the early 1940s, continuous detection to monitor the column effluent by refractive index measurement was added as a possibility. Another innovation was the consecutive use of different solvents or solvent mixtures with increasing developing and eluting power. This in effect may be considered to be a precursor to gradient elution chromatography which was introduced by Alm et al. in 1952. Classical liquid chromatography based on adsorption desorption was essentially a nonlinear process where the time of retardation (today the retention time) and the quantitative response depended on the position on the adsorption isotherm. Liquid partition chromatography developed by A.J.R Martin and R.L.M. Synge was first described in 1941.

The achievements of Martin and Synge revolutionized not only chromatography but also the way biochemical investigations were carried out. In addition, within a decade, the application of partition chromatography for the analysis of gaseous and volatile compounds completely changed the way of industrial analytical processes. Thus, it is not surprising that in 1952, Martin and Synge were honored with the Nobel Prize for Chemistry.

TLC did not represent a new invention but only an extension of existing techniques. The first report is in 1938 by N.A. Izmailov and M.S. Shraiber, who while investigating complex natural samples that form part of pharmaceutical solutions, used Twett's classical column chromatography, but found it to be too slow. Therefore, they tried thin longitudinal layers of the adsorbent on small glass plates, adding only a drop of the sample solution and developing it by the drop-wise addition of the solvent. The chromatograms obtained consisted of concentric circles of the separated substances. Izmailov and Shraiber called this technique Spot chromatography. About 10 years later J.E. Meinhard and N.F. Hall at the University of Wisconsin, also carried out chromatography on adsorbent-coated glass plates and were the first to use a binder in preparing the adsorbent layer; they called this technique Surface chromatography. However, real thin-layer chromatography started in 1951 by Justus G. Kirchner, U.S. Department of Agriculture, Pasadena, CA. He used glass plates coated with silicic acid (called chromatostrips) for the analysis of terpenes in essential oils.

In the following years, Kirchner widely demonstrated the usefulness of the technique and he also described a number of possibilities, such as carrying out reactions on the plate or utilizing two-dimensional chromatography, similar to the way already known in paper chromatography. Yet, in the public opinion, not Kirchner but Egon Stahl is usually considered as the originator of thin-layer chromatography. Stahl published his first paper on Diinnschicht chromatographie (thin-layer chromatography) in 1956. However, the technique won a wide interest only after his second paper (published 2 years later) which described the equipment used for the coating of glass plates with a thin layer of special, fine-grain adsorbent. Such equipment as well as standardized adsorbents soon became commercially available and this fact greatly helped in the wide acceptance of the technique. In 1962, Stahl also published a very useful and highly popular handbook of TLC which was translated into a number of languages. He was also instrumental in the standardization of the procedure and its materials.

The main advantage of TLC is that it is a simple method that permits the parallel analysis of a number of samples. In the last 30 years, the technique has undergone a number of improvements, expanding its scope and permitting automation. It is a living method and will continue to have a role in the routine analysis of large numbers of samples (Poole 2003). Undoubtedly TLC is a modern analytical separation method with extensive versatility, much already utilized, but still with great potential for future development into areas where research apparently is only just beginning. By the mid 1970s, it was recognized that HPTLC added a new dimension to TLC as it was demonstrated that precision could be improved tenfold, analysis time could be reduced by a similar factor, less mobile phase was required, and the development distances on the layers could be reduced (Zlatkis and Kaiser 1977). The technique could now be made fully instrumental to give accuracy comparable with HPLC. Commercially the plates were first called nano-TLC plates by the manufacturer (Merck), but this was soon changed to the designation "HPTLC." In 1977, the first major HPTLC publication appeared, as *HPTLC high performance thin-layer chromatography* edited by Zlatkis and Kaiser (1977). In this volume Halpaap and Ripphahn described their comparative results with the new 5 × 5 cm HPTLC plates versus conventional TLC for a series of lipophilic dyes. Bonded phases then followed in quick succession. Reversed-phase HPTLC was reported by Halpaap et al. 1980 and this soon became commercially available as precoated plates (Hauck et al. 1982). Hauck and Jost in 1985 reported an amino (NH2-)-modified HPTLC plate, which was soon followed by cyano-bonded (Hauck et al. 1987) and diol-bonded phases (Günter 1984). The decade of 1980s has also seen improvements in spectrodensitometric scanners with fully automated system, including options for peak purity and the measurement of full UV/visible spectra for all separated components. Automated multiple development (AMD) made its appearance in 1984 due to the pioneering work of Burger (Mack et al. 1988). This improvement enabled a marked increase in number and resolution of the separated components. In recent years, TLC/HPTLC research has entered the chiral separation field using a number of chiral selectors and chiral stationary phases. Only one type of chiral precoated plate is presently commercially available, which is based on a ligand-exchange principle and is produced as either a TLC or HPTLC plate. Günter has reported results with amino acids and derivatives on the TLC plate (Tames et al. 1999a) and Mack and Hauck similarly with their HPTLC equivalent. Currently, all steps of the TLC processes can be computer controlled. The use of highly sensitive charge-coupled device (CCD) cameras has enabled the chromatographer to electronically store images of chromatograms for future use (identity or stability testing) and for direct entry into reports at a later date.

Commercially available HPTLC plates coated with pure 4–5 μm spherical silica gel have added further capabilities to the technique. Background interference has been reduced and resolution further improved, which has enabled HPTLC to be hyphenated effectively with Raman spectroscopy, MALDI, mass spectrometry, and FTIR.

Hyphenation

TLC/HPTLC is an analytical method in its own right. It is also complimentary to other chromatographic techniques and spectroscopic procedures. Results obtained with TLC/HPTLC can often be transferred to HPLC or vice versa with some adjustment in eluting solvent conditions. For multicomponent samples (e.g., pesticides in water), selected fractions from an HPLC separation can be collected and subsequently re-chromatographed on HPTLC to give a "fine-tuned" separation of the components of the fractions.

Planar chromatography on high-performance thin-layer chromatography (HPTLC) plates is well known cost-effective offline separation technique. In recent years, the entire system reproducibility, from application to detection, has been enhanced through automation of the equipment making it useful and reliable for quantitative analysis. Extraordinarily helpful and at the same time an underestimated feature is the fact, that planar chromatography shows what is left at the start, which would never be seen in column chromatography, and gives a more comprehensive sample image when unknowns have to be analyzed. The system versatility, i.e., parallel chromatography of up to 36 runs, wide range of application volumes (nl to ml), automated single or multiple development and multiple detection by fluorescence (FLD), UV, and Vis, together with less sample preparation than other chromatographic techniques, and the simplicity of carrying out in situ pre- and post-chromatographic derivatization, provide planar chromatography with a still treasured position among the analytical tools for high-throughput analysis.

Thin-layer chromatography has been successfully hyphenated with high-performance liquid chromatography (HPLC), mass spectroscopy (MS), Fourier transform infra-red (FTIR), and Raman spectroscopy to give far more detailed analytical data on separated compounds. Even the UV/visible diode array technique has been utilized in TLC/HPTLC to determine peak purity or the presence of unresolved analytes.

However, coupling with mass spectrometry (MS) has been a weakness of this technique in comparison with the advances achieved for HPLC, gas chromatography (GC), and capillary electrophoresis (CE). Due to the selective detection offered by MS, several approaches for TLC/HPTLC/MS hyphenation have been proposed during the extensive research done in the last decade, e.g., FAB (Orinak et al. 2006), secondary ion mass spectrometry (SIMS) (Parent et al. 2006a; Nakamura et al. 2006), MALDI (Dreisewerd et al. 2005; Peng et al. 2005), continuous wave diode LD/atmospheric pressure chemical ionization (APCI) (Kertesz et al. 2005) extraction using an automated surface sampling probe via liquid microjunction with electrospray ionization (ESI) (Van Berkel et al. 2007), automated sampling with desorption ESI (DESI) (Morlock and Ueda 2007), direct analysis in real time (DART) (Luftmann 2004), and plunger-based extractors (Prosek et al. 2004; Hauck and Schulz 2002). Considering the advantages of ultra thin-layer chromatography (UTLC) (Salo et al. 2007), i.e., increased detection limits, reduced solvent consumption, and increased speed, some authors have developed UTLC/MALDI (Kauppila et al. 2006), UTLC/DESI (Crecelius et al. 2004), and UTLC/SIMS. All these approaches have the great advantage that the planar chromatographic separation can be performed regardless of MS-suitable solvents. All HPTLC/MS systems were first assayed as proof of concept, and then after successful qualitative results, several of these systems were also evaluated for quantitative determinations with promising results. However, besides requiring internal standard (IS) to carry out the quantification, each system has its own limitations. MALDI (Brown and Busch 1991), FAB (Banno et al. 1991), and LSIMS (Ford et al. 2005) required matrix incorporation after chromatography via spotting or electrospray causing ion suppression and band spreading, and LD/APCI were only suitable for a limited plate size.

The automated sampling via a liquid microjunction was useful only with reversed-phase plates (Aranda and Morlock 2007). The automated sampling DESI (Dreisewerd et al. 2005) and the LD/APCI systems required modifications in the mass spectrometer and manual positioning of plunger-based extractor for producing some level of dispersion (Morlock and Schwack 2006). A robust manually positioned plunger-based device to couple normal phases with ESI–MS was successful for qualitative (Alpmann and Morlock 2006) and quantitative (Jautz and Morlock 2006) evaluation of analytes in food and pharmaceuticals at nanogram levels per band on the plate. As the interface elutes the complete substance band with its whole depth profile, detection ability ranges in the low-pg per band. However, a fully automated interface to couple HPTLC with ESI–MS for evaluating quantitative analysis was used for the first time by Heinrich Luftmann et al. (2007) in 2007. A new fully automated online interface for HPTLC/MS hyphenation has several features that make it highly attractive as a coupling tool for several systems and in different laboratories. The validation results obtained for caffeine quantification in energy drinks and pharmaceutical samples, without internal standard, proved the reliability of the interface and its usefulness for quantitative analysis. Interface designed by Luftmann et al. (2001) can be considered as one of the most reliable and universal interfaces for HPTLC/MS hyphenation (Poole and Dias 2000). It is generally agreed that HPTLC is most effective for the low-cost analysis of samples requiring minimal sample cleanup or where it allows a reduction in the number of sample preparation steps (e.g., the analysis of samples containing components that remain adsorbed to the stationary phase at the point of application). Thin-layer chromatography is also preferred for the analysis of substances with poor detection characteristics requiring post-chromatographic treatment for detection. Since all sample components are located in the chromatogram, thin-layer chromatography is the most suitable technique for surveying sample properties. The common method of development in HPTLC employs capillary forces to transport the mobile phase through the layer. These weak forces arise from the decrease in free energy of the solvent as it enters the porous structure of the layer. For fine particle layers, capillary forces are unable to generate sufficient flow to minimize the main sources of band broadening. In addition:

A: The mobile phase velocity varies with function of time (and migration distance)

B: The mobile phase velocity is established by the system variables

Multiple developments provide a general approach to increase the zone capacity by taking advantage of a focusing mechanism that occurs each time the solvent front passes over the stationary zone on the layer. This affords perhaps a doubling of the zone capacity at the expense of longer separation times. Multiple developments with an incremental increase in the development length and a decreasing solvent strength gradient are the basis of separations by AMDs.

The combination of chromatographic separation and mass spectrometric detection is considered as a tool for problem solving in analytical chemistry and increasingly for routine analytical methods. Mass spectrometric detection brings an added level of information, complementary to the chromatographic process that involves the certainty of identification and the specificity of detection. Mass spectral information can generally be obtained from sample size typical of common analytical methods. Robust and affordable coupled instruments for GC–MS and LC–MS are available in many laboratories, and CE–MS is not widely available. By comparison, HPTLC–MS is mainly a research tool available to a very small number of research groups, which have mainly constructed their own interface. The evolution of TLC–MS has been slow compared to LC–MS. In practice, mass spectra of samples separated by TLC are generally obtained by slow and labor-intensive manual methods (e.g., solvent extraction of excised zones with conventional liquid introduction into the mass spectrometer) earlier. The problems of interfacing thin-layer chromatography to mass spectrometry (TLC–MS) are largely different from those experienced in LC–MS and other column separation techniques. At the completion of the separation, the chromatogram is fixed in time and space with major portion of the mobile phase eliminated by evaporation. The thin-layer plate can be considered as a storage device effectively decoupling the separation requirements from those of measuring the mass spectra. Immobilized separation is also generally compatible with the vacuum requirement of the mass spectrometer. On the other hand, the sample is embedded within the stationary phase from which it must be extracted into gas phase and ionized for mass analysis. It is not surprising that the principal methods of instrumental HPTLC–MS are based on surface deposition and ionization techniques. Commercial interfaces provide limited automation capabilities till now, but this is likely to change soon. As an alternative to these general approaches, probes for direct solvent extraction of individual separated zones followed by ESI of the extracted sample have been described by Anderson and Busch (1998) and Berkel et al. in (2002). Each system has its own limitations; FAB, MALDI, and LC-ESI-MS requires matrix incorporation after chromatography via spotting or electrospray causing ion suppression and band spreading. To improve the sensitivity of the FAB method, a matrix solution or phase transition matrix is applied to the layer, and spectral acquisition and time averaging of the ion signal are employed. Because ions are removed from the surface only, the ionization efficiency will be low unless a mechanism is provided to the surface in a continuous manner. Impregnating the chromatogram with a viscous liquid or low melting point solid (phase transition matrix) fulfills this role without destroying the integrity of the chromatogram through zone broadening. For reliable mass spectra in the scan mode, sample amounts in the tens of nanograms to microgram range are required. Since FAB ionization produces largely molecular ion adducts with limited fragmentation, tandem mass spectrometry and collision-induced dissociation are used for identification purposes (Nyiredy 2001; Butz and Stan 1995; Wagner et al. 1984; Renger 1993, 2001; Szepesi and Nyiredy 1992; Galand et al. 2002; Mincsovics et al. 2002; Morovjan 2002; Somsen et al. 1995; Oka et al. 1992; Wilson 1999; Parent et al. 2006b; Wilson and Modern 1996; Tames et al. 1999b, 2000; Crecelius et al. 2000).

Methods relying on MALDI requires application of an extraction solvent to the layer to move sample to the layer surface followed by cocrystallization with a MALDI matrix. TLC/HPTLC–MALDI direct coupling methodologies use one of the four methods of layer treatment. A MALDI matrix solution is deposited directly on the layer or, better still, applied to the layer by electrospray (Crecelius et al. 2002), and the solvent evaporated with crystallization of the matrix. The main problem with this approach is loss of separation integrity due to zone broadening caused by convection driven by matrix crystallization. The pressing method attempts to reduce sample spreading by separating the extraction and crystallization steps. A MALDI matrix layer is formed in the smooth inert substrate, which is separate from the TLC plate and is transferred to the surface of the separation layer by pressing the matrix layer and TLC plate face to face in the presence of a sprayed-on extraction solvent. The spatial resolution and detection limits are largely determined by the selection of the extraction solvent, the extraction time, the pressure, time used for the pressing steps, and the thickness of the stationary phase. Even for optimum conditions, the sample recovery remains low because of the poor extraction efficiency of the sample from the sorbent pores as well as some lateral broadening of sample zones. The extraction efficiency is virtually complete using a hybrid HPTLC–MALDI plate, in which two juxtaposed layers, a HPTLC layer and a MALDI matrix layer, are formed on a common support. The separation is performed in one direction on the TLC plate layer, the mobile phase is evaporated, and the MALDI matrix is applied to the MALDI zone, followed by elution of the sample in the direction of the MALDI matrix layer. The MALDI matrix is used to acquire the mass spectra. The most recent approach uses a suspension of graphite particles in ethylene glycol and ethanol sprayed onto the layer as an energy transfer matrix. This approach improves the sensitivity and reduces the contribution of matrix ions to the mass spectral background. Optimization of all four methods is critical to their success with good-quality mass spectra consisting of mainly molecule–ion adducts with a few fragmentation ions obtained from nanogram amounts of sample (Anderson and Busch 1998; Van Berkel et al. 2002).

The capabilities of LC–MS can be used to assess the level of satisfaction for HPTLC–MS, given that user desires are similar for both techniques. It is quite obvious that HPTLC–MS falls well below the satisfaction level of LC–MS, as it stands today. HPTLC–MS is generally used by experimenting analysts for obtaining spectra only and is not considered reliable for quantitative analysis. The possibilities of chromatogram reconstruction from stored ion intensities or base peak ion intensities, use of mass chromatograms for target compound's location, and automated identification by library searching are very limited in HPTLC–MS. A major advantage of liquid chromatography is the active transport mechanism of the mobile phase that delivers the whole sample to the mass spectrometer in a time defined by the peak widths. It is necessary for HPTLC–MS to provide data in an automated fashion and in near real time if it is to become accepted alongside LC–MS.

Scanning of TLC/HPTLC PLATE for Quantification or Finger Printing

Slit Scanning

Slit-scanning densitometry is the dominant method of recording thin-layer separations for interpretation and quantification. This technology is now relatively mature and although limited to absorption and fluorescence detection in the UV–visible range, it has adequately served the needs of thin-layer chromatography for the last three decades. Evolution of slit-scanning densitometry is now largely progressive and has incorporated major changes in its operation and performance as seen in the development of scanners employing a fiber optic bundle for illumination of sample zones and collection of reflected light (or fluorescence). Photodiode-array detector is used for simultaneous multiwave length detection and spectral recording. This approach simplifies data acquisition for some applications of modern chemometric approaches for data analysis. This may improve the quality of available data from thin-layer separations but does not overcome the principal limitations of slit-scanning densitometry. Major improvement may be realized through image analysis, also known as video densitometry. However, practical working difficulties and commercial aspects have practically abandoned this concept.

Video Scanning

For video densitometry, optical scanning takes place electronically, using a computer with video digitizer, light source, monochromators, and appropriate optics to illuminate the plate and focus the image onto a charge-coupled device (CCD) video camera. The main attractions of video densitometry for detection in thin-layer chromatography are fast and simultaneous data acquisition from the whole plate. It is a simple instrument designed without moving parts and compatibility with data analysis of two-dimensional chromatograms. There is increasing confidence that software for post-chromatographic data processing can be used to evaluate and minimize chromatographic and scanning errors resulting in improved data quality. This could be the potential use of two-dimensional separations for thin-layer chromatography with image analysis. Video densitometry has proven popular in the development of field-portable instruments and as a replacement for photographic documentation of thin-layer separations. Modern instruments provide attractive options for searching and comparing sample images, as well as integration of peak area. The ease of storing raw images into documents is responsible for a significant amount of the commercial success of video densitometers in current production.

As things stand today, video densitometry competes with slit-scanning densitometry in terms of sensitivity, resolution, and available wavelength measuring range. A major difference in the two techniques is that while slit-scanning densitometry is a stable technology, image analysis is rapidly evolving, driven by broad applicability to many analytical techniques, of which thin-layer chromatography is an obvious application. The problems affecting thin-layer chromatography are inhomogeneous illumination of the layer during acquisition of the image, which is the main source of scanning error. Proper lighting can increase the image contrast and resolution, thereby improving the overall performance of the system. At present, there is no system that affords uniform illumination of standard sized layers with the required spatial homogeneity for high-resolution and high-sensitivity measurements. One persistent problem in the video scanning is insensitivity of CCD cameras in the UV spectral region. Back-thining CCD cameras provide much higher efficiency in the UV spectral range but are currently very expensive.

HPTLC–MS

Over the last two decades, mass spectrometry has become one of the central techniques in analytical chemistry and the analysis of biological (macro) molecules in particular. Its importance is now comparable to that of the traditional electrophoresis and liquid separations techniques, and it is often used in conjunction with them as so-called "hyphenated" techniques, such as LC–MS, TLC/HPTLC–MS, and CE–MS.

This development was originally triggered by the discovery of novel techniques to generate stable ions of the molecules of interest and the development of associated ion sources. Such a technique needs to meet two basic requirements: first the molecules, usually existing in the liquid or solid condensed state, have to be transferred into the gas phase and eventually into the vacuum of a mass analyzer; and second, the neutral molecules have to acquire one or several charges to separated and detectable in the mass analyzer. Both steps had traditionally been prone to internal excitation of the molecules leading to fragmentation and loss of analytical information. The two techniques that evolved as the front runners and nowadays dominate mass spectrometry are ESI and MALDI. Although these two techniques solve the problem of transfer from the condensed to gas phase, as well as the ionization in different ways, they were developed completely independently. It is also important to realize that both ESI and MALDI make use of principles developed in the years before, such as field desorption and desorption by particle beams, as well as chemical ionization in the gas phase. The novel ionization mechanisms have early on induced the revival of some mass analyzer principles, such as the (axial ion extraction) time-of-flight (TOF) instruments, which had been written off as having too low a performance earlier. More recently, a whole plethora of new mass spectrometers have been marketed, combining both ESI and MALDI with high-performance spectrometers, such as the orthogonal extraction TOFs, Fourier transform ion cyclotron (FTICR), and orbitrap instruments. These developments have been largely introduced by the instrument manufacturers in recent years.

Type of MS Techniques Used for TLC/HPTLC–MS Analysis

Electrospray Ionization

ESI produces highly charged droplets by nebulizing liquids in a strong electrostatic field. The highly charged droplets are generally formed in a dry bath gas at atmospheric pressure by evaporation of neutral solvent until the charge repulsion-"Coulombic explosion." The mechanism of ion formation itself is a volatile topic and has been treated in several reports. Whatever the ultimate process of ion formation, it is clear that ESI produces multiple charged molecules from solution under mild conditions. These ions generally arise by attachment of protons, alkali cations, or ammonium ions for positive ion formation.

Electrospray ionization (ESI) was first introduced by Dole and co-workers in 1968 and coupled with MS in 1984 by Yamashita and Fenn. These workers laid the foundation of ESI, demonstrating its ultimate utility for the analysis of macromolecules and as a potential interface for the combination of liquid chromatography with mass spectrometry. The production of molecules bearing multiple charges accesses higher molecular weights by extending the mass range for mass-to-charge (m/z) limited mass spectrometers. The multiple charging phenomenon has been demonstrated to apply to molecules of over 150 kDa, and it has permitted the measurement of relative mass with precision of better than 0.05%.

In ESI, the sample is dissolved in a polar, volatile solvent and transported through a needle placed at high positive or negative potential (relative to a nozzle surface). The high electric potential (1–4 kV) between the needle and the nozzle causes the fluid to form a Taylor cone, which is enriched with positive or negative ions at the tip. A spray of charged droplets is ejected from the Taylor cone by the electric field. The droplets shrink through evaporation, assisted by a warm flow of nitrogen gas passing across the front of the ionization source. Ions are formed at atmospheric pressure and pass through a cone-shaped orifice into an intermediate vacuum region and from there through a small aperture into the high vacuum of the mass analyzer. ESI has been used in conjunction with all common mass analyzers. The exact mechanism of ion formation from charged droplets has still not been fully elucidated.

ESI is an extremely gentle ionization method, accompanied by very little fragmentation of the formed molecular ions. Consequently, weak bonds are often preserved and analysis of intact posttranslationally modified peptides/proteins and noncovalently bound complexes, such as protein–ligand complexes, can be successfully performed with ESI–MS. Even though fragments are seldom produced in ESI, the ions generated are especially favorable for collision-induced dissociation (CID), because the high-charge state of the molecular ions increases the energy available for the collision event. Analyte signal suppression caused by charge competition between electrolytes and other analytes is a major problem in ESI. It may, in practice, prevent thorough analysis of complex mixtures if chromatographic prefractionation is not applied. These charge competition phenomena and analyte signal's strong dependence on experimental conditions (pH, solvent composition, and salt concentration) thereby making it risky to draw quantitative conclusions from ESI–MS data.

However, as in MALDI–MS, quantification can be achieved, within a limited concentration range, using a carefully chosen internal calibrant of known quantity and close chemical resemblance to the peptide/protein of interest. The combination of ionization at atmospheric pressure and the continuous flow of solvent in ESI allows direct coupling with separation techniques, such as liquid chromatography and capillary electrophoresis.

The sensitivity of ESI–MS is good, with low femtomole or attomole detection levels for many peptides. However, the sensitivity of the method is a function of the concentration of the injected sample. High flow rates (1–1,000 μl/min) in conventional ESI–MS result in high sample consumption. It is therefore advantageous to use the lowest possible flow rate. Nano-ESI (or nanospray) is a low flow rate method (20–200 nl/min) in comparison to ESI, with lower sample consumption and considerably higher sensitivity. Nano-ESI has also been shown to be more tolerant to salts than conventional ESI. Sub-attomole levels of analyte have been detected by coupling capillary electrophoresis (CE) to ESI. ESI–MS can be used for analysis of polar molecules ranging from less than 100 Da. An important feature of ESI is the capability to generate a distribution of multiple charged ions, which allows the analysis of very large proteins using mass analyzers with limited *m*/*z* range. Relatively small changes in analysis conditions, such as pH, solvent composition, salt concentration, and partial denaturation of the analyte molecules, can alter the charge state distribution of a large molecule.

Matrix-Assisted Laser Desorption Ionization

Koichi Tanaka of Shimadzu Corp. (Japan) shared the 2002 Nobel Prize in Chemistry with Fenn "for their development of soft desorption ionization methods for mass spectrometric analyses of biological macromolecules". Fenn's contributed on ES, while Tanaka was recognized for his laser desorption method of protein ionization.

Laser desorption has been studied for many years as another method to volatilize large biomolecules for mass spectrometry. However, mass spectra for larger molecular weight compounds were frequently dominated by fragment ions with few if any intact molecular ions. At nearly the same time ESI was demonstrated, matrix-assisted laser desorption ionization (MALDI) was developed as a method for analysis of proteins. The key of MALDI process is the incorporation of the analyte into a matrix that absorbs radiation from the ultraviolet (or infrared) laser. Transfer of energy from the matrix to the sample leads to desorption of the solid to the gas phase for analysis by the mass spectrometer. Again, as with ESI, the mechanism of desorption and ion formation is complicated and unresolved. However, the formation of ions to nearly 10^6 Da and the high sensitivity of MALDI (sub-femtomole) have led to the explosive commercialization of MALDI–MS. The dominant feature of the sample spectrum is the appearance of an abundant singly charged molecular ion. In some cases, multiple charged ions of relatively low charge (2+ to 5+) are observed, depending on the sample matrix, the wavelength type of matrix and power of the laser pulse. In comparison to ESI mass spectra, the appearance of only singly and lower charged ions generally allow quicker and simpler interpretation of the data. MALDI mass spectra of complex mixtures of macromolecules are much easier to interpret than ESI spectra (without further computer support), where a suite of multiple charged ions representing each component is formed. The mass range of the MALDI–MS method is comparable to ESI–MS.

The relative simplicity of a MALDI mass spectrum compared to an electrospray mass spectrum is mirrored by the simplicity of the experiment. MALDI mass spectra are typically acquired on a TOF mass spectrometer. The instrument consists of a laser source, a flight tube, and a detector. The ions line up at the starting point (sample plate). When the gun (laser source) is fired, the ions run in a straight line to the finish line (detector). The smaller ions are quicker runners and reach the finish line before the larger, slower runners. Similarly, the smaller mass ions impact the detector surface before the larger mass ions. The instrument measures the flight time of each ion, which thus provides a measurement of mass. In a typical experiment, biochemical samples are prepared by placing 1 μl of an aqueous solution of the analyte of interest and an appropriate matrix solution, a saturated solution of the matrix in 1:2 acetontrile–0.1% TFA aq. on a sample target. The sample/matrix solution is then allowed to air dry at room temperature before insertion into the mass spectrometer. The key idea of the method is to embed the analyte macromolecule in a suitable matrix of molecules having a strong absorption at the laser wavelength and is present in high molar excess over the analyte. This induces an efficient transfer of the laser pulse energy to the analyte and results in a soft desorption process, i.e., little or no fragmentation. The matrix may also play a role in the subsequent ionization of the analyte molecules. Popular MALDI matrices for peptide and protein analysis are α-cyano-4-hydroxy cinnamic acid (αHCCA), 1,5-dihydroxybenzoic acid (DHB), and 3,5-dimethoxy-4-hydroxy-cinnamic acid (sinapinic acid) and for oligonucleotides, 3-hydroxypicolinic acid. Because of the low cost and small size of the nitrogen laser (N_2, 337 nm), it is by far the most common radiation source for the desorption process on commercial systems. In some cases, the UV region of the electromagnetic spectrum can be reached by more powerful lasers, such as Nd:YAG lasers [with the appropriate tripling (355 nm) or quadrupling (266 nm) of laser frequency]. The infrared region can be reached with a Er:YAG laser (2.3 nm). Although the power density of a nitrogen laser is more than adequate for most MALDI applications, the advantage of a YAG laser is the ability to access different wavelength regions. The use of the fundamental infrared frequency (1.06 nm) of a YAG laser has been shown to be advantageous for oligonucleotides analysis.

Desorption Electrospray Ionization

Desorption ESI (DESI) was introduced by Takatz et al. The phenomenon actually was observed earlier but was discarded as a nuisance (an analyte or calibration mixture that coated the entrance of the transfer capillary and contributed to undesired peaks in the spectra). The idea of using the electrospray for desorption is as clever as it is simple. The method is sensitive and large species such as proteins can be detected. The ions observed are more or less the same as with regular ESI. A DESI source consists of a spray capillary and a coaxial capillary providing the nebulizer gas. High voltage is applied to the spray needle, which is directed toward the target surface. Sample species are then desorbed and subsequently enter the orifice to the mass spectrometer. Normal distances between the spray needle, sample, and orifice range from some millimeters to several centimeters. The optimum geometry depends on the sample and on the size of the desired sampling area. The advantages with DESI are that the target can be, in principle, any type of surface and that the analysis time often can be very short (in seconds). This means that rapid analyses can be performed without the need for sample preparation. A sample, such as a dollar bill, tomato, or tablet, can be placed close to the spectrometer inlet and after a few seconds of spraying a spectrum is recorded. There are numerous applications, including high-throughput analysis, screening for trace levels of drugs, explosives, pesticides, and contaminations. DESI also has a potential for chemical imaging.

There are some variants that have emerged in the wake of DESI. By replacing the electrospray emitter by a metal needle and allowing solvent vapor into the coaxial gas flow, desorption APCI (DAPCI) can be performed. Other versions are atmospheric pressure solids analysis probe (ASAP) where a heated gas jet desorbs the analyte, which is subsequently ionized by a corona discharge, and electrospray assisted laser desorption/ionization (ELDI).

Direct Analysis in Real Time

DART was introduced in 2005. Similar to DESI, samples can be analyzed directly without preparation. This is a versatile new ion source which is working in open air under ambient conditions. Ionization is performed by reaction of electronic or vibronic excited state species with reagent molecules and analytes. Without requiring sample preparation hundreds of polar and nonpolar analytes, such as toxic chemicals, pharmaceutics, drugs, explosives, chemical warfare agents in liquids, gasses, and solids, were directly detected on various surfaces. The efficacy of DART–TOF–MS was demonstrated in numerous application fields such as food analysis, clinical applications, and forensics. Functioning of DART was shown on different surfaces, such as concrete, human skin, currency, documents, airline boarding passes, business etc and the reduced functioning and detectability on silica gel glass plates which are mostly used for quantitative HPTLC. Being a continuous source, DART is suitable for coupling to the mass analyzers like quadrupoles, magnetic sectors, Qq-TOFs, and trapping analyzers.

The zones of interest on HPTLC can be directed into the excited gas stream of DART. Helium was used as gas and gas heater was used for heating the spot. DART–TOF–MS does not have any chromatographic systems, and the data acquired by the DART–TOF–MS system was similar to the data acquired by LC–TOF systems. When the HPTLC plate was introduced into the DART gas stream, the peaks of the target compound appeared on the TIC or mass chromatograms. The peak intensity normally depends on the amount of target compound introduced in the DART gas stream. During the measurement, several peaks could be detected for a single zone on the HPTLC plate due to the position change of the plate in the DART gas stream to find a suitable measurement position. Therefore, the largest peak during the corresponding time interval, when the plate was kept in the DART gas stream, was integrated to calculate the peak area. The new ion source DART has decisive advantages for coupling with planar chromatography. The advantages of the system are the following first, an electrospray is not necessary which can cause physical damage to the plate or liquid spreading effects obviously leading to lower signal intensities on normal phases compared to nonpolar ones and, second, in situ analysis directly from surfaces is possible and mass spectrometric results can be obtained within seconds.

The zone of interest has to be exposed into the DART gas stream (excited state helium atoms) and presumably charged water clusters (originating from water molecules in ambient air) allow proton transfer. These charged and desorbed sample molecules were guided into the MS vacuum system and mass spectra were obtained instantaneously. However, most important thing is proper positioning of the zone of interest in the excited gas stream. Therefore, the plate is to be cut within the track (y-axis) or within the substance assignment window (x-axis) to ensure that the analyte is positioned directly on the plate edge. The plate cut and thus analyte's edge position are crucial. If the plate position is too close, the gas flow will be blocked and if it is too far away no signal will be obtained. Desorption and ionization of substances from different stationary phases, such as silica gel, amino phases, or RP-18 phases, are possible. The analytical response regarding DART is dependent on distance and angle of the plate related to the DART gas stream. Employment of DART in the field of planar chromatography was successfully demonstrated for the first time by Luftmann.

The feasibility to obtain instantaneous mass spectra from substance zones highly contributes to the progress of planar chromatography. Within seconds mass spectra of substances adsorbed on a HPTLC plate were obtained directly without any additional step (extraction, addition of matrix, etc.). Spatial resolution of DART was better than 3 mm. Repeatability, analytical response, and detectability were heavily influenced by manual positioning. This drawback is overcome by employment of stable isotope-labeled standards. This way, analytical response and repeatability were impressively improved. However, their availability and cost poses a major hurdle (Morlock and Ueda 2007; Ekman et al. 2009; Loo 1995; Takats et al. 2005; Cody et al. 2005; Williams et al. 2006; Thomson 1910).

Secondary Ion Mass Spectrometry

SIMS has quite a long historical record although the term SIMS was not coined until 1970. Ejection of neutrals and ions from a surface as a result of ion bombardment was first observed by J.J. Thomson as early as 1910. In 1949, Herzog and Viehbock constructed an ion source to a secondary ion mass spectrometer, while the first practical instrument was developed by Honig in the 1950s. Further development in the area was carried out by Liebl, who designed an ion microprobe in 1967. A direct combination of TLC/HPTLC with SIMS provides a method for quantitative analysis of thermally unstable compounds or compounds of low volatility. In the mid-1970s, Benninghoven et al. obtained the first secondary ion mass spectra of amino acids. Since then, the technique continued to develop with new primary ion sources and refined components into today's powerful chemical imaging, depth profiling, and surface analysis technique. It is even possible to analyze compounds up to about 10 kDa, although the sensitivity is normally poor for 1 kDa mass. For most applications, the mass range of interest is rather 300 Da. Larger species can nevertheless be probed by fragment signatures, which often provide sufficient information. The primary ion impact is believed to start a cascade of collisions between the impacting particle and the atomic nuclei in the sample, resulting in ejection of neutral molecules and ions through so-called sputtering (McDonnell and Heeren 2007). The SIMS technique can be divided into two major areas based on its application: (1) static SIMS where the objective is to analyze the surface of a sample and (2) dynamic SIMS where the sample depth profile is analyzed. In static SIMS, the irradiation of the sample is kept at a low level so that surface damage can be neglected. The TOF analyzer has proven to be well suited for static SIMS, especially for imaging, a technique that can offer submicrometer spatial resolution (Herzog and Viehbock 1949; Honig 1958, 1985, 1995; Castaing and Slodzian 1962; Liebl 1967; Benninghoven et al. 1976; Sigmund and Claussen 1981; McDonnell and Heeren 2007; Mc Phail 2006).

Dynamic SIMS, on the other hand, exploits the fact that material is desorbed only from the surface. Here, a pit is literary dug in the sample to analyze the material as a function of depth. In dynamic SIMS, most molecular information is lost, but atomic and small molecular species can be detected. Since dynamic SIMS is considerably faster with a continuous primary ion beam than with a pulsed ion beam, quadrupoles and magnetic sectors are the analyzers of choice. Major applications of static SIMS include microelectronics, materials science, polymers, particles, and life sciences, while applications for dynamic SIMS include semiconductors, metallurgy, geochronology, and biology. In SIMS, the analyte ions (i.e., the secondary ions) are created by letting a pulse or continuous beam of primary ions impinge on the sample surface Dynamic SIMS is used for depth profile analysis of mainly inorganic samples. The objective is to measure the distribution of certain compound as a function of depth. At best the resolution in this direction is 0.1 nm, that is, considerably better than the lateral resolution. Depth profiling of semiconductors is used, for example, to monitor trace level elements or to measure the sharpness of the interface between two layers of different composition. For glass it is of interest to investigate slow processes, such as corrosion, and small particle analyses including environmental samples contaminated by radioisotopes and isotope characterization in extraterrestrial dust.

Finally, secondary neutral mass spectrometry (SNMS) deserves mentioning. SNMS was introduced in the early 1970s by Oechsner and Gerhard. In SIMS, the ionization is very selective so the signal typically does not reflect the true sample composition. Most ejected species in a sputtering event, however, are neutral. Postionization of these with an appropriate technique allows better quantitative analysis. This can be achieved in different ways, for example, using an electron beam, plasma, or laser for ionization. The typical application of SNMS is depth profiling of inorganic samples, such as semiconductors, but trace element analysis of biological samples has also been performed (Sodhi 2004; Hofmann 2004; Jones et al. 2006; Oechsner and Gerhard 1972; Barber et al. 1981).

Fast Atom Bombardment

A technique very closely related to SIMS is FAB, where a liquid sample is bombarded with energetic atoms (typically Ar or Xe atoms of approximately 10 keV kinetic energy) instead of ions (Beranova-Giorgianni and Desiderio 1997; Benninghoven and Sichtermann 1978). There is actually a technique named liquid SIMS (LSIMS) where the liquid sample is bombarded with energy (Barber et al. 1982). In principle, there is no difference in the sputtering mechanism whether the primary particles are ions or atoms. The original reason to use a neutral beam was to avoid influence from the comparatively high accelerating voltage in the source of the magnetic sector of the instrument where it was introduced (Heine et al. 1989). Charging of insulating targets is also significantly reduced. The FAB ion source can be combined with many different mass analyzers, but is most widely used together with quadrupole and magnetic sector instruments. A disadvantage with FAB is the rapid contamination of the ion source region, so frequent cleaning is required. In order to keep the sample in the liquid state when it enters the high vacuum ion source, the sample is usually dissolved in a viscous solvent with low vapor pressure and freezing point, such as glycerol. The matrix also shields the sample molecules from damage caused by the impinging high-energy particles. In continuous flow FAB (CF-FAB), sample solution is continuously delivered to the target, thereby making it possible to provide FAB–MS with online coupling to LC and CE. Less organic matrix (0.5% glycerol instead of 0.90% glycerol) is necessary to keep the sample liquid in CF-FAB, resulting in greatly reduced chemical background. The reduced background and the constant refreshment of the surface layers make CF-FAB more sensitive than conventional FAB.

FAB mass spectra are dominated by singly charged molecular ions, although doubly charged molecular ions and dimers are also occasionally observed. Prompt fragmentation of the analyte ions often gives partial sequence information but is usually not of sufficient intensity to fully sequence a peptide of unknown structure. However, fragmentation readily occurs with CID and provides further structural information. One major disadvantage with FAB is the intense chemical background due to matrix cluster ions and matrix fragment ions. The matrix can also react directly with the sample molecules forming radical anions or causing reduction of the analyte. The FAB ion source is not as "soft" and sensitive as ESI or MALDI (Oka et al. 1994; Lin et al. 2000; Sweetman and Blair 1988; Mallis et al. 1989; Dass 1996; Moseley et al. 1991; Caprioli et al. 1986; Caprioli and Moore 1990; Shu-Yao et al. 2007).

Electrospray-Assisted Laser Desorption Ionization/Fused Droplet

Elcctrospray-assisted laser desorption/ionization (ELDI), a new ionization method that combines some of the features of ESI and LD, allows direct, sensitive, and rapid characterization of small organic and large biological compounds in solids under ambient conditions. Tedious post-separation and sample pretreatment procedures are often avoided when using this technique for direct solid sample analysis. The ionization processes of ELDI are suggested to be similar to those of fused-droplet ESI (FD-ESI, or two-step ESI). In an FD-ESI source, gaseous analyte molecules or neutral droplets containing the analyte molecules are conducted to the tip of an electrosprayer by nitrogen where the analyte molecules are ionized through fusion or reactions with the charged species (including charged solvent droplets, hydronium ions, or protonated solvent species) in the ESI plume. One of the advantages of using FD-ESI for sample analysis is that the ionization and nebulization processes are separate events; this feature provides independent control over the conditions of the sample solution and the composition of the ESI solvent. By varying the methods of introducing the sample into a FD-ESI source, a number of unique applications have been demonstrated for the analyses of liquid, gas, and solid samples. The ELDI source uses laser irradiation to produce gaseous analyte molecules; the desorbed analyte molecules then join into the electrospray plume, where they are post-ionized through ESI. Because a large amount of energy can be introduced to the analyte through a laser pulse, analytes on solid surfaces are often efficiently desorbed through ELDI. In addition, the high spatial resolution and scanning capability of the laser beam makes ELDI a useful technique for the rapid and continuous characterization of chemical entities directly from the surfaces of HPTLC plates. Ionization of the desorbed molecules through ESI processes may also greatly increase the mass accuracy of the detected ions (relative to MALDI).

ELDI–MS can be used directly to characterize molecules separated on either reversed-phase or normal-phase silica gel HPTLC plates when employing MS or MS/MS in positive- or negative-ion mode. Since the use of organic or inorganic matrices is unnecessary for ELDI analysis, this feature greatly simplifies the sample preparation procedure. Because a large amount of energy can be transferred to the analyte from the UV laser pulse, the molecules on solids having hard surfaces (such as TLC/HPTLC plates) can be efficiently desorbed when using ELDI. Presently high spatial resolution and scanning capability of the laser beam makes ELDI a useful technique for the rapid and continuous characterization of molecules adsorbed onto the surfaces of TLC/HPTLC plates. The desorption and ionization processes of ELDI are performed at atmospheric pressure, and thus it is readily compatible with almost all mass analyzers.

Tandem Mass Spectrometry

Tandem mass spectrometry (MS/MS) is a technique where structural information on sample molecules is obtained using multiple stages of mass selection and mass separation. A prerequisite is that the sample molecules can be transferred into gas phase and ionized intact and that they can be induced to fall apart in some predictable and controllable fashion after the first mass selection step. Multistage MS/MS, or MS^n, can be performed by first selecting and isolating a precursor ion (MS2), fragmenting it, isolating a primary fragment ion (MS3), fragmenting it, isolating a secondary fragment (MS4), and so on, as long as one can obtain meaningful information or the fragment ion signal is detectable. A variety of imaginative modes of tandem MS is described in the literature, but the four most common modes are (a) precursor ion mode (b) product ion mode (also referred to as multi-reaction monitoring, MRM) (c) neutral loss mode, and (d) single reaction monitoring. The product ion mode is the most explorative mode and is typically utilized to get structural information of as many different sample molecules as possible. In precursor ion mode and neutral loss mode, the objective is to detect a certain class of molecules with a common functional group, such as phosphopeptides or different classes of lipids. Selected reaction monitoring (SRM) is universally utilized in combination with online chromatographic separation to quantify a specific compound in a complex (often biological) matrix (Loo 1995; Griffiths 2008).

Present Challenge for Drug Discoveries

The present challenge in drug discovery is to synthesize new compounds efficiently in minimal time with true identification. The trend is toward carefully designed and well-characterized compound libraries because fast and effective synthesis methods easily produce thousands of new compounds. The need for rapid and reliable analysis methods is increased at the same time. The principles of classical planar chromatographic separation combined with ultraviolet (UV) and mass spectrometric (MS) detection methods have resulted in powerful, rapid, easy, low-cost, alternative tools and techniques for qualitative and quantitative analysis of small drug or "drug-like" molecules. High-performance thin-layer chromatography (HPTLC) was introduced and evaluated for fast semiquantitative assessment of the purity of synthetic target compounds. HPTLC methods were compared with the liquid chromatography (LC) methods. Electrospray ionization mass spectrometry (ESI–MS) and atmospheric pressure matrix-assisted laser desorption/ionization MS (AP–MALDI–MS) were used to identify and confirm the product zones on the plate. AP–MALDI–MS was rapid and easy to carry out directly on the plate without scraping. The thinner adsorbent layer, the monolithic UTLC plate provided 20–100 times better sensitivity in MALDI analysis than HPTLC plates. The limits of detection (LODs) went down to as low as picomole range for UTLC–AP–MALDI- and UTLC–DESI–MS. In a comparison of AP–MALDI and vacuum MALDI–MS detection for UTLC plates, desorption from the irregular surface of the plates with the combination of an external AP–MALDI ion source and an ion trap instrument provided less variation in mass accuracy than the vacuum MALDI–TOF instrument.

Notes

High-performance thin-layer chromatography has been an all-time favorite chromatographic technique of phyto/biochemists. Direct interfacing of TLC/HPTLC with MS has always been difficult, but nowadays this has become possible due to TLC/HPTLC–MS coupling. Because of versatility and affordability, TLC/HPTLC is routinely used in the chemical and life sciences. The technology related to TLC/HPTLC has been changed significantly with considerable attention given to detection techniques capable of identifying the components present in separated bands. Coupling of TLC/HPTLC with mass spectrometry (MS) become a very simple and robust separation method with a detector with an ability to detect selectively a wide variety of analytes ranging from trace to ultra-trace levels. Highly specific quantitative TLC/HPTLC is an additional advantage made possible with the coupling of TLC/HPTLC to MS detection. Several quantitative TLC/HPTLC–MS methods reported in the literature have used different methods for producing gas-phase analyte ions directly from the surface of a TLC/HPTLC plate.

References

Alpmann A and Morlock G. Improved online coupling of planar chromatography with electrospray mass spectrometry: extraction of zones from glass plates. Anal Bioanal Chem 2006; 386, 1543–1551.

Anderson RM and Busch KL. Thin-layer chromatography coupled with mass spectrometry: interface to electrospray ionization. J Planar Chromatogr 1998; 11, 336–441.

Aranda M and Morlock G. New method for caffeine quantification by planar chromatography coupled with electrospray ionization mass spectrometry using stable isotope dilution analysis. Rapid Commun Mass Spectrom 2007; 21, 1297–1303.

Banno K, Matsuoka M and Takahashi R. Quantitative analysis by thin-layer chromatography with secondary ion mass spectrometry. Chromatographia 1991; 32, 179–181.

Barber M, Bordoli RSR, Sedgewick D and Tyler AN. Fast atom bombardment of solids (F. A. B.): a new ion source for mass spectrometry. J Chem Soc Chem Commun 1981; 7, 325–327.

Barber M, Bordoli RS, Elliott GJ, Sedgwick RD and Tyler AN. Fast atom bombardment mass spectrometry. Anal Chem 1982; 54, 645A–657A.

Benninghoven A and Sichtermann WK. Detection, identification, and structural investigation of biologically important compounds by secondary ion mass spectrometry. Anal Chem 1978; 50, 1180–1184.

Benninghoven A, Jaspers D and Sichtermann D. Secondary-ion emission of amino acids. Appl Phys 1976; 11, 35–39.

Beranova-Giorgianni S and Desiderio DM (1997). Fast Atom Bombardment Mass Spectrometry of Synthetic Peptides. In: Methods in Enzymology: Solid-Phase Peptide Synthesis Methods in Enzymology 289, published by Academic Press, San Diego.

Brown SM and Busch KL. Direct identification and quantitation of diuretic drugs by fast atom bombardment mass spectrometry following separation by thin-layer chromatography. J Planar Chromatogr 1991; 4, 189–193.

Butz S and Stan HJ. Screening of 256 pesticides in water by thin-layer chromatography with automated multiple developments. Anal Chem 1995; 67, 620–630.

Caprioli RM and Moore WT (1990). Continuous-Flow Fast Atom Bombardment Mass Spectrometry. In Methods in Enzymology: Mass Spectrometry-Methods in Enzymology 193, published by Academic Press, San Diego.

Caprioli, RM, Fan T and Cottrell JS. Continuous-flow sample probe for fast atom bombardment mass spectrometry. Anal Chem 1986; 58, 2949–2954.

Castaing R and Slodzian GJ. Optique corpusculaire – premiers essais de microanalyse par emission ionique secondaire. J Microsc 1962; 1, 395–399.

Cody RB, Laramee JA and Durst HD. Versatile new ion source for the analysis of materials in open air under ambient conditions. Anal Chem 2005; 77, 2297–2302.

Crecelius A, Clench MR, Richards DS, Mather J and Parr V. Analysis of UK-224,671 and related substances by thin-layer chromatography/matrix assisted laser desorption ionisation/time of flight mass spectrometry of small molecules. J Planar Chromatogr 2000; 13, 76–81.

Crecelius A, Clench MR, Richards DS, Evason D and Parr V. Thin-layer chromatography-postsource-decay matrix assisted laser desorption ionisation time-of-flight mass spectrometry. J Chromatogr Sci 2002; 40, 538–549.

Crecelius A, Clench MR, Richards DS and Parr V. Quantitative determination of piroxicam by TLC-MALDI TOFMS. J Pharmaceut Biomed Anal 2004; 35, 31–39.

Dass C. The role of a liquid matrix in controlling FAB-induced fragmentation. J Mass Spectrom 1996; 31, 77–82.

Dreisewerd K, Muthing J, Rohlfing A, Meisen I, Vukelic Z, Peter-Katalinic J, Hillenkamp F and Berkenkamp S. Analysis of gangliosides directly from thin-layer chromatography plates by infrared matrix-assisted laser desorption/ionization orthogonal time-of-flight mass spectrometry with a glycerol matrix. Anal Chem 2005; 77, 4098–4107.

Ekman R, Silberring J, Brinkmalm Ann W and Kraj A (2009). Mass Spectrometry (Instrumentation, Interpretation and Applications), published by John Wiley and Sons, New Jersey.

Ford MJ, Deibel MA, Tomkins BA and Van Berkel GJ. Quantitative thin-layer chromatography/mass spectrometry analysis of caffeine using a surface sampling probe electrospray ionization tandem mass spectrometry system. Anal Chem 2005; 77, 4385.

Galand N, Pothier J and Viel C. Plant drug analysis by planar chromatography. J Chromatogr Sci 2002; 40, 585–597.

Griffiths J. A brief history of mass spectrometry. Anal Chem 2008; 80, 5678–5683.

Günter K. Thin-layer chromatographic enantiomeric resolution via ligand exchange. J Chromatogr 1984; 448, 11–30.

Halpaap H, Krebs KF and Hauck HE. Thin-layer chromatographic and high performance thin-layer chromatographic ready-for-use layers with lipophilic modifications. HRC CC J High Resolut Chromatogr Chromatogr Commun. 1980; 3, 215–240.

Hauck HE and Schulz M. Ultrathin-layer chromatography. J Chromatogr Sci 2002; 40, 550.

Hauck HE and Jost W. Instrumental High Performance, Thin-Layer Chromatography, Proceedings of 2nd International Symposium, Interlaken, Switzerland, 1982; 25–37.

Hauck HE and Jost W. Instrumental High Performance Thin-Layer Chromatography, Proceedings of 3rd International Symposium, Wurzburg, Germany, 1985; 83–91.

Hauck HE and Jost W. Instrumental High Performance Thin-Layer Chromatography, Proceedings of 4th International Symposium, Selvino/Bergamo, Italy, 1987; 241–253.

Heine CE, Holland JF and Watson JT. Influence of the ratio of matrix to analyte on the fast atom bombardment mass spectrometric response of peptides sampled from aqueous glycerol. Anal Chem 1989; 61, 2674–2682.

Herzog RFK, Viehbock FP. Ion source for mass spectrography. Phys Rev 1949; 76, 855–856.

Hofmann S. Sputter-d depth profiling for thin-film analysis. Philos Trans R Soc A 2004; 362, 55–75.

Honig RE. Sputtering of surfaces by positive ion beams of low energy. J Appl Phys 1958; 29, 549–555.

Honig RE. The development of secondary ion mass spectrometry (SIMS): a retrospective. Int J Mass Spectrom Ion Process 1985; 66, 31–54.

Honig RE. Stone-age mass spectrometry: the beginnings of "SIMS" at RCA laboratories, Princeton. Int J Mass Spectrom Ion Process 1995; 143, 1–10.

Jautz U and Morlock G. Efficacy of planar chromatography coupled to (tandem) mass spectrometry for employment in trace analysis. J Chromatogr A 2006; 1128, 244–250.

Jones EA, Fletcher JS, Thompson CE, Jackson DA, Lockyer NP and Vickerman JC. ToF-SIMS analysis of bio-systems: are polyatomic primary ions the solution? Appl Surf Sci 2006; 252, 6844–6854.

Kauppila TJ, Talaty N, Salo PK, Kotiah T, Kostiainen R and Cooks RG. New surface for desorption electrospray ionization mass spectrometry: porous silicon and ultra-thin layer chromatography plates. Rapid Commun Mass Spectrom 2006; 20, 2143–2150.

Kertesz V, Ford MJ and Van Berkel GJ. Automation of a surface sampling probe/electrospray mass spectrometry system. Anal Chem 2005; 77, 7183–7189.

Liebl H. Ion microprobe mass analyzer. J Appl Phys 1967; 38, 5277–5283.

Lin HY, Gonyea G, Killeen S and Chowdhury SK. Negative fast atom bombardment ionization of aromatic sulfonic acids: unusual sample–matrix interaction. Rapid Commun Mass spectrum 2000; 14, 520–522.

Loo JA. Bioanalytical mass spectrometry: many flavour to choose. Bioconjug Chem 1995; 6, 644–665.

Luftmann H. A simple device for the extraction of TLC spots: direct coupling with electrospray mass spectrometer. Anal Bioanal Chem 2004; 378, 964–968.

Mack M, Hauck HE and Herbert H. Enantiomeric separation in TLC with the new HPTLC precoated plate CHIR with concentrating zone. J Planar Chromatogr 1988; 1, 304–308.

Mallis. LM, Wang HM, Loganathan D and Linhardt RJ. Sequence Analysis of Highly Sulfated, Heparin-Derived Oligosaccharides Using Fast Atom Bombardment Mass Spectrometry. Anal Chem 1989; 61, 1453–1458.

Mc Phail D. Applications of secondary ion mass spectrometry (SIMS) in materials science. J Mater Sci 2006; 41, 873–903.

McDonnell LA and Heeren RMA. Imaging mass spectrometry. Mass Spectrom Rev 2007; 26, 606–643.

Mincsovics E, Kiss BD, Morovjan G, Nemes KB and Klebovich I. A new tool in metabolism research-combination of OPLC on-line radioactivity detection with HPLC-RD technique. J Planar Chromatogr 2002; 14, 312–317.

Morlock G and Schwack W. Determination of isopropylthioxanthone (ITX) in milk, yoghurt and fat by HPTLC-FID, HPTLC-ESI/MS and HPTLC-DAT/MS. Anal Bioanal Chem 2006; 385, 586–595.

Morlock G and Ueda Y. New coupling of planar chromatography with direct analysis in real time spectrometry. J Chromatogr A 2007; 1143, 243–251.

Morovjan G, Dalmadi-Kiss Klebovich BI and Mincsovics E. Metabolite analysis, isolation and purity assessment using various liquid chromatographic techniques combined with radioactivity detection. J Chromatogr Sci 2002; 40, 603–608.

Moseley MA, Deterding LJ, Tomer KB and Jorgenson JW. Nanoscale packed-capillary liquid chromatography coupled with mass spectrometry using a coaxial continuous-flow fast atom bombardment interface. Anal Chem 1991; 63, 1467–1473.

Nakamura K, Suzuki Y, Goto-Inoue N, Yoshida-Noro C and Suzuki A. Structural characterization of neutral glycospingolipids by thin-layer chromatography coupled to matrix-assisted laser desorption/ionization quadrupole ion trap time-of-flight MS/MS. Anal Chem 2006; 78, 5736–5743.

Nyiredy Sz (2001). Planar Chromatography. A Retrospective View for the Third Millennium, published by Springer Scientific, Budapest, Hungary.

Oechsner H and Gerhard WA. Method for surface analysis by sputtered neutrals. Phys Lett 1972; 40, 211–212.

Oka H, Ikai Y, Kawamura N, Hayakawa J, Masuda K, Harada K.-I and Suzuki M. Development of a condensation technique for thin-layer chromatography/fast-atom bombardment mass spectrometry of non-visible compounds. Rapid Commun Mass Spectrom 1992; 6, 89–94.

Oka H, Ikai Y, Ohno T, Kawamura N, Hayakawa J, Harada K and Suzuki M. Identification of unlawful food dyes by thin-layer chromatography-fast atom bombardment mass spectrometry. J Chromatogr A 1994; 674, 301–307.

Orinak A, Orinakova R, Arlinghaus HF, Vering and Hellweg G. Post-chromatographic TOF-SIMS identification of diterpenes. Surf Interface Anal 2006; 38, 599–603.

Parent A, Anderson TM, Michaelis DJ, Jiang G, Savage PB and Linford MR. Analysis of organic halides and amines on TLC plates. Appl Surf Sci 2006; 252, 6746–6749.

Parent AA, Anderson TM, Michaelis DJ, Jiang G, Savage PB and Linford MR. Direct ToF-SIMS analysis of organic halides and amines on TLC plates. Appl Surf Sci 2006; 252, 6746–6749.

Peng S, Edler M, Ahlmann N, Hoffmann and Franzke TJ. A new interface to couple thin-layer chromatography with laser desorption/atmospheric pressure chemical ionization mass spectrometry for plate scanning. Rapid Commun Mass Spectrom 2005; 19, 2789–2793.

Poole CF. Thin-layer chromatography: challenges and opportunities. J Chromatogr A 2003; 1000, 963.

Poole CF and Dias NC. Practitioners guide to method development in thin-layer chromatography. J Chromatogr A 2000; 892, 123–142.

Prosek M, Milivojevic L, Krizman M and Fir M. On-line TLC-MS. J Planar Chromatogr 2004; 17, 420.

Renger B. Quantitative planar chromatography as a tool in pharmaceutical analysis. J AOAC Int 1993; 76, 7–13.
Renger B. Planar chromatographic procedures in pharmaceutical analysis. J AOAC Int 2001; 84, 1217.
Salo PK, Vilmunen S, Salomies H, Ketola RA and Kostiainen R. Two-dimensional ultra thin-layer chromatography and atmospheric pressure matrix-assisted laser desorption/ionization mass spectrometry in bioanalysis. Anal Chem 2007; 79, 2101–2108.
Shu-Yao L, Min-Zong H, Chang HC and Shiea J. Using electrospray-assisted laser desorption/ ionization mass spectrometry. Anal Chem 2007; 79, 8789–8795.
Sigmund P and Claussen C. Sputtering from elastic-collision spikes in heavy-ion-bombarded metals. J Appl Phys 1981; 52, 990–993.
Sodhi RNS. Time-of-flight secondary ion mass spectrometry (TOF-SIMS): versatility in chemical and imaging surface analysis. Analyst 2004; 129, 483–487.
Somsen GW, Morden W and Wilson ID. Planar chromatography coupled with spectroscopic techniques. J Chromatogr A 1995; 703, 613–665.
Sweetman BJ and Blair IA. 3-Nitrobenzyl alcohol has wide applicability as a matrix for FAB-MS. Biomed Environ Mass Spectrom 1988; 17, 337–340.
Szepesi G and Nyiredy Sz. Planar chromatography: Current status and future perspectives in pharmaceutical analysis-I. Applicability, quantitation and validation. J Pharm Biomed Anal 1992; 10, 1007–1015.
Takats Z, Cotte-Rodrıguez I, Talaty N, Chen H and Cooks RG. Direct, Trace level detection of explosives on ambient surfaces by desorption electrospray ionization mass spectrometry. Chem Commun 2005; 15, 1950–1952.
Tames F, Watson ID, Morden W and Wilson ID. Detection and identification of morphine in urine extracts using thin-layer chromatography and tandem mass spectrometry. J Chromatogr B 1999a; 729, 341–346.
Tames F, Watson ID, Morden W and Wilson ID. Detection and identification of morphine in urine extracts using thin-layer chromatography tandem mass spectrometry. J Chromatogr B Biomed Sci Appl 1999b; 729, 341–346.
Tames F, Watson ID, Morden WE and Wilson ID. Analysis of benzodiazepines by thin-layer chromatography and tandem mass spectrometry. J Planar Chromatogr 2000; 13, 432–436.
Thomson JJ. On rays of positive electricity. Philos Mag Ser 6 1910; 20, 752–767.
Van Berkel GJ, Sanchez AD and Quirke JME. Thin-layer chromatography and electrospray mass spectrometry coupled using a surface sampling probe. Anal Chem 2002; 74, 6216–6223.
Van Berkel GJ, Tomkins BA and Kertesz V. Thin-layer chromatography/desorption electrospray ionization mass spectrometry: Investigation of goldseal alkaloids. Anal Chem 2007; 79, 2778–2789.
Wagner H, Bladt S and Zgainski EM (1984). Plant Drug Analysis, published by Springer-Verlag, Berlin.
Williams JP, Patel VJ, Holland R and Scrivens JH. The use of recently described ionisation techniques for the rapid analysis of some common drugs and samples of biological origin. Rapid Commun Mass Spectrom 2006; 20, 1447–1456.
Wilson ID. The state of the art in thin-layer chromatography-mass spectrometry: a critical appraisal. J Chromatogr A 1999; 856, 429–442.
Wilson ID and Modern W. Advances and applications in the use of HPTLC-MS-MS. J Planar Chromatogr Mod-TLC 1996; 9, 84–91.
Zlatkis A and Kaiser RE (1977). HPTLC high performance thin-layer chromatography, published by Elsevier, Amsterdam, Netherlands.

Chapter 16
TLC/HPTLC with Direct Mass Spectrometric Detection: A Review of the Progress Achieved in the Last 5 Years

Jurgen Schiller, Beate Fuchs, Kristin Teuber, Ariane Nimptsch, Kathrin Nimptsch, and Rosmarie Süß

Abstract Thin-layer chromatography (TLC) is a widely used, fast and inexpensive method. Even minor constituents of complex mixtures can be often characterized. Unfortunately, the unambiguous assignment of the spots to defined compounds is often difficult because common staining methods normally reveal only (at best) a substance class, not a specific compound. For instance, in the case of phospholipids, the lipid class, but not the detailed fatty acyl composition, can be determined (normal phase TLC). Nowadays mass spectrometry (MS) seems to be the most suitable method for analyte characterization due to its high sensitivity and mass accuracy. This became possible by the invention of soft ionization methods such as matrix-assisted laser desorption and ionization time-of-flight (MALDI–TOF) MS and electrospray ionization (ESI) although many further desorption techniques are available nowadays. This review summarizes the so far available knowledge about direct TLC/MS couplings and gives an overview about selected compounds that could be successfully analyzed.

Chromatographic Methods and Product Identification

Liquid chromatographic methods, especially high-performance liquid chromatography (HPLC) and planar chromatography [in particular thin-layer (TLC) or high-performance thin-layer chromatography (HPTLC)], are indispensable tools of modern analytical chemistry (Hahn-Deinstrop 2006). Many different qualitative as well as quantitative applications are illustrated in detail in this book. While HPLC can be used for the analysis of large biomolecules such as proteins or polysaccharides that cannot be appropriately characterized by TLC, TLC is a very

J. Schiller (✉), B. Fuchs, K. Teuber, A. Nimptsch, K. Nimptsch, and R. Süß
University of Leipzig, Medical Department, Institute of Medical Physics and Biophysics, Härtelstr, 16/18, 04107 Leipzig, Germany

MM. Srivastava (ed.), *High-Performance Thin-Layer Chromatography (HPTLC)*,
DOI 10.1007/978-3-642-14025-9_16, © Springer-Verlag Berlin Heidelberg 2011

common separation technique in (synthetic) organic chemistry and natural product chemistry, where normally only small amounts of the samples are available. This holds because TLC offers in addition to the simple performance and the relatively low costs, many important advantages in comparison to HPLC (Hahn-Deinstrop 2006; Fuchs et al. 2009a):

- TLC is a very robust system and to a much lesser extent affected by impurities potentially present in the sample. Thus, TLC may also be used for the analysis of "suspicious" samples (for instance, crude tissue or food extracts) that might easily damage or plug the column of an HPLC system.
- HPLC columns are normally rather expensive and are, thus, many times reused subsequent to careful washing to remove previous analytes. However, it can never be completely ruled out that there is some residual material on the column. Thus, it is often more difficult to certify an HPLC assay than a TLC assay because TLC does not provide any "memory" effects as a completely new stationary phase is used in all cases.
- The consumption of solvents is much lower than in the case of HPLC. This makes TLC less expensive and particularly less harmful and more environment friendly.
- Although nowadays there are also "multiplexing" HPLC systems available (Woodbury et al. 1995) that allow the simultaneous analysis of several samples in parallel, the simple analysis of a multitude of samples is also a clear advantage of TLC.

There is hardly a class of small molecules that have not yet been analyzed by TLC, and thus this review will not deal with improvements of the achievable separation quality, neither by optimizing the stationary phase nor the solvent system nor sample application.

However, separation is only one (unequivocally very important) step of successful analysis, and the detection of the different analytes on the TLC plate and their structural elucidation is also very important. This particularly holds because the retardation factor (R_f value) by which a compound is characterized is not very well reproducible but is influenced by many parameters such as the solvent composition, the used TLC plates, the relative humidity, and the temperature. Analyte identification is a particular problem if no reliable reference compounds are available (Hahn-Deinstrop 2006), and this is often very difficult if unknown natural products have to be analyzed (Marderosian and Chao 1974). Identification of TLC-separated spots can be basically performed in the following ways (Geiss 1987).

Directly on the TLC Plate with and Without Any Derivatization

This method can be only applied to as such colored compounds (e.g., plant or synthetic dyes) or molecules that are characterized by an intense UV absorption or that give rise to fluorescence. This approach has the considerable advantage that the analyte can be directly visualized and is not destroyed by any harsh staining process and can be subsequently reanalyzed by any suitable analytical method. Obviously, however, the majority of compounds cannot be detected by this most simple approach because they are not colored or lack UV absorbance. Finally, the achievable sensitivity by this approach is normally quite low.

The use of characteristic dyes to monitor specific compounds is a commonly used approach particularly as in that way information on the presence of defined functional chemical groups of the analyte is easily available (Nakamura and Pisano 1976). For instance, diphenylhydrazine (DNPH) is a well-known and useful derivatization agent of carbonyl groups present in aldehydes or ketones, whereas ninhydrin reacts specifically with amino groups. A commonly used, but completely unspecific approach is the use of moderately concentrated sulfuric acid and subsequent heating (Hahn-Deinstrop 2006). Under these conditions, virtually all organic compounds are destroyed and the resulting black spots allow the identification of colorless compounds directly on the TLC plate. This is particularly used in the field of lipid or fatty acid analysis, i.e., if compounds with relatively high carbon contents are of interest (Klein et al. 1998).

Analysis by Reelution from the TLC Plate

Although this approach is applicable to all compounds and seems, thus, most straightforward, this method is obviously tedious and time consuming and bears the particular risk of losing large amounts of the analytes. On the other hand, this approach offers the considerable advantage that all common spectroscopic methods [such as nuclear magnetic resonance spectroscopy (NMR) or mass spectrometry (MS)] can be applied. A very useful tool that helps to extract samples from a TLC plate in a very convenient and highly reproducible manner is commercially available from CAMAG (http://www.camag.com/v/products/tlc-ms) and based on a recently described method (Luftmann et al. 2007).

Spectroscopic Analysis Directly on the TLC Plate

Clearly this approach does not suffer from potential losses upon the reextraction process of the analyte and retains the achieved separation quality completely (components with strongly different R_f values cannot be very accurately re-eluted and analyzed by the method described above). Several different spectroscopic techniques have been already successfully applied and will be shortly discussed here.

NMR Spectroscopy

NMR spectroscopy is a powerful (but unfortunately not very sensitive) analytical tool. As ^1H is the most sensitive nucleus (Berger and Braun 2004), nearly all the so far available data are based on ^1H NMR. There are only very few papers dealing with this topic: Organic reaction products could be identified directly (without the need of previous elution with solvents) from separated TLC spots by high-resolution magic angle spinning (HR MAS) NMR spectroscopy (Bradley and McLaughlin 2007; Wilson et al. 1997). HR MAS uses fast sample rotation (a few thousand Hertz) in the magic angle (54.7°) to reduce the line-widths of the individual resonances but does not require the complex pulse sequences of solid state NMR, i.e., the less complex pulse sequences of high-resolution NMR can be used (Schiller et al. 2004a). The spot of interest is scraped from the TLC plate, transferred to an HR MAS sample rotor (that are now available from nearly all manufacturer of NMR spectrometers) and suspended in a small amount of deuterated solvent. Particularly, the use of D_2O instead of H_2O helps significantly to suppress the solvent signal. There are nowadays rotors with special inserts available that require only about 20 μl of sample. One problem is that the stationary phase, i.e., the –OH groups of the silica gel, gives a broad background signal near about 4.6 ppm. If the spectra are recorded with little care, this broad resonance dominates the spectra and reduces the sensitivity to detect resonances of interest. However, the intensities of broad resonances can be considerably reduced by the so-called Carr–Purcell–Meiboom–Gill (CPMG) pulse sequence that eliminates compounds with short relaxation times that would lead to broad resonances (Berger and Braun 2004). One-dimensional ^1H NMR spectra with a good signal-to-noise ratio (*S/N*) can be obtained in a few minutes, while the required instrument times are of the order of several hours if 2D-NMR is required. According to our best knowledge, all TLC/NMR applications were so far exclusively based on ^1H NMR and no attempts were so far made to record NMR spectra of less sensitive nuclei such as ^{13}C (Berger and Braun 2004).

IR and Raman Spectroscopy

Although infrared (IR) is normally much more common (Günzler and Gremlich 2002) than Raman spectroscopy, both techniques clearly offer complementary information: While IR is particularly suitable to detect polar functional groups such as C=O or O–H, Raman spectroscopy is more useful for the characterization of relatively apolar groups such as C–H. In comparison to UV light that is very common for monitoring spots of aromatic or unsaturated compounds on a TLC plate, IR irradiation has a higher wavelength, i.e., a lower energy and thus penetrates deeper into the TLC sample plate. This clearly reduces the risk of "loosing" compounds at the inner of the silica gel layer. Recently it could be shown that amino acids such as L-lysine can be quantitatively determined by near IR (NIR) spectroscopy directly from the TLC plate. Commercially available apple juice served as the sample and cellulose TLC with *n*-butanol/acetic acid/water (8:4:2, v/v/v) as mobile phase was used. Under these conditions, amino acids could be determined with a detection level of 10 μg/ml (Heigl et al. 2006).

As already stated above, IR detects primarily polar compounds, while Raman spectroscopy is more suitable for apolar compounds. Thus, the ingredients from the small berry fruit of *Evodia rutaecarpa* (well known as "Wu-Zhu-Yu" in Chinese traditional medicine with anti-inflammatory activity) were analyzed at the microgram level by HPTLC and surface enhanced Raman spectrometry (SERS). It is particularly remarkable that no standard samples were needed (Zhang et al. 2007a) but the achieved absorptions could be directly used (Günzler and Gremlich 2002).

Mass Spectrometry

The application of the spectroscopic approaches mentioned above is quite limited because they do not provide high sensitivity and can be only used if the analyte is available in huge amounts. In contrast, mass spectrometric (MS) techniques offer much higher sensitivities and are, thus, more often used for the evaluation of TLC-separated compounds. There are basically two different approaches. In the first technique, the sample is re-eluted from the TLC plate and later characterized by MS. Electrospray ionization (ESI) MS is normally used because it seems the most versatile ionization technique, i.e., it enables the ionization of nearly all compounds. The second approach is based on MS desorption methods where the ions can be directly generated from a solid surface.

MS Methods Based on Previous Extraction of the Analytes

A plunger-based extraction interface (now commercially available as the "ChromXtract" from the CAMAG company) used with an HPLC pump was shown to provide good results for quantitative TLC/ESI MS on HPTLC silica gel plates regarding repeatability of the MS spectra and the achievable limit of detection (LOD) and the limit of quantification (LOQ). This was shown for harmane, a heterocyclic aromatic amine, as selected model compound (Jautz and Morlock 2006). Modifications of the device enabled complete, highly reproducible extraction of analytes from glass-backed as well as aluminum-backed TLC and HPTLC plates, layers with thickness up to 100 μm and different stationary phases (Alpmann and Morlock 2006). The "ChromXtract" was also used in the simultaneous determination of caffeine, ergotamine, and metamizole in a solid pharmaceutical formulation by HPTLC (silica gel 60F layer, ethyl acetate/methanol/ammonia (90:15:1, v/v/v) as the mobile phase) with UV absorption densitometry for caffeine and metamizole and fluorescence densitometry for ergotamine. Additional mass confirmation was performed by single quadrupole [a very common mass analyzer (Hillenkamp and Peter-Katalinić 2007; Fuchs et al. 2008a)] MS in the positive ESI full scan mode for caffeine and ergotamine, and negative ion mode mass spectra were used for metamizol (Aranda and Morlock 2007). A caffeine standard with (deuterated) caffeine-d_3 was used for correction of the ChromXtractor plunger positioning, and this resulted in significantly improved data accuracy. Although this is a very promising approach, our focus will be on the use of desorption techniques. Readers primarily interested in TLC/MS by reextraction of the analyte are referred to the excellent papers by Gerda Morlock and her group members (Morlock and Sherma 2009; Sherma and Morlock 2008).

MS Desorption Methods

Although there are nowadays a lot of desorption MS techniques, only two of them will be considered in this review, namely desorption electrospray (DESI) (Miao and Chen 2009) MS and matrix-assisted laser desorption ionization (MALDI) (Hillenkamp and Peter-Katalinić 2007) MS. Although MALDI MS will be discussed in more detail, a detailed discussion of the basics and applications of DESI MS is beyond the scope of this review. The reader primarily interested in this technique is referred to the excellent papers by Zoltan Takats (Chen et al. 2005). It should be particularly noted that this technique is not only suitable to characterize TLC plates but may also be used to characterize native tissue samples (Schäfer et al. 2009). Due to the rapid performance of this technique [direct analysis in real time (DART)], it seems likely that applications "in the operating room" will be developed in the near future. For instance, this method would be very helpful for the differentiation between normal and cancer tissue. A new HPTLC/DART–TOF MS coupling method (that is already commercially available from IonSence: http://www.ionsense.com/dart-svp.php) was shown to be successful for identification and qualitative purposes with detection limits in the low ng/zone range for isopropylthioxanthone (ITX) as a selected example. ITX is commonly used as photoinitiator in the offset print and was previously found as an impurity in baby milk products. Quantitative analysis could be significantly improved using isotope labeled standards, and spatial resolution using an in-house plate holder system was better than 3 mm (Morlock and Ueda 2007).

TLC/MALDI Combination

Although combined TLC/MALDI is already commercially available from the Bruker Daltonics Company (http://www.bdal.de/products/system-solutions/tlc-maldi.html), only a few applications have been described so far (Fuchs et al. 2009a). These will be summarized in this review because this topic is currently under intense research and one "hot topic" in the TLC community. This interest is among other reasons also coming from the fact that MALDI MS is increasingly used as an "imaging" technique for thin tissue slices, and much research work was performed to improve the quality and reproducibility of the achievable images (Cornett et al. 2007). As a thin tissue slide (particularly subsequent to dehydration), where the analyte of interest is dispersed in a "matrix" of other molecules and a TLC plate (where the analyte is dispersed in the stationary phase of silica gel) are similar samples, some of the methods developed for tissue imaging may also be used to analyze a lane on a TLC plate and to obtain a "scan" of the TLC surface. Directly combined TLC/MALDI MS confers several important advantages as already discussed above: (a) the mass spectrum of a given compound is much more indicative than the corresponding, rather difficult to reproduce R_f value and (b) significant spatial resolution may be obtained by MALDI MS as the laser beam has a focal spot of the order of 100 μm or even smaller (Hillenkamp et al. 1975). This allows the identification of spots that cannot be differentiated by visual inspection of the TLC plate or by scraping off the spots of interest. Although a further decrease in the laser spot size would help to increase the achievable resolution, one must also consider that the spot size correlates with the achievable sensitivity. It seems that 100 μm is a good compromise between resolution and sensitivity.

The direct combination of TLC with MS detection has been pursued for about 30 years and is driven by the aim of obtaining a detailed TLC "image" without compromising TLC resolution and MS sensitivity. Two reviews that summarize the specifics of the coupling between MALDI and planar as well as column chromatography are available in (Gusev 2000; Busch 2004).

Several different approaches to combine TLC with MS detection have been suggested:

- Scraping the spot of the analyte from the TLC plate and extracting the compound into a suitable solvent. The extracted analyte can be later analyzed by MS in the conventional manner. Although this method is very useful and provides the desired information, this approach is obviously tedious and time-consuming. Additionally, compounds with very similar R_f values can be hardly differentiated under these conditions.
- Analysis of a small piece of the TLC plate cut from the whole plate. MS analysis of the spot of interest is subsequently performed in the presence of the stationary phase – silica gel in the majority of cases. This method has the advantage that irregularities of the TLC plate play the smaller roles the smaller the size of the TLC plate.
- Placing the intact total TLC plate into the mass spectrometer for direct analysis of the entire TLC plate so that two-dimensional scanning can be achieved.

Independent of the sample size, the (most common UV) MALDI MS method requires two further important requirements: (a) the UV-absorbing matrix must be applied onto the TLC plate and (b) as the UV laser does not penetrate deeply into the TLC plate, the analyte must be extracted from the inner of the sample plate to the TLC surface. Both conditions must be fulfilled without major spreading of the analyte during the addition of the solvents, i.e., without compromising the achieved quality of chromatographic separation. Both important prerequisites were discussed in more detail in (Mehl et al. 1997).

There are basically five different methods how the MALDI matrix solution may be applied onto the TLC plate (Wilson 1999).

- Adding small droplets of the matrix solution onto the (previously marked) spots of interest by a common pipette.
- Using a solvent or a solvent mixture with a relatively high surface tension resulting in rather small spots when applied to the TLC plate (mixtures of water and acetonitrile were already successfully used in this context) (Nakamura et al. 2006).
- Spraying the matrix solution onto the TLC plate (e.g., using an electrospray equipment). Instruments based on this approach are nowadays already commercially available in the context of MS imaging (for instance, see the "ImagePrep" device from Bruker Daltonics: http://www.bdal.com/products/robotics-hplcs-api-sources/imageprep.html).
- Brushing the TLC plate with a "supersaturated" solution of the matrix (Mowthorpe et al. 1999).
- "Pressing" matrix crystallized on a special surface (e.g., a stainless steel block) onto the TLC plate.
- In order to overcome problems with surface irregularities of the TLC plate (and the resulting mass inaccuracies), the use of a "hybrid plate" was also suggested (Mehl and Hercules 2000): After TLC separation, the analyte spots are eluted from the silica layer to the MALDI layer. Although this is a rather indirect approach, it was shown that the detection limits (in the femtomole range) could be significantly improved under these conditions (Mehl and Hercules 2000).

Despite some sophisticated approaches that make use of very special TLC plates, the most common method is to fix a common TLC plate with conductive adhesive tape onto a standard MALDI target and to load it directly into the mass spectrometer. Note that the back of the TLC plate is very important. As electric conductivity is required to allow desorption of ions from the TLC plate, glass backs may not be used as they are not conductive. The simplest way to overcome this problem is to use commercially available TLC plates with aluminum backs.

How Does MALDI–TOF MS Work?

Although other soft-ionization methods [in particular electrospray (ESI)] are also available, this chapter focuses exclusively on the MALDI method. Although there is no absolute need to combine the MALDI ion generation (the "ion source") with a time-of-flight (TOF) mass analyzer (Fuchs et al. 2008a), we will focus nearly exclusively on this combination because it is most often used and the majority of commercially available MALDI MS devices are equipped with this type of mass analyzer.

MALDI MS was developed at the end of the 80s of the last century and is nowadays widely used. A detailed survey of the different applications of MALDI–TOF MS is provided in the timely book by Hillenkamp and Peter-Katalinić (2007). A scheme of ion generation and mass analysis by MALDI–TOF MS is provided in Fig. 16.1. MALDI–TOF MS is based on the utilization of an (in most cases) ultraviolet-absorbing matrix. Although there are also IR lasers in use nowadays that require completely different matrix compounds (e.g., glycerol due to its intense absorption in the IR range) (Fuchs and Schiller 2009a), UV matrices will be primarily discussed here because nearly all commercially available MALDI devices are equipped so far with N_2 lasers emitting at $\lambda = 337$. The matrix fulfills two important tasks: (a) it absorbs the energy emitted by the laser and (b) prevents the aggregation of the analyte molecules that would otherwise result in cluster ion formation and unnecessarily complicate the spectra (Schiller et al. 2007a).

When the pulsed laser beam hits the sample (normally cocrystals between the matrix and the analyte), its energy is absorbed by the matrix that is present in a typically 50–1,000-fold excess over the analyte of interest. Consequently, the matrix is vaporized, carrying intact analyte molecules into the vapor phase. During the expanding process of this gas cloud, ions (e.g., H^+ and Na^+ or other alkali metal ions) are exchanged between the matrix and the analyte leading to the formation of charged analyte molecules. These ions are called "adducts" or "quasimolecular ions".

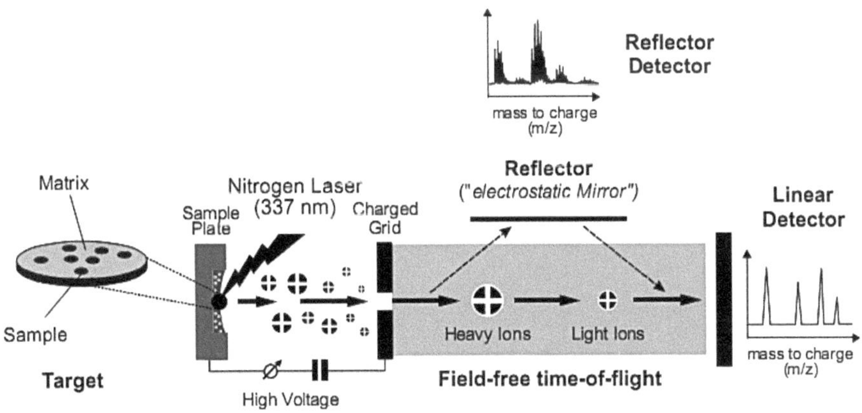

Fig. 16.1 Schematic representation of events occurring in a conventional MALDI–TOF mass spectrometer. The analyte–matrix mixture is evaporated and ionized by laser irradiation. These ions are accelerated in an electric field and then allowed to drift in a field-free pathway. Over this distance, a separation between low-mass and high-mass ions occurs. The "time of flight" (t) depends on the flight path (L), the mass of the ion (m), its energy (e U), and the number of charges (z). Reprinted with modifications and with permission from Elsevier (Schiller et al. 2004b)

After being formed, ions are accelerated in a strong electric field (typically of the order of 20 kV). After passing a charged grid, the ions are drifting freely over a field-free space where mass separation is achieved. Low mass ions arrive at the detector in a shorter time than high-mass ions (Schiller et al. 2004b). There is no absolute need to combine a MALDI ion source with a "TOF" detector. However, as MALDI is often used for the detection of large molecules, the TOF detector is very popular because it has a nearly unlimited mass range. An additional reason is the pulsed ion generation of MALDI that is most suitable for the TOF detector (Hillenkamp and Peter-Katalinić 2007).

Besides cation generation, anions may also be generated by abstracting H^+ or Na^+ from the analyte. The ratio between cations and anions is determined by the (gas phase) acidities of the analyte and the matrix (Fuchs and Schiller 2009a). Positive and negative ions can be simply differentiated by inverting the direction of the applied electric field.

Although MALDI–TOF MS is primarily used for the analysis of analytes with higher masses in the range of several kilodaltons, it may also be used for small molecules that will be the focus of the present review. There is, however, one problem in the low mass range: The matrix itself tends to undergo photochemical reactions and the resulting products may interfere with the analytes of interest (Peterson 2007). Therefore, some approaches were suggested making no use of conventional (small organic molecule) matrices. However, these approaches are currently neither widely accepted nor widely used. For a review see Peterson (2007). The chemical structures of some selected important UV MALDI matrix compounds that are mentioned in this chapter are given in Fig. 16.2.

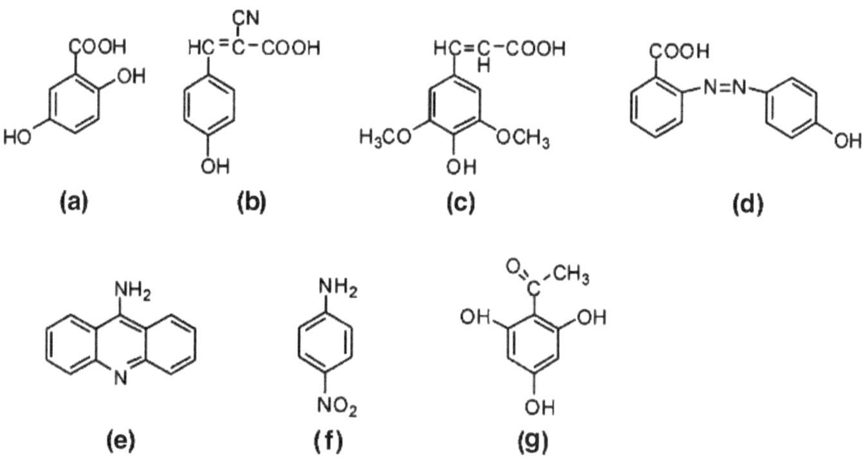

Fig. 16.2 Chemical structures of important, frequently used MALDI UV matrices: (**a**) 2,5-dihydroxybenzoic acid (DHB); (**b**) α-cyano-4-hydroxycinnamic acid (CHA); (**c**) sinapinic acid (SA); (**d**) 2-(4-hydroxyphenylazo)benzoic acid (HABA); (**e**) 9-aminoacridine (9-AA); (**f**) *p*-nitroaniline (PNA); (**g**) 2,4,6-trihydroxyacetophenone (THA). Note that 9-AA, THA, and PNA are particularly useful for recording negative ion mass spectra

Quantitative MALDI–TOF MS

The accuracy of quantitative information available from MALDI–TOF mass spectra is often questioned due to the irregularities of the cocrystals between the matrix and the analyte. Regarding quantitative data analysis, chromatographic and mass spectrometric methods ("LC/MS") that make use of solutions provide obviously more accurate data. By definition, solutions always show a homogeneous distribution of the analyte and there are no inhomogeneities. Nevertheless, there are an increasing number of papers showing that carefully recorded MALDI mass spectra may be throughout quantitatively analyzed (Duncan et al. 2008). It is, however, rather difficult to provide absolute data as the detection limits of different compounds may differ significantly. For instance, the detection limit of lipids is of the order of 20 pg (Gellermann et al. 2006) corresponding to about 25 fmol if 800 g/mol is assumed as a typical molecular mass of a lipid.

These data hold for the use of 2,5-dihydroxybenzoic acid (DHB) as matrix and an isolated lipid sample in pure (nearly salt-free and in the absence of impurities) organic solvents. The presence of salts, detergents, other lipids, etc. may significantly increase the detection levels (Zschörnig et al. 2006). Note that these considerations do not only hold for lipids but for other substances, too. Another problem of mixture analysis is caused by the different detectabilities of the individual compounds (Schiller et al. 2007b) that depend significantly on the molecular structure. Therefore, the previous separation of a mixture into the individual compounds is normally required in order to obtain reliable data (Schiller et al. 2007b).

Characterization Using Direct TLC/MALDI Coupling

This chapter gives an overview which compounds were recently investigated by combined TLC/MALDI. Of course, the focus of this survey will be on smaller molecules because TLC is particularly useful for the separation of such molecules (Sherma 2008). It is not the aim of this chapter to give a complete survey but to provide some selected examples for the illustration of the capabilities and limitations of this approach.

Small Proteins and Peptides

As the traditional focus of MALDI is the analysis of proteins and peptides ("proteomics") (Yates et al. 2009), peptides were historically used as the first components to evaluate the capabilities and limitations of combined TLC/MALDI. Already Gusev et al. (1995a) investigated the suitability of various MALDI matrices for the detection of smaller proteins (bradykinin, angiotensin, and enkephalin that possess molecular masses between about 500 and 1,500 Da) on silica gel and cellulose TLC plates. These peptides are particularly important because they are widely used as calibrants for MALDI–TOF mass spectra (Hillenkamp and Peter-Katalinić 2007).

Absolute detection limits between about 2–4 ng (on the TLC plate) could be established using 2-(4-hydroxyphenylazo)benzoic acid (HABA) as matrix that provided the very best homogeneity with the analytes, whereas ferulic acid and sinapinic acid (cf. Fig. 16.2 for chemical structures) were less sensitive but showed a higher salt tolerance. Analyte fragmentation was comparable to standard MALDI (on a stainless steel target). The problem of analyte spreading upon applying the matrix solution to the TLC plate was solved by the same authors (Gusev et al. 1995b) in an interesting way. A matrix layer was first generated on a stainless steel target, from which it was transferred to the TLC plate of interest. The transfer was accomplished by pressing the matrix onto the previously wetted TLC plate. Subsequent to this preparation, the TLC plate was directly inserted into the MS device. Under these conditions, spatial resolutions of about 250–500 μm and absolute detection limits in the picogram range could be obtained even for larger peptides (Gusev et al. 1995b). Although this is remarkable, TLC/MALDI has experienced very little interest in the field of peptide analysis. This is most probably caused by the high establishment of LC/MS methods in this field (Mueller et al. 2007).

The direct investigation of TLC-separated tryptic digests of phosphopeptide mixtures was also described (Kochin et al. 2006). Detection limits in the femtomolar range and a superior separation quality in comparison to the established (electrophoretic) protocols could be obtained. Regarding the analysis of phosphopeptides, a combined two-dimensional planar electrochromatography/TLC with MALDI–TOF mass spectrometric detection was suggested as a new approach (Panchagnula et al. 2007). Phosphopeptides migrate more slowly in the first dimension, based on their anionic phosphate residues and the resulting charge, and certain predominantly acidic phosphopeptides even migrate in the opposite direction, relative to the bulk of the peptides. Phosphopeptides were further distinguished based on their hydrophilicities in the second dimension. This permits a restricted region of the plate to be directly interrogated for the presence of phosphopeptides by MALDI–TOF MS without the interference of residual proteins.

Home-made TLC plates are sometimes more advantageously used in certain applications than commercially available ones. For instance, it could be shown that a monolithic porous polymer layer significantly improves both the TLC separation of larger peptides in comparison to commercially available precoated TLC plates and their detection by MALDI–TOF MS (Bakry et al. 2007). Fifty to two hundred micrometer thin poly (butyl methacrylate-*co*-ethylene dimethacrylate) layers were prepared in situ using UV-initiated polymerization. These monolithic thin layers could be also prepared directly on the stainless steel MALDI targets.

DNA and Nucleic Acids

Charged compounds are much more refractive to MS analysis than neutral analogs with a comparable molecular weight (Hillenkamp and Peter-Katalinić 2007; Fuchs et al. 2008a). This results in a much lower sensitivity and, thus, much higher amounts of samples are needed. Therefore, the MS analysis of DNA is a rather challenging task due to the phosphate backbone and the resulting significant negative charge density (Douthwaite and Kirpekar 2007). Although this problem might be solved by performing the MS analysis under sufficiently acidic conditions to protonate the phosphate group and to screen its charge in that way, this attempt should be regarded with great caution as cleavage of the phosphate residues might easily occur. The problem of the significant negative charge density cannot be simply overcome in the most obvious way by recording negative ion spectra. In the case of phosphorylated inositols, it has been shown that each phosphate residue aggravates the MS detection. This particularly holds not only for positive ion detection but also – to a slightly minor extent – for negative ion detection (Müller et al. 2001).

Nevertheless, the detection of smaller oligonucleotides (obtained by enzymatic digestion of native DNA prior to TLC) is possible without major problems (Isbell et al. 1999). Using 2,4,6-trihydroxyacetophenone as matrix and ammonium hydroxide/methanol as solvent system, detection limits of small oligonucleotides in the range of about 10 pg were reported (Isbell et al. 1999).

Carbohydrates

Carbohydrates are generally less frequently investigated by MALDI–TOF MS than proteins or peptides because they are much more refractive to the ionization process (Fuchs et al. 2008a) and give generally much smaller ion yields, i.e., they are only detectable if present in relatively huge amounts. This particularly holds for acidic oligosaccharides that are still more difficult to detect than neutral ones (Harvey 2009).

However, Dreisewerd and coworkers have shown using oligosaccharides isolated from human and elephant milk (Dreisewerd et al. 2006) that amounts of the order of 10 pmol can be still easily analyzed by direct TLC/MALDI coupling. Since these authors used an IR laser, simple glycerol could be applied as MALDI matrix resulting in a very homogeneous wetting of the silica gel and a high reproducibility of the mass spectra. The used orthogonal MALDI device (Guilhaus et al. 2000) that enables the decoupling of the ion source from the mass analyzer provided an excellent mass accuracy independent of potential irregularities of the TLC silica gel surface. Data obtained by the IR laser (Dreisewerd et al. 2006) were also compared with data obtained with an UV laser on the same instrument. For the UV laser, α-cyano-4-hydroxycinnamic acid was used as matrix. Compared to the IR–MALDI mode, the sensitivity of the UV–MALDI was about one order of magnitude lower, whereas unspecific analyte ion fragmentation, as well as adduct formation, was significantly reduced upon UV excitation (Dreisewerd et al. 2006). This particularly holds for glycerol adducts that were observed with significant intensities using the IR laser.

Glycosaminoglycans (GAG) such as hyaluronan or chondroitin sulfate are of particular interest because they are very abundant in the human body and possess a lot of biomedical applications that are currently investigated in more detail. Unfortunately, their high polarity causes problems and highly polar solvent systems have to be applied if successful separation by TLC is planned. It has been recently shown that a solvent system that contains significant amounts of formic acid helps to separate the individual oligosaccharides that can be easily obtained by enzymatic digestion of the native polysaccharides (Zhang et al. 2007b). Although not yet published, we could recently show that the direct characterization of these highly acidic oligosaccharides is possible directly on the TLC plate by MALDI–TOF MS, and some selected spectra are shown in Fig. 16.3.

Unfortunately, there are two problems that complicate the situation: The first problem is coming from the use of high amounts of formic acid. Under these conditions, esterification occurs (detectable by the mass shift of 28) and, thus, a certain part of the GAG is not detectable as such but only as the corresponding formyl ester. This is, however, not a major problem because the extent of formylation is only moderate. A second much more serious problem is the loss of the sulfate residue. This does, however, occur under all so far tested MS conditions (Tissot

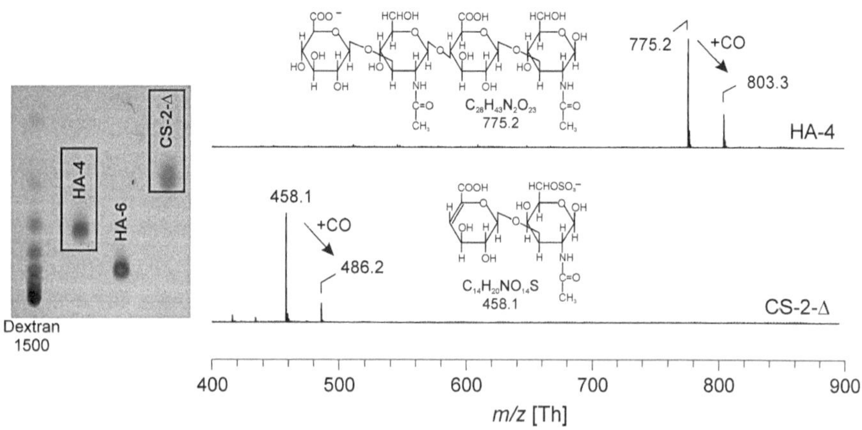

Fig. 16.3 Negative ion MALDI-TOF mass spectra of selected oligosaccharides obtained by enzymatic digestion of hyaluronan (HA) and chondroitin sulfate (CS). Mass spectra were recorded directly from the TLC plate (shown at the left). The molecular structures and the *m/z* ratios are provided along with the mass spectra. Please note that "+ CO" indicates the formation of the formyl ester because the separation of the carbohydrates was performed in the presence of formic acid. Dextran 1500 as well as HA-6 are only shown for comparative purposes on the TLC plate (left) but no mass spectra of these compounds are provided

et al. 2007) and is only slightly enhanced if measurements are performed directly on the TLC plate.

Pharmaceuticals

This is a very important aspect because many companies need to certify their products. As HPLC always bears the risk that there is some residual material on the column, while TLC always uses a completely new stationary phase, a TLC-based separation method is more easily certifiable than a HPLC-based method. In a pioneering study by Crecelius et al. (2000), it could be shown that (known) impurities of certain pharmaceuticals can be directly investigated by combined TLC/MALDI. α-Cyano-4-hydroxycinnamic acid was used as the matrix and was electrosprayed onto the TLC plate to provide maximum homogeneity. This approach gave relatively stable signals and a good reproducibility. However, these authors also noted that irregularities of the used TLC plate led to significant deviations of mass accuracy although standard MALDI–TOF MS is capable of providing mass accuracies of the order of 20 ppm (Crecelius et al. 2000). Therefore, a program was newly developed that recalibrates each mass spectrum acquired during a run using a matrix ion or other selected ion as "lock masses." The same authors have also shown that it is possible to record "post source decay" (PSD) spectra [a special type of MS/MS spectra typical of TOF devices, for details see (Hillenkamp and Peter-Katalinić 2007)] directly from the TLC sample plate. As in a typical PSD experiment several individual mass spectra must be recorded; this is an additional confirmation of the high reproducibility of the suggested approach (Crecelius et al. 2002a). Nearly at the same time, it was also shown that the application of a particle suspension matrix may help to further improve the mass spectral quality of smaller molecules (Crecelius et al. 2002b): Particles of different materials and sizes (TiN, TiO_2, silicone and graphite) were applied as suspensions. Among them, graphite showed the lowest background and the highest peak intensities of the analyte. Using this approach it could be shown that the antibiotics tetracycline and chlortetracycline can be easily differentiated by TLC/MALDI although they gave only an unresolved spot on the TLC (Crecelius et al. 2002b).

Quantitative analysis by TLC/MALDI often yields comparable results to those obtained with stainless steel targets (the standard MALDI sample plate) (Pan et al. 2007). However, the achievable detection limits normally differ for both approaches. For instance, 60 pg of cocaine were the detection limit on the TLC plate, whereas only about 10 pg were required on the stainless steel plate. This difference might be caused by (a) irregularities of the surface of the TLC plate or (b) the inability of bringing all the analyte to the surface of the TLC plate. As already indicated above, UV lasers penetrate not very deeply into the sample, and therefore the analyte on the surface of the TLC plate is primarily detected. This is a major drawback of the application of UV lasers. It is very important to note that an internal reference compound of known structure and concentration is mandatory if absolute (but not only relative intensity ratios) amounts are to be determined. An internal standard was also used when Crecelius and coworkers determined the absolute amounts of the drug Piroxicam by combined TLC/MALDI (Crecelius et al. 2004). Four hundred to eight hundred nanograms of the drug was still detectable by this approach.

Comparable sensitivities as reported above were also found upon the analysis of siderophores (Hayen and Volmer 2005) that are low molecular weight Fe (III)-specific binding molecules that typically occur in bacteria. These authors additionally reported that the sensitivity, directly, on the sample plate is about 4–24 times lower in comparison to conventional stainless steel MALDI targets. These differences were assumed to be caused by the incomplete transport of the trapped analyte molecules from the deeper layers of the TLC gel to the surface (Hayen and Volmer 2005). Thus, it seems reasonable to assume that a reduced thickness of the stationary phase would enhance the detectability of the complete analyte. Thus, standard TLC plates were replaced by ultra-thin-layer chromatography (UTLC) plates (Salo et al. 2005) and it could be shown that due to the thinner adsorbent layer the monolithic UTLC plates provide 10–100 times better sensitivities. Although the majority of commercially available MALDI devices use high vacuum conditions, atmospheric pressure (AP) MALDI is increasingly used for two reasons: (a) these devices are less accident-sensitive and (b) provide the capability to record collisionally induced (CI) fragment ion spectra without the need of an extra collision cell. Unfortunately, AP–MALDI possesses normally lower sensitivity than common high vacuum MALDI. Nevertheless, it was shown that AP–MALDI in combination with UTLC is sufficiently sensitive to detect benzodiazepines directly in human urine without the need of a major workup of the sample (Salo et al. 2007). One interesting application is the investigation of changes of dyes during photo-aging, i.e., upon the light-induced bleaching process. Paintings of old masters that are a few hundred years old usually contain di- and triterpenoids as varnish resins and undergo photooxidation – particularly in the presence of atmospheric O_2. Using such compounds on a cellulosic surface it could be shown that different di- and triterpenoids as well as their characteristic oxidation products containing up to six oxygen molecules could be easily detected by direct TLC/MALDI (Scalarone et al. 2005). Another related work focused on the question if the age of an ink entry from a questioned document that is an important issue in forensic sciences is possible by direct MS investigations. Therefore, the age-induced changes of methyl and ethyl violet (that are both commonly used as ballpoint dyes) under the influence of light were investigated by LDI and MALDI MS directly on a layer of common paper (Weyermann et al. 2006). Many different parameters (including light, wavelength of light, heat, humidity, etc.) could be identified to be responsible for dye aging and could be clearly differentiated by MS. Due to the relatively small masses of the dyes, LDI is superior in comparison to MALDI because there is no matrix background seen upon LDI. Very recently different quaternary protoberberine alkaloids (a pharmaceutically important class of isoquinoline alkaloid) were also investigated by TLC and subsequent LDI MS

(Shariatgorji et al. 2009). Successful separation of berberine and palmatine from Berberis barandana as well as quantitative data analysis could be accomplished. Metabolites could be detected very sensitively due to the presence of the quaternary ammonia group that lends the molecules a permanent positive charge.

Glycolipids

Glycolipids are nowadays regarded as important molecules that are involved in the pathogenesis of different diseases such as cancer. Therefore, their analysis is of increasing interest. A short survey of the structures of selected sphingo- and phospholipids is given in Fig. 16.4.

In a rather early attempt (Guittard et al. 1999), native glycosphingolipids were separated on a conventional TLC plate and subsequently heat-transferred to a different membrane to get rid of impurities that are potentially coming from the

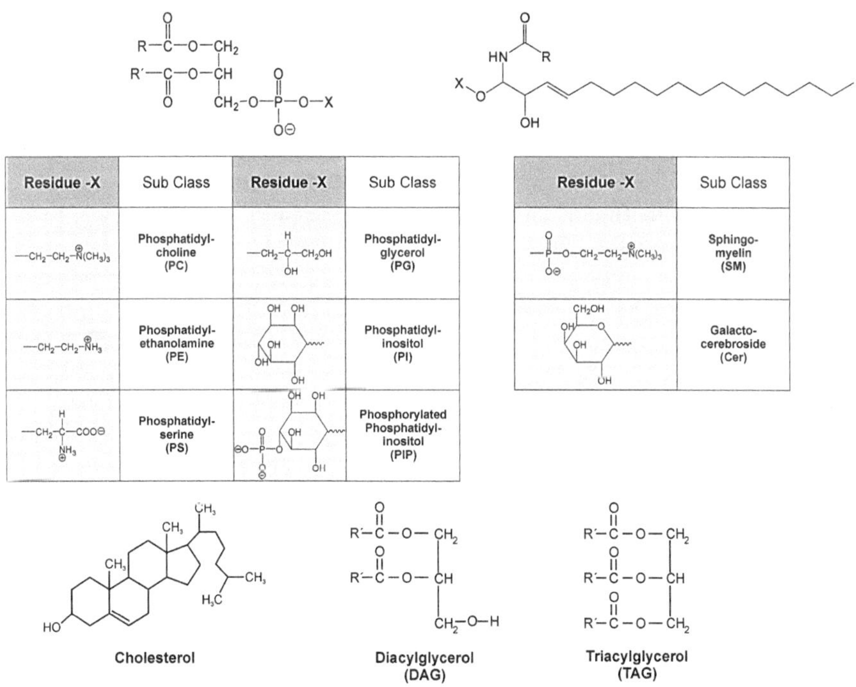

Fig. 16.4 Survey of lipids occurring in significant amounts in cellular membranes. The head group structures of the relevant phospholipids are shown in the left table, whereas the structures of sphingosin-derived lipids are shown at the right. The sinuous lines indicate that stereochemical aspects of carbohydrate chemistry will not be considered. Please note that only PC and PE represent neutral (zwitterionic) phospholipids, while all others are acidic phospholipids. R and R′ represent varying fatty acyl residues. At the bottom, some apolar lipids that are also abundant constituents of cellular membranes are additionally shown

TLC material and because spectral quality and particularly the achievable sensitivity can be highly improved under these conditions. Polyvinylidene difluoride (PVDF) as membrane provided the very best results (Guittard et al. 1999). Lipopolysaccharides that experience currently significant biochemical interest were so far primarily analyzed by the "classical" approach, i.e., by scraping off the individual bands from the TLC plate and their subsequent MALDI analysis. Using more sophisticated equipment (Fourier Transform (FT) MS) (O'Connor et al. 2004), however, it could be shown that the analysis of gangliosides (that primarily occur in brain tissues and are assumed to indicative of Alzheimer disease) is also possible by direct TLC/MALDI without the induction of major fragmentation. In order to enhance the stability of the generated ions, these authors used a relatively high gas pressure to allow collisional cooling of the generated ions (O'Connor et al. 2004). Recently, it was also shown that glycolipids from brain can also be analyzed using a very simple equipment, i.e., a commercially available MALDI–TOF MS device and a normal nitrogen laser (Fuchs et al. 2008b). The majority of the so far performed TLC/MALDI studies of glycolipids used UV lasers. However, there are also a few studies where infrared lasers were applied. These are primarily available on home-built MALDI devices but commercially only on special request. Erbium-doped yttrium–aluminum–Garret (Er:YAG) IR lasers are normally used and have the considerable advantage that IR radiation penetrates deeper into the sample than UV radiation. Therefore, bringing the complete analyte from the inner of the plate to the TLC surface is less important. Beside an IR laser, Dreisewerd et al. (2005) used an orthogonal but not an axial system for glycolipid analysis. The orthogonal configuration has the significant advantage that irregularities of the sample do not significantly influence the achievable spectral mass accuracy. It was found that even minor gangliosides from a complex mixture extracted from cultured Chinese hamster ovary cells can be characterized. These authors (Dreisewerd et al. 2005) also provided convincing evidence that the fluorescent dye "primuline" that is widely used in lipid research as it binds noncovalently to the fatty acyl residues (White et al. 1998) may be used as a nondestructive, MALDI-compatible staining agent. Similar data were independently obtained by another group that additionally introduced "vibrational" cooling (Ivleva et al. 2005). The term "vibrational cooling" describes the desorption process in the pressure range, where "cooling" of the excess energy of the generated ions is achieved. Under these conditions, fragmentations of labile bonds of the analytes can be minimized. This is normally performed by introducing an inert gas (e.g., N_2) under a moderate pressure. In a recent work, DHB in acetonitrile/water (1:1, v/v) was found very useful as matrix for the analysis of glycosphingolipids because in this solvent mixture DHB is highly soluble and this solvent mixture has a relatively high surface tension leading to reduced analyte spreading. Sensitivities between 25 and 50 pmol could be obtained under these conditions (Nakamura et al. 2006). It should be noted that the application of highly concentrated matrix solutions is very important because a larger excess of matrix in comparison to standard MALDI is required for the direct TLC/MALDI coupling to minimize fragmentation of the analyte (Fuchs et al. 2008b, 2009a). A combination between MALDI, TLC, and antibody detection was also recently suggested (Distler et al. 2008; Souady et al. 2009). The authors used the following workflow: (a) TLC separation of cancer-associated glycosphingolipids (GSLs) from human hepatocellular and pancreatic tumors, (b) their detection with oligosaccharide-specific proteins, and (c) the in situ MS analysis of the previously protein-detected GSLs. Detection limits of less than 1 ng of immunostained GSLs could be obtained under these conditions. It is a particular advantage of this approach that only crude lipid extracts of biological sources are required for TLC–IR–MALDI–MS, and no laborious previous GSL purification is needed. A "TLC-Blot-MALDI-Imaging" method was recently also suggested (Goto-Inoue et al. 2009) to visualize whole lipids and individual molecular species. This method provides higher sensitivity than common staining methods and allows for visualization of all lipids within a linear range of approximately one order of magnitude. The important field of glycolipid has been recently comprehensively reviewed (Müthing and Distler 2010).

One important problem particularly related to MS imaging of biological tissues is stemming from the interference of matrix peaks with the biological molecules of interest. In order to overcome this problem, it was suggested to use graphite-assisted laser desorption/ionization (GALDI) MS and this technique was already successfully applied for the analysis of cerebrosides in a complex total brain lipid extract (Cha and Yeung 2007). Beside the lack of typical "matrix" peaks it could also be shown that GALDI suffers by far less from suppression effects, for instance, the suppression of cerebrosides by other dominant phospholipids, particularly PC. In addition to positive ion detection, (negatively charged) sulfatides could also be easily detected using a thin layer of graphite. Thus, graphite enables the recording of not only positive ion spectra but also negative ion spectra.

Phospholipids

Although phospholipids (PL) are important constituents of all tissues and particularly cells, these molecules were not considered to be very important but were primarily regarded to represent the "packaging material" for other much more important molecules such as proteins. This situation has, however, changed significantly as soon as PLs were discovered to represent important second messenger molecules and to be involved in different pathologies (Fernandis and Wenk 2007). Consequently, the term "lipidomics" became very popular (Fuchs and Schiller 2009b). Additionally, MALDI–TOF MS analysis of lipids (Schiller et al. 1999; Harvey 1995) became also popular since MALDI imaging has provided evidence that these molecules ionize particularly well and are, thus, easily detectable under MALDI conditions.

One serious problem is, however, coming from the strongly varying detectabilities of the individual PL classes. In the presence of compounds with quaternary ammonia groups (e.g., PC, SM, LPC), further lipids that are only detectable with

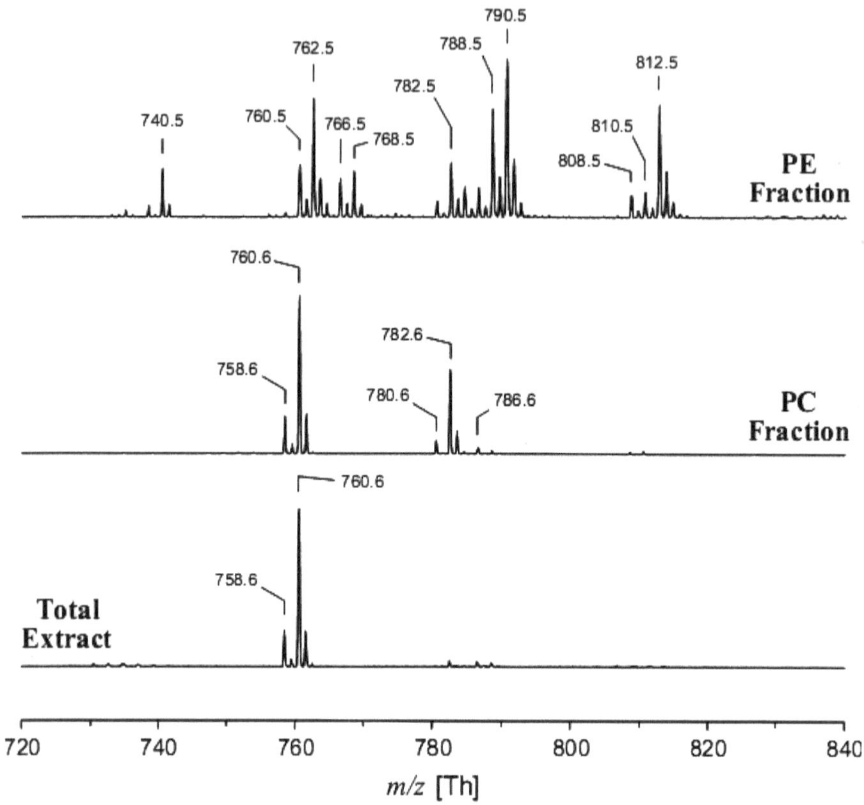

Fig. 16.5 MALDI–TOF mass spectra of an organic extract of hen egg yolk. The total extract, as well as the phosphatidylcholine (PC) and the phosphatidylethanolamine (PE) fractions are shown. Separation of lipids was performed by TLC before MS analysis. Although some further TLC fractions could be obtained, only the most relevant lipid classes are shown here. Note that the spectra of the total extract and the isolated PC fraction are virtually identical, while there are not even small peaks of the most abundant PE species (m/z = 790.5 and 812.5) detectable. This indicates marked differences regarding detection limits

lower sensitivities may be suppressed (Petković et al. 2001). This is illustrated in Fig. 16.5 on the selected example of an extract from a hen egg yolk.

Of course, this may also be considered as an advantage because only some selected classes are detected and, thus, spectra are less complex. Nevertheless, chromatographic separation of the total lipid extracts is normally required if a detailed analysis of the mixture is needed. As TLC is a very common and traditional method of lipid analysis (Touchstone 1995), there are many reports where MALDI MS analysis was performed subsequent to TLC separation accompanied by scraping off the spots of interest and the reelution of the related lipids. The lipid composition of body fluids (Sommerer et al. 2004), cells (Leßig et al. 2004), and tissues (Schiller et al. 2003a) could be successfully analyzed by this approach. It is necessary to mention that (in addition to diayl- and alkyl-acyl compounds) there are also alkenyl-acyl lipids. These "plasmalogens" are not formed by the reaction of the glycerol backbone with a fatty acid but with a fatty aldehyde and a comprehensive review is available in (Leßig and Fuchs 2009). The alkenyl ether group is extremely sensitive toward even traces of acids and this is reflected by the finding that plasmalogen PL fractions do often contain the corresponding lysophospholipids (Schiller et al. 2003b). Thus, it is reasonable to assume that the plasmalogens are even sensitive against the acidic properties of the silica gel and are already hydrolyzed under these conditions into the corresponding lyso compounds. This is particularly important if spermatozoa or stem cells are investigated because these cells possess extraordinarily high plasmalogen content (Leßig and Fuchs 2009; Schiller et al. 2003b).

The direct analysis of lipids separated on a TLC plate by MALDI–TOF MS was, however, only recently attempted by two different methods. One approach is based on the use of an IR laser and glycerol as matrix (Rohlfing et al. 2007). This approach has the significant advantage that IR radiation penetrates deeper into the TLC plate, but simultaneously confers the disadvantage that abundant glycerol adducts (and to a minor extent even NaCl adducts) with the PLs of interest are detected. Another problem is that IR lasers are commercially barely available so far. Therefore, another approach used a readily available N_2 laser and standard DHB as matrix (Fuchs et al. 2007).

A special adapter device was used for the introduction of the TLC plate into the mass spectrometer (Fig. 16.6), whereby the TLC plate was simply mounted onto the target by conductive adhesive tapes (as mechanical fixation was not yet possible) (Fuchs et al. 2007, 2009b). One selected TLC lane of a separated hen egg yolk extract and some selected mass spectra are shown in Fig. 16.7.

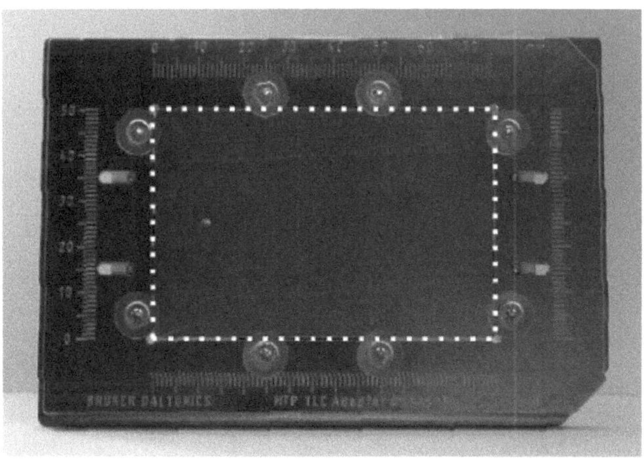

Fig. 16.6 Photograph of the MALDI–TOF adapter target prototype used in this study (Bruker Daltonics, Bremen, Germany). The TLC plate is inserted into the center of the target (along the *dotted white rectangle*) and can then be fixed mechanically. Adhesive tape is no longer required

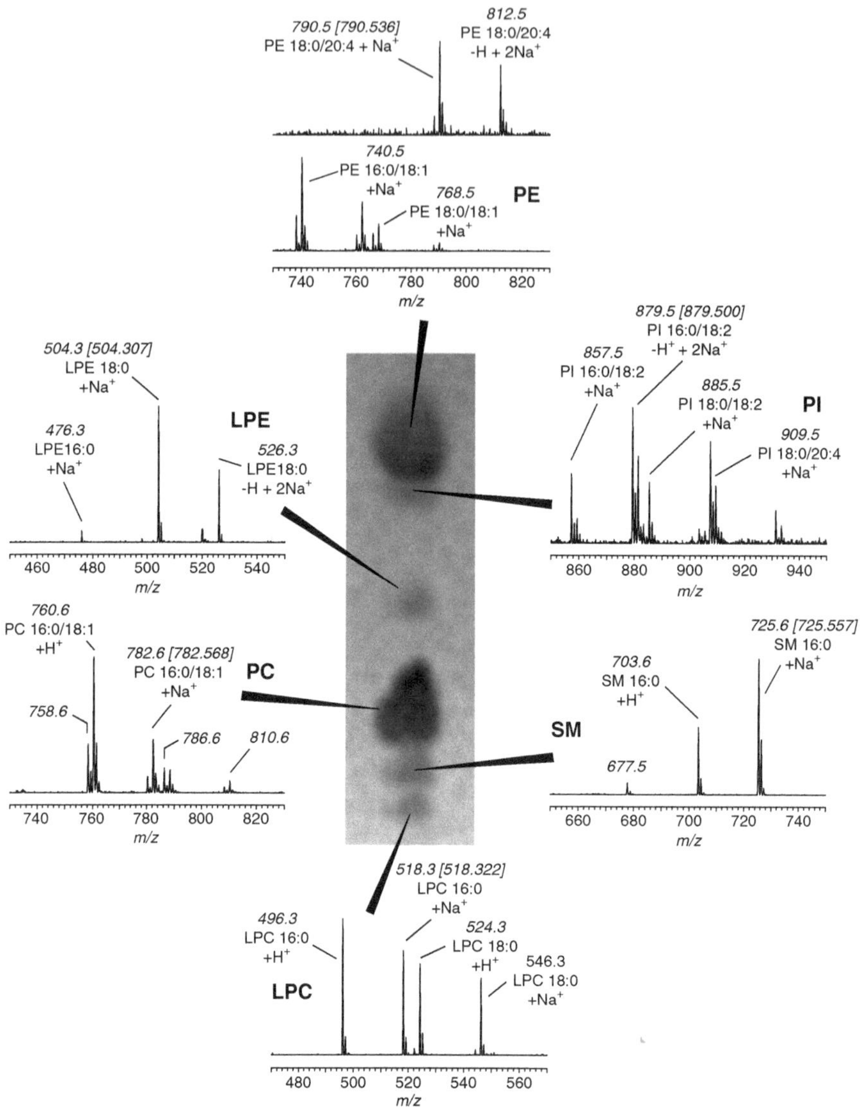

Fig. 16.7 Expanded region of a TLC-separated egg yolk extract and the corresponding positive ion MALDI-TOF mass spectra recorded directly from the indicated positions on the plate. Only the relevant mass regions of each PL class are shown and assignments are provided directly in the individual traces. Data given in parentheses correspond to theoretical masses and were introduced to enable comparison with the experimental data in selected cases. Please also note that the PE fraction provides different spectra, depending on the position where the laser hits the PE spot. The only marked fragmentation is the loss of the head group of SM ($m/z = 677.5$). Reprinted from [95] with permission

Two facts are particularly remarkable. First, even rather minor fractions of e.g., phosphatidylinositol (PI) can be easily analyzed and give spectra with a convincing signal-to-noise (*S/N*) ratio. As PI makes out only about 0.5% of the total PLs of egg yolk, this clearly proves the significant dynamic range of more than two orders of magnitude. Second, the lower and the upper part of the PE spot provide significantly different mass spectra, whereby PEs with longer fatty acyl (in particular arachidonoyl) residues are found in the upper and PEs with shorter acyl residues in the lower part of the spot. This is a clear indication that even the separation quality obtained under normal phase conditions is sufficient to allow the differentiation of fatty acyl residues if MS is used for subsequent analysis (Fuchs et al. 2009b).

Regarding detection limits, both approaches (Rohlfing et al. 2007; Fuchs et al. 2007) provided comparable results (about 400 pmol) and, therefore, both approaches might be useful for routine lipid analysis. This particularly holds as surprisingly good mass accuracies (about 100 ppm) and mass resolutions (about 3,000) could be obtained that are absolutely sufficient for the differentiation of different phospholipids. Very recently, an automated routine became available and was already successfully applied for the investigation of the lipid composition of human mesenchymal stem cells (Fuchs et al. 2008c) as well as plasma PL signatures associated with respiratory disease severity in cystic fibrosis patients (Guerrera et al. 2009). This is a clear indication that MALDI mass spectra of lipids can be quantitatively analyzed without major problems although further attempts are clearly necessary to establish more useful matrix compounds – particularly for negative ion detection (Fuchs et al. 2009c).

Notes

TLC is a simple, inexpensive, and convenient analytical method for many (particularly small) organic molecules. Although there are many dyes that enable the assignment of the individual analytes, this method often fails if unknown molecules have to be analyzed and – especially – if no standards are available. In these cases an additional detection system is necessary. In combination with liquid chromatography, MS analysis is widely established, but there are so far much lesser attempts to combine TLC with MS. Although there were already different reports about successfully combining TLC with MS desorption techniques such as DESI MS, the clear focus of this chapter was on the combination between TLC and MALDI. As the MALDI method anyway makes use of a solid sample, this technique seems to be the ionization method of choice for combining TLC and MS. The most important advantages are the simple coupling (in the simplest way the developed TLC plate is simply mounted onto a commercially available MALDI target) between commercially available MALDI devices and TLC as well as the simple to interpret mass spectra because primarily singly charged ions are generated by MALDI, and complex adduct patterns can be also easily avoided. Additional advantages are the high spatial resolution provided by the small laser beam spot as well as the high detection sensitivities.

This chapter has given a survey of the different substance classes that were already investigated by TLC/MALDI. Hopefully, it became evident that small, relatively apolar molecules such as lipids can be most successfully analyzed by this method, while highly polar molecules such as carbohydrates (especially if they contain additional charged functional groups such as phosphate or sulfate) are at present much more difficult. Although the authors hope that they were able to show the significant progress that was achieved in the last years, there are still some problems one must try to overcome:

- It is so far not known how much of the analyte is actually detected if UV lasers (the standard on commercially available devices) are used because UV does not penetrate deeply into the TLC layer but remains primarily on the surface. Therefore, major attempts are necessary to shed light on the quantitative aspects of TLC/MALDI.
- The search for useful TLC/MALDI matrices has to be going on because it is not yet known whether the so far used matrices do already provide the optimum results.
- It would be good to overcome the reduced mass accuracy obtained by the combined TLC/MALDI approach in comparison to standard MALDI due to irregularities of the TLC surface. It is, however, expected that this problem could be significantly reduced by adding a suitable calibration standard, using UTLC plates or MALDI devices with an orthogonal geometry.

Despite these problems, however, the illustrated examples clearly confirmed the significant future potential of the direct combination between TLC and MALDI and its wide applicability.

References

Alpmann A and Morlock G. Improved online coupling of planar chromatography with electrospray mass spectrometry: extraction of zones from glass plates. Anal Bioanal Chem 2006; 386 (5), 1543–1551.

Aranda M and Morlock G. Simultaneous determination of caffeine, ergotamine, and metamizol in solid pharmaceutical formulation by HPTLC-UV-FLD with mass confirmation by online HPTLC-ESI-MS. J Chromatogr Sci 2007; 45 (5), 251–255.

Bakry R, Bonn GK, Mair D and Svec F. Monolithic porous polymer layer for the separation of peptides and proteins using thin-layer chromatography coupled with MALDI-TOF-MS. Anal Chem 2007; 79 (2), 486–493.

Berger S and Braun S (2004). 200 and More NMR Experiments – A Practical Course, first edition, published by Wiley-VCH, Weinheim.

Bradley SA and McLaughlin RL. High-resolution magic-angle spinning NMR for the identification of reaction products directly from thin-layer chromatography spots. Magn Reson Chem 2007; 45 (10), 814–818.

Busch KL. Planar separations and mass spectrometric detection. J Planar Chromatogr 2004; 17 (6), 398–403.

Cha S and Yeung ES. Colloidal graphite-assisted laser desorption/ionization mass spectrometry and MSn of small molecules. 1. Imaging of cerebrosides directly from rat brain tissue. Anal Chem 2007; 79 (6), 2373–2385.

Chen H, Talaty NN, Takáts Z and Cooks RG. Desorption electrospray ionization mass spectrometry for high-throughput analysis of pharmaceutical samples in the ambient environment. Anal Chem 2005; 77 (21), 6915–6927.

Cornett DS, Reyzer ML, Chaurand P and Caprioli RM. MALDI imaging mass spectrometry: molecular snapshots of biochemical systems. Nat Methods 2007; 4 (10), 828–833.

Crecelius A, Clench MR, Richards DS and Mather J, Parr V. Analysis of UK-224,671 and related substances by thin-layer chromatography-matrix-assisted laser-desorption ionization-time of flight mass spectrometry. Planar Chromatogr 2000; 13 (2), 76–81.

Crecelius A, Clench MR, Richards DS, Evason D and Parr V. Thin-layer chromatography-postsource-decay matrix-assisted laser desorption/ionization time-of-flight mass spectrometry of small drug molecules. J Chromatogr Sci 2002a; 40 (10), 614–620.

Crecelius A, Clench MR, Richards DS and Parr V. Thin-layer chromatography-matrix-assisted laser desorption ionisation-time-of-flight mass spectrometry using particle suspension matrices. J Chromatogr A 2002b; 958 (1–2), 249–260.

Crecelius A, Clench MR, Richards DS and Parr V. Quantitative determination of Piroxicam by TLC-MALDI TOF MS. J Pharm Biomed Anal 2004; 35 (1), 31–39.

Distler U, Hülsewig M, Souady J, Dreisewerd K, Haier J, Senninger N, Friedrich AW, Karch H, Hillenkamp F, Berkenkamp S, Peter-Katalinić J and Müthing J. Matching IR-MALDI-o-TOF mass spectrometry with the TLC overlay binding assay and its clinical application for tracing tumor-associated glycosphingolipids in hepatocellular and pancreatic cancer. Anal Chem 2008; 80 (6), 1835–1846.

Douthwaite S and Kirpekar F. Identifying modifications in RNA by MALDI mass spectrometry. Methods Enzymol 2007; 425, 1–20.

Dreisewerd K, Müthing J, Rohlfing A, Meisen I, Vukelić Z, Peter-Katalinić J, Hillenkamp F and Berkenkamp S. Analysis of gangliosides directly from thin-layer chromatography plates by infrared matrix-assisted laser desorption/ionization orthogonal time-of-flight mass spectrometry with a glycerol matrix. Anal Chem 2005; 77 (13), 4098–4107.

Dreisewerd K, Kölbl S, Peter-Katalinić J, Berkenkamp S and Pohlentz G. Analysis of native milk oligosaccharides directly from thin-layer chromatography plates by matrix-assisted laser desorption/ionization orthogonal-time-of-flight mass spectrometry with a glycerol matrix. J Am Soc Mass Spectrom 2006; 17 (2), 139–150.

Duncan MW, Roder H and Hunsucker SW. Quantitative matrix-assisted laser desorption/ionization mass spectrometry. Brief Funct Genomic Proteomic 2008; 7 (5), 355–370.

Fernandis AZ and Wenk MR. Membrane lipids as signaling molecules. Curr Opin Lipidol 2007; 18 (2), 121–128.

Fuchs B and Schiller J. Recent developments of useful MALDI matrices for the mass spectrometric characterization of apolar compounds. Curr Org Chem 2009a; 13 (16), 1664–1681.

Fuchs B, Schiller J. Application of MALDI-TOF mass spectrometry in lipidomics. Eur J Lipid Sci Technol 2009b; 111 (1), 83–98.

Fuchs B, Schiller J, Süß R, Schürenberg M and Suckau D. A direct and simple method of coupling matrix-assisted laser desorption and ionization time-of-flight mass spectrometry (MALDI-TOF MS) to thin-layer chromatography (TLC) for the analysis of phospholipids from egg yolk. Anal Bioanal Chem 2007; 389 (3), 827–834.

Fuchs B, Arnold K and Schiller J (2008a). Mass Spectrometry of Biological Molecules, In: Encyclopedia of Analytical Chemistry (ed. by Meyers RA), first edition, published by John Wiley & Sons, Chichester, pp. 1–39.

Fuchs B, Nimptsch A, Süß R and Schiller J. Analysis of brain lipids by directly coupled matrix-assisted laser desorption ionization time-of-flight mass spectrometry and high-performance thin-layer chromatography. J AOAC Int 2008b; 91 (5), 1227–1236.

Fuchs B, Schiller J, Süß R, Zscharnack M, Bader A, Müller P, Schürenberg M, Becker M and Suckau D. Analysis of stem cell lipids by offline HPTLC-MALDI-TOF MS. Anal Bioanal Chem 2008c; 392 (5), 849–860.

Fuchs B, Süß R, Nimptsch A and Schiller J. Matrix-assisted laser desorption and ionization time-of-flight mass spectrometry (MALDI-TOF MS) directly combined with thin-layer chromatography (TLC) – a review of the current state. Chromatographia 2009a; 69 (Suppl.), 95–105.

Fuchs B, Schiller J, Süß R, Nimptsch A, Schürenberg M and Suckau D. Capabilities and disadvantages of combined matrix-assisted laserdesorption/ionization time-of-flight mass spectrometry (MALDI-TOF MS) and high-performance thin-layer chromatography (HPTLC): analysis of egg yolk lipids. J Planar Chromatogr 2009b; 22 (1), 35–42.

Fuchs B, Bischoff A, Süß R, Teuber K, Schürenberg M, Suckau D and Schiller J. Phosphatidylcholines and -ethanolamines can be easily mistaken in phospholipid mixtures: a negative ion MALDI-TOF MS study with 9-aminoacridine as matrix and egg yolk as selected example. Anal Bioanal Chem 2009c; 395 (8), 2479–2487.

Geiss F (1987). Fundamentals of Thin Layer Chromatography, first edition, published by Hüthig, Heidelberg.

Gellermann GP, Appel TR, Davies P and Diekmann S. Paired helical filaments contain small amounts of cholesterol, phosphatidylcholine and sphingolipids. Biol Chem 2006; 387 (9), 1267–1274.

Goto-Inoue N, Hayasaka T, Taki T, Gonzalez TV and Setou M. A new lipidomics approach by thin-layer chromatography-blot-matrix-assisted laser desorption/ionization imaging mass spectrometry for analyzing detailed patterns of phospholipid molecular species. J Chromatogr A 2009; 1216 (42), 7096–7101.

Guerrera IC, Astarita G, Jais JP, Sands D, Nowakowska A, Colas J, Sermet-Gaudelus I, Schuerenberg M, Piomelli D, Edelman A and Ollero M. A novel lipidomic strategy reveals plasma phospholipid signatures associated with respiratory disease severity in cystic fibrosis patients. PLoS One 2009; 4 (11), e7735.

Guilhaus M, Selby D and Mlynski V. Orthogonal acceleration time-of-flight mass spectrometry. Mass Spectrom Rev 2000; 19 (2), 65–107.

Guittard J, Hronowski XPL and Costello CE. Direct matrix-assisted laser desorption/ionization mass spectrometric analysis of glycosphingolipids on thin layer chromatographic plates and transfer membranes. Rapid Commun Mass Spectrom 1999; 13 (18), 1838–1849.

Günzler H and Gremlich H-U (2002). IR-Spectroscopy – An Introduction, first edition, published by Wiley-VCH, Weinheim.

Gusev AI. Interfacing matrix-assisted laser desorption/ionization mass spectrometry with column and planar separations. Fresenius J Anal Chem 2000; 366 (6–7), 691–700.

Gusev AI, Proctor A, Rabinovich YI and Hercules DM. Thin-layer chromatography combined with matrix-assisted laser-desorption ionization mass spectrometry. Anal Chem 1995a; 67 (11), 1805–1814.

Gusev AI, Vasseur OJ, Proctor A, Sharkey AG and Hercules DM. Imaging of thin-layer chromatograms with matrix-assisted laser-desorption ionization mass spectrometry. Anal Chem 1995b; 67 (24), 4565–4570.

Hahn-Deinstrop E (2006). Applied Thin-Layer Chromatography – Best Practice and Avoidance of Mistakes, second edition, published by Wiley-VCH, Weinheim.

Harvey DJ. Matrix-assisted laser desorption/ionization mass spectrometry of phospholipids. J Mass Spectrom 1995; 30 (9), 1333–1346.

Harvey DJ. Analysis of carbohydrates and glycoconjugates by matrix-assisted laser desorption/ionization mass spectrometry: an update for 2003–2004. Mass Spectrom Rev 2009; 28 (2), 27–361.

Hayen H and Volmer DA. Rapid identification of siderophores by combined thin-layer chromatography/matrix-assisted laser desorption/ionization mass spectrometry. Rapid Commun Mass Spectrom 2005; 19 (5), 711–720.

Heigl N, Huck CW, Rainer M, Najam-Ul-Haq M and Bonn GK. Near infrared spectroscopy, cluster and multivariate analysis hyphenated to thin layer chromatography for the analysis of amino acids. Amino Acids 2006; 31 (1), 45–53.

Hillenkamp F and Peter-Katalinić J (2007). MALDI MS – A Practical Guide to Instrumentation, Methods and Application, first edition, published by Wiley-VCH, Weinheim.

Hillenkamp F, Unsöld E, Kaufmann R and Nitsche R. Laser microprobe mass analysis of organic materials. Nature 1975; 256 (5513), 119–120.

Isbell DT, Gusev AI, Taranenko NI, Chen CH and Hercules DM. Analysis of nucleotides directly from TLC plates using MALDI-MS detection. Fresenius J Anal Chem 1999; 365 (7), 625–630.

Ivleva VB, Sapp LM, O'Connor PB and Costello CE. Ganglioside analysis by thin-layer chromatography matrix-assisted laser desorption/ionization orthogonal time-of-flight mass spectrometry. J Am Soc Mass Spectrom 2005; 16 (9), 1552–1560.

Jautz U and Morlock G. Efficacy of planar chromatography coupled to (tandem) mass spectrometry for employment in trace analysis. J Chromatogr A 2006; 1128 (1–2), 244–250.

Klein TR, Kirsch D, Kaufmann R and Riesner D. Prion rods contain small amounts of two host sphingolipids as revealed by thin-layer chromatography and mass spectrometry. Biol Chem 1998; 379 (6), 655–66.

Kochin V, Imanishi SY and Eriksson JE. Fast track to a phosphoprotein sketch – MALDI-TOF characterization of TLC-based tryptic phosphopeptide maps at femtomolar detection sensitivity. Proteomics 2006; 6 (21), 5676–5682.

Leßig J and Fuchs B. Plasmalogens in biological systems: their role in oxidative processes in biological membranes, their contribution to pathological processes and aging and plasmalogen analysis. Curr Med Chem 2009; 16 (16), 2021–2041.

Leßig J, Gey C, Süß R, Schiller J, Glander HJ and Arnhold J. Analysis of the lipid composition of human and boar spermatozoa by MALDI-TOF mass spectrometry, thin layer chromatography and ^{31}P NMR spectroscopy. Comp Biochem Physiol B Biochem Mol Biol 2004; 137 (2), 265–277.

Luftmann H, Aranda M and Morlock GE. Automated interface for hyphenation of planar chromatography with mass spectrometry. Rapid Commun Mass Spectrom 2007; 21 (23), 3772–3776.

Marderosian AD and Chao JM. Identification of active drug principles in natural products. J Chromatogr Sci 1974; 12 (5), 285–292.

Mehl JT and Hercules DM. Direct TLC-MALDI coupling using a hybrid plate. Anal Chem 2000; 72 (1), 68–73.

Mehl JT, Gusev AI and Hercules DM. Coupling protocol for thin layer chromatography matrix-assisted laser desorption ionization. Chromatographia 1997; 46 (7–8), 358–364.

Miao Z and Chen H. Direct analysis of liquid samples by desorption electrospray ionization-mass spectrometry (DESI-MS). J Am Soc Mass Spectrom 2009; 20 (1), 10–19.

Morlock G and Sherma J. New and improved liquid chromatographic methods for food analysis. J AOAC Int 2009; 92 (3), 689–690.

Morlock G and Ueda Y. New coupling of planar chromatography with direct analysis in real time mass spectrometry. J Chromatogr A 2007; 1143 (1–2), 243–251.

Mowthorpe S, Clench MR, Cricelius A, Richards DS, Parr V and Tetler LW. Matrix-assisted laser desorption/ionisation time-of-flight thin layer chromatography mass spectrometry – A rapid method for impurity testing. Rapid Commun Mass Spectrom 1999; 13 (4), 264–270.

Mueller DR, Voshol H, Waldt A, Wiedmann B and Van Oostrum J. LC-MALDI MS and MS/MS – an efficient tool in proteome analysis. Subcell Biochem 2007; 43, 355–380.

Müller M, Schiller J, Petković M, Oehrl W, Heinze R, Wetzker R, Arnold K and Arnhold J. Limits for the detection of (poly-)phosphoinositides by matrix-assisted laser desorption and ionization time-of-flight mass spectrometry (MALDI-TOF MS). Chem Phys Lipids 2001; 110 (2), 151–164.

Müthing J and Distler U. Advances on the compositional analysis of glycosphingolipids combining thin-layer chromatography with mass spectrometry. Mass Spectrom Rev 2010; 24 (7), 1032–1038.

Nakamura H and Pisano JJ. Derivatization of compounds at the origin of thin-layer plates with fluorescamine. J Chromatogr 1976; 121 (1), 33–40.

Nakamura K, Suzuki Y, Goto-Inoue N, Yoshida-Noro C and Suzuki A. Structural characterization of neutral glycosphingolipids by thin-layer chromatography coupled to matrix-assisted laser desorption/ionization quadrupole ion trap time-of-flight MS/MS. Anal Chem 2006; 78 (16), 5736–5743.

O'Connor PB, Budnik BA, Ivleva VB, Kaur P, Moyer SC, Pittman JL and Costello CE. A high pressure matrix-assisted laser desorption ion source for Fourier transform mass spectrometry designed to accommodate large targets with diverse surfaces. J Am Soc Mass Spectrom 2004; 15 (1), 128–132.

Pan C, Xu S, Zhou H, Fu Y, Ye M and Zou H. Recent developments in methods and technology for analysis of biological samples by MALDI-TOF-MS. Anal Bioanal Chem 2007; 387 (1), 193–204.

Panchagnula V, Mikulskis A, Song L, Wang Y, Wang M, Knubovets T, Scrivener E, Golenko E, Krull IS, Schulz M, Hauck H-E and Patton WF. Phosphopeptide analysis by directly coupling two dimensional planar electrochromatography/thin-layer chromatography with matrix-assisted laser desorption/ionization time-of-flight mass spectrometry. J Chromatogr A 2007; 1155 (1), 112–123.

Peterson DS. Matrix-free methods for laser desorption/ionization mass spectrometry. Mass Spectrom Rev 2007; 26 (1), 19–34.

Petković M, Schiller J, Müller M, Benard S, Reichl S, Arnold K and Arnhold J. Detection of individual phospholipids in lipid mixtures by matrix-assisted laser desorption/ionization time-of-flight mass spectrometry: phosphatidylcholine prevents the detection of further species. Anal Biochem 2001; 289 (2), 202–216.

Rohlfing A, Müthing J, Pohlentz G, Distler U, Peter-Katalinić J, Berkenkamp S and Dreisewerd K. IR-MALDI-MS analysis of HPTLC-separated phospholipid mixtures directly from the TLC plate. Anal Chem 2007; 79 (15), 5793–5808.

Salo PK, Salomies H, Harju K, Ketola RA, Kotiaho T, Yli-Kauhaluoma J and Kostiainen R. Analysis of small molecules by ultra thin-layer chromatography-atmospheric pressure matrix-assisted laser desorption/ionization mass spectrometry. J Am Soc Mass Spectrom 2005; 16 (6), 906–915.

Salo PK, Vilmunen S, Salomies H, Ketola RA and Kostiainen R. Two-dimensional ultra-thin-layer chromatography and atmospheric pressure matrix-assisted laser desorption/ionization mass spectrometry in bioanalysis. Anal Chem 2007; 79 (5), 2101–2108.

Scalarone D, Duursma MC, Boon JJ and Chiantore O. MALDI-TOF mass spectrometry on cellulosic surfaces of fresh and photo-aged di- and triterpenoid varnish resins. J Mass Spectrom 2005; 40 (12), 1527–1535.

Schäfer KC, Dénes J, Albrecht K, Szaniszló T, Balog J, Skoumal R, Katona M, Tóth M, Balogh L and Takáts Z. In vivo, in situ tissue analysis using rapid evaporative ionization mass spectrometry. Angew Chem Int Ed Engl 2009; 48 (44), 8240–8242.

Schiller J, Arnhold J, Benard S, Müller M, Reichl S and Arnold K. Lipid analysis by matrix-assisted laser desorption and ionization mass spectrometry: a methodological approach. Anal Biochem 1999; 267 (1), 46–56.

Schiller J, Süß R, Fuchs B, Müller M, Zschörnig O and Arnold K. Combined application of TLC and matrix-assisted laser desorption and ionisation time-of-flight mass spectrometry (MALDI-TOF MS) to phospholipid analysis of brain. Chromatographia 2003a; 57 (Suppl 1), 297–302.

Schiller J, Müller K, Süß R, Arnhold J, Gey C, Herrmann A, Leßig J, Arnold K and Müller P. Analysis of the lipid composition of bull spermatozoa by MALDI-TOF mass spectrometry a cautionary note. Chem Phys Lipids 2003b; 126 (1), 85–94.

Schiller J, Huster D, Fuchs B, Naji L, Kaufmann J and Arnold K. Evaluation of cartilage composition and degradation by high-resolution magic-angle spinning nuclear magnetic resonance. Methods Mol Med 2004a; 101, 267–85.

Schiller J, Süß R, Arnhold J, Fuchs B, Leßig J, Müller M, Petković M, Spalteholz H, Zschörnig O and Arnold K. Matrix-assisted laser desorption and ionization time-of-flight (MALDI-TOF) mass spectrometry in lipid and phospholipid research. Prog Lipid Res 2004b; 43 (5), 449–488.

Schiller J, Süß R, Fuchs B, Müller M, Petković M, Zschörnig O and Waschipky H. The suitability of different DHB isomers as matrices for the MALDI-TOF MS analysis of phospholipids: which isomer for what purpose? Eur Biophys J 2007a; 36 (4–5), 517–527.

Schiller J, Süß R, Fuchs B, Müller M, Zschörnig O and Arnold K. MALDI-TOF MS in lipidomics. Front Biosci 2007b; 12, 2568–2579.

Shariatgorji M, Spacil Z, Maddalo G, Cardenas LB and Ilag LL. Matrix-free thin-layer chromatography/laser desorption ionization mass spectrometry for facile separation and identification of medicinal alkaloids. Rapid Commun Mass Spectrom 2009; 23 (23), 3655–3660.

Sherma J. Planar chromatography. Anal Chem 2008; 80 (12), 4253–4267.

Sherma J and Morlock G. Chronology of thin-layer chromatography focusing on instrumental progress. J Planar Chromatogr 2008; 21 (6), 471–477.

Sommerer D, Süß R, Hammerschmidt S, Wirtz H, Arnold K and Schiller J. Analysis of the phospholipid composition of bronchoalveolar lavage (BAL) fluid from man and minipig by MALDI-TOF mass spectrometry in combination with TLC. J Pharm Biomed Anal 2004; 35 (1), 199–206.

Souady J, Soltwisch J, Dreisewerd K, Haier J, Peter-Katalinić J and Müthing J. Structural profiling of individual glycosphingolipids in a single thin-layer chromatogram by multiple sequential immunodetection matched with direct IR-MALDI-o-TOF mass spectrometry. Anal Chem 2009; 81 (22), 9481–9492.

Tissot B, Gasiunas N, Powell AK, Ahmed Y, Zhi ZL, Haslam SM, Morris HR, Turnbull JE, Gallagher JT and Dell A. Towards GAG glycomics: analysis of highly sulfated heparins by MALDI-TOF mass spectrometry. Glycobiology 2007; 17 (9), 972–982.

Touchstone JC. Thin-layer chromatographic procedures for lipid separation. J Chromatogr B Biomed Appl 1995; 671 (1–2), 169–195.

Weyermann C, Kirsch D, Costa-Vera C and Spengler B. Photofading of ballpoint dyes studied on paper by LDI and MALDI MS. J Am Soc Mass Spectrom 2006; 17 (3), 297–306.

White T, Bursten S, Frederighi D, Lewis RA and Nudelman E. High-resolution separation and quantification of neutral lipid and phospholipid species in mammalian cells and sera by multi-one-dimensional thin-layer chromatography. Anal Biochem 1998; 258 (1), 109–117.

Wilson ID. The state of the art in thin-layer chromatography-mass spectrometry: a critical appraisal. J Chromatogr A 1999; 856 (1–2), 429–442.

Wilson ID, Spraul M and Humpfer E. Thin-layer chromatography combined with high-resolution solid-state NMR for compound identification without substance elution: preliminary results. J Planar Chromatogr 1997; 10 (3), 217–219.

Woodbury CP Jr, Fitzloff JF and Vincent SS. Sample multiplexing for greater throughput in HPLC and related methods. Anal Chem 1995; 67 (5), 885–890.

Yates JR, Ruse CI and Nakorchevsky A. Proteomics by mass spectrometry: approaches, advances, and applications. Annu Rev Biomed Eng 2009; 11, 49–79.

Zhang JZ, Wang Y, Chen H and Shao HB. TLC-SERS study on evodiamine in evodia rutaecarpa. Spectrosc Spectral Anal 2007a; 27 (5), 944–947.

Zhang Z, Xie J, Zhang F and Linhardt RJ. Thin-layer chromatography for the analysis of glycosaminoglycan oligosaccharides. Anal Biochem 2007b; 371 (1), 118–120.

Zschörnig O, Richter V, Rassoul F, Süß R, Arnold K and Schiller J. Analysis of human blood plasma by MALDI-TOF MS – Evaluation of critical parameters. Anal Lett 2006; 39 (6), 1101–1113.

Chapter 17
Scanning Diode Laser Desorption Thin-Layer Chromatography Coupled with Atmospheric Pressure Chemical Ionization Mass Spectrometry

Song Peng, Norman Ahlmann, Michael Edler, and Joachim Franzke

Abstract Continuous wave diode laser is applied for desorption of an analyte from a porous surface of a thin-layer plate covered with a graphite suspension. The thermally desorbed analyte molecules are ionized in the gas phase by a corona discharge at atmospheric pressure. Therefore, both essential processes – the desorption and the ionization of analyte molecules, which are often performed in one step – are separated. Reserpine was chosen as model analyte, which is often used for specification of mass spectrometers. No fragmentation was observed because of efficient collisional cooling under atmospheric pressure. The influence of diode laser power and the composition of the graphite suspension were investigated, and a primary optimization was performed. An interface to allow online qualitative and quantitative full plate detection and analysis of compounds separated by thin-layer chromatography is presented.

Thin-layer chromatography (TLC), as an economic and handy chromatographic separation technique, has been extensively used for many years. In contrast to column chromatography, such as high-performance liquid chromatography, TLC allows many samples to be run simultaneously, which makes it a useful tool for rapid screening in many fields of applications. Mostly, the separated analytes are visually detected by staining techniques, UV absorption and fluorescence extinction. An automated and more convenient detection is based on UV/Vis absorption or fluorescence imaging allowing the quantification of the analytes. The major inherent drawback of all these types of detection is the inability to identify the analytes. Recently, mass spectrometry has become the method of choice for identification of compounds in conjunction with separation techniques. In the last few decades, a series of attempts have been made to couple TLC and MS (Somsen et al. 1995; Wilson 1999). Mainly, there are two different approaches to realize TLC/MS. First, the stationary phase containing the separated analytes is removed

S. Peng (✉), N. Ahlmann, M. Edler, and J. Franzke
ISAS – Institute for Analytical Sciences, Bunsen-Kirchhoff-Straße 11, 44139 Dortmund, Germany
ISAS – Institute for Analytical Sciences, Otto-Hahn-Straße 6b, 44227 Dortmund, Germany

MM. Srivastava (ed.), *High-Performance Thin-Layer Chromatography (HPTLC)*,
DOI 10.1007/978-3-642-14025-9_17, © Springer-Verlag Berlin Heidelberg 2011

from the plate by scratching, followed by liquid extraction of the compounds of interest from the silica gel and subsequent MS analysis. An improvement in this technique is the invention of a device enabling liquid extraction directly from the TLC plate (Luftmann 2004). Nevertheless, these techniques are destructive, time consuming, and not suited for scanning. Second, an alternative approach is to directly transfer the analytes from the intact TLC plate to the mass spectrometer. This technique has received much attention in recent years as it offers the possibility of fast and automatic scanning and imaging of TLC plates.

Laser desorption (LD) allows sampling in a small and defined area and therefore guarantees higher spatial resolution than the TLC separation itself. This property offers the possibility to characterize partially overlapping components after separation on TLC plates. The initial works on LD-TLC/MS were reported by Hercules (1983), Novak and Hercules (1985), followed by a number of applications of this technique (Somsen et al. 1995). To desorb analytes from a commercially available TLC plate, the power density of the radiation should be approximately 10^6 W/cm^2 and should not exceed 10^7 W/cm^2 (Cotter 1981). This power density is valid for desorption of analytes from a white surface. However, conventional LDTLC/MS based on a UV laser suffers from extensive fragmentation due to the high laser power, which is required for the ionization of the analytes. Consequently, infrared (IR) laser desorption from TLC plates, followed by a post-ionization step, was developed and applied by several groups (Li and Lubman 1989; Roger et al. 1993; Fanibanda et al. 1994; Krutchinsky et al. 1995).

The combinations of matrix-assisted laser desorption/ionization (MALDI) and TLC was introduced (Gusev et al. 1995a, b; Oka et al. 1992; Mehl et al. 1997; Mehl and Hercules 2000). TLC/MALDI was realized by using a solvent to enrich the sample molecules onto the surface of the plate, followed by a cocrystallization step of the analyte molecules with the matrix to form a structure like conventional MALDI. Nevertheless, because an "extraction solvent" is used in TLC/MALDI, a certain loss of chromatographic resolution because of lateral diffusion of separated compounds is inevitable. Different optimized protocols have been developed to reduce the problem (Oka et al. 1992; Gusev et al. 1995b).

Another method called surface-assisted laser desorption/ionization (SALDI) was developed recently. Here, particles or particle suspensions are used to couple the laser energy and assist desorption/ionization of sample molecules (Therisod et al. 2001; Crecelius et al. 2002; Chen et al. 1998; Wu and Chen 2002). The mechanism involves heating of particles and achieving heat conduction to the surface or surrounding liquid, i.e., a thermal desorption/ionization process. It is advantageous that the application of a particle powder supplies almost wavelength-independent absorption ability (Dale et al. 1996). This feature opens the possibility of a wider choice of laser source, including diode lasers that emit in the visible and near IR wavelength range. Crecelius et al. (2002) investigated a series of particle suspensions of different materials and sizes by applying them to TLC plates. The results from the graphite particle with a diameter of 2 mm showed the lowest background noise and the highest peak intensities. Graphite has the advantage that not only the absorption coefficient is relatively high but also the heat of conductivity is high.

Desorption from a TLC Plate by a Continuous Wave Near Infrared Diode Laser

The first examination of the combination of TLC and MS supported by a continuous wave near infrared diode laser is presented in (Peng et al. 2004), motivated by the low costs, the ease of use, and the small size of diode lasers. Certainly, a diode laser with low power has to be considered to represent a weak source for desorption, which is not able to realize desorption/ionization in TLC/MS coupling directly. Today, there are diode lasers with power up to 50 W commercially available. Diode lasers with power lower than 4 W are applied in our experiments. However, when the radiation with a power of 4 W is focused on a round spot with a diameter of 100 μm, the power density is at least one order of magnitude lower than the required power density for desorption. To overcome that problem and achieve efficient desorption, we applied a graphite suspension on the surface of the plate, which absorbs the power of the diode laser radiation. At the same time, a corona discharge was used to chemically ionize the desorbed molecules before they enter the ion trap mass spectrometer, hence representing an atmospheric pressure laser desorption/ chemical ionization (AP-LD/CI) technique (Schurrenberg et al. 1999). This technique offers the possibility to use diode lasers in TLC/MS for the first time. The decoupling of desorption and ionization allows individual optimization for both processes.

Experimental Arrangement for Laser Desorption

Figure 17.1 shows a sketch of the ion source used for the following experiments. A stainless steel target with a diameter of 4 mm was mounted on a xy-stage (DS40-xy, Newport, USA) to support a small piece of a TLC plate. The distance from the target to the front of the heated capillary inlet (transfer capillary) of a LCQ Classic ion trap mass spectrometer (ThermoFinnigan, USA) and the position in reference to the axis could be changed by the xy-stage. A potential of 1.5 kV was applied to the target by an additional power supply in order to improve the ion transmission efficiency. The corona needle was positioned on axis of the transfer capillary and above the rear edge (in reference to the capillary inlet) of the target. The needle could also be moved along the axis, therefore allowing an adjustment of the needle-to-plate distance for optimal APCI conditions. The standard APCI power supply of the LCQ system was connected to the needle to generate the corona discharge. Diode lasers (GBL981000G from Roithner Lasertechnik and SLD326B-24 from SONY) with wavelengths of 985 and 807 nm and maximum power outputs of 1 and 4 W, respectively, were used to desorb the analyte molecules. For all experiments, the laser beam was focused on the surface with a calculated spot diameter of approximately 0.1 mm, resulting in a maximum power density in the order of 10^4 W/cm^2.

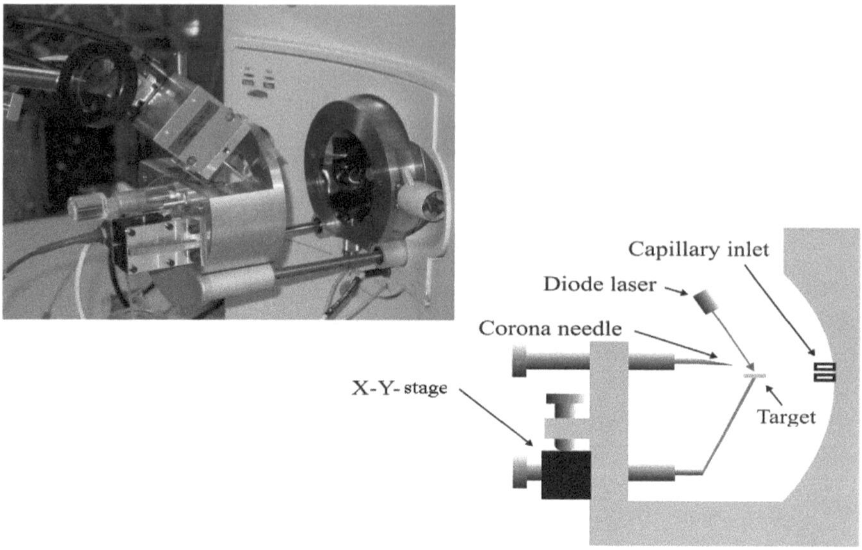

Fig. 17.1 Schematic diagram and photo of the ion source

Reserpine (RSP) was used as a model compound for all investigations. The capillary voltage was set to 6 V, the tube lens voltage offset was set to 57 V, and the capillary temperature was maintained at 150°C. These parameters were all optimized for RSP molecules. The ion injection was set to AGC (automatic gain control) mode because synchronization between laser and mass spectrometer is not necessary because of the use of a continuous diode laser system.

Glycerol (99.5%, photometric grade) was obtained from Aldrich, methanol and acetic acid from Merck (Germany), graphite powder (99.5%, 2 mm) from Fluka, and RSP from Sigma. No further treatment or purification was performed for the materials used in these experiments.

The sample solutions (5×10^{-4} M) were prepared by dissolving RSP in an aqueous solution of 50% methanol with 0.1% acetic acid. Graphite suspension was prepared by diluting glycerol in methanol and afterward adding graphite powder. To obtain a well-mixed suspension, the mixture was then mechanically shaken for 30 min.

Evaluation with Reserpine as Sample

The purpose of the present study is to evaluate the applicability of desorption from a TLC plate by continuous diode laser radiation. To avoid the construction and evaluation of a system to transport the desorbed molecules to the mass spectrometer, which is necessary when a whole TLC plate is to be used, the samples were deposited on small pieces of TLC plates without actual separation. The pieces were obtained by cutting a commercial TLC plate (glass, 20 × 20 cm^2, Merck Art. 5721, Germany) to a size of 3 × 3 mm^2. Then, a piece of the TLC plate was fixed onto the sample holder using double-sided adhesive tape, 5 μl of sample solution (2.5 nmol) was deposited on the TLC piece, and subsequently dried for 5 min at room temperature. Next, 2 μl of graphite/methanol/glycerol mixture was placed on top of the TLC piece and left in ambient air for 10 min to obtain a dry and flat surface. Finally, the AP-LD/CI experiment was started, and the protonated RSP was detected by the LCQ Classic ion trap mass spectrometer.

The evaluation was accomplished with the calmative and antihypertensive drug RSP as analyte (for structure, see Fig. 17.2). The sample contained 2.5 nmol of RSP per analysis. Initial experiments were carried out by using the 1-W diode laser with maximal power output. The total ion counts in the *m*/*z* range 150–700 were measured by the mass spectrometer.

The first step of the following experiments was to find the best position of the corona discharge in reference to the entrance of the transfer capillary in order to realize ionization of desorbed molecules and an efficient transportation by the electric field. A short distance between the target and the inlet of the mass spectrometer is helpful to improve the transmission efficiency. However, if the target is too close to the capillary inlet, an electric arc will appear in between. A distance of 3 mm from the front edge of the TLC plate to the inlet was finally evaluated as the optimized parameter, whereas the target was 3 mm underneath the axis of the transfer capillary. This ensured a cloud from desorbed materials formed on the axis. It was found that the target can influence the corona discharge if it is too close to the axis of the electric field. The tip of the corona needle was positioned on axis and above the rear edge of the TLC plate, 1–2 mm offset from the spot of the laser beam, which ensured that the electric field contained the whole region of the desorbed molecules in the gas phase. It should be noted that when the corona needle is moved above the front part of the TLC plate, only a very weak sample signal can be measured, and when the position is beyond the front edge, no signal can be obtained. Such a geometric layout assures that a well-built electric field is set to form the steady ion current to carry ionized sample molecules into the mass spectrometer. In other publications about AP-LD/CI (Coon et al. 2002; Coon and Harrison 2002), a long distance between each part of the source (1.5 or 3 cm from heated capillary inlet to the tip of corona needle) was applied and for efficient ionization in such a distance, a high discharge voltage (5.3 or 8.1 kV) was used.

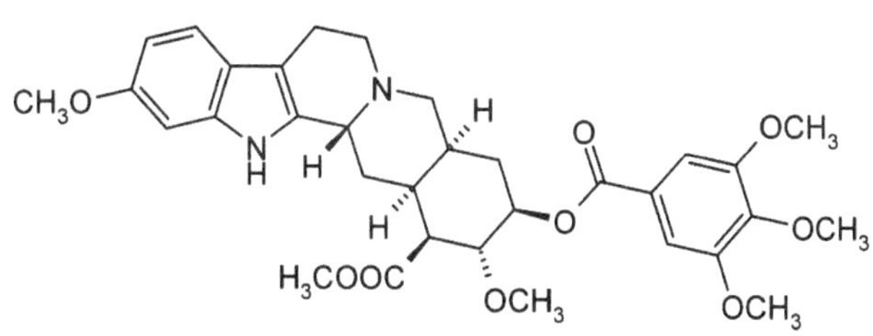

Fig. 17.2 Structure of reserpine

In our case, a compact source layout was used to improve the transmission efficiency, and the discharge voltage was dropped to around 3 kV, which results in a current of 0.5 A. Higher discharge currents were also used, but no obvious enhancement of the signal intensity could be observed, and the risk of a plasma breakthrough between corona needle and target is high. A potential of 1.5 kV on the target was found to enhance desorption and also improve ion transmission efficiency.

Separated Process of Desorption and Ionization

The following experiment proves that both essential processes – desorption and ionization of analyte molecules, which are often performed in one step – are separated. The surface of the TLC plate where the RSP spot was located was treated with 2 μl of graphite mixture. The graphite mixture was prepared by diluting glycerol in methanol in a ratio of 60:40, and thereafter adding 14 mg/ml of graphite powder.

As shown in Fig. 17.3, no signal from RSP is measured when only the corona discharge is in operation (a), obviously because RSP is not evaporated from the surface of the TLC plate. After approximately 1 min, the diode laser is also switched on (b), and an intensive signal appears. Finally, after approximately 2 min when a steady-state detector signal is reached (the height of the signal is dependent on the applied laser power and the duration of the radiation), the corona discharge is switched off (c) and the signal intensity decreases immediately to zero.

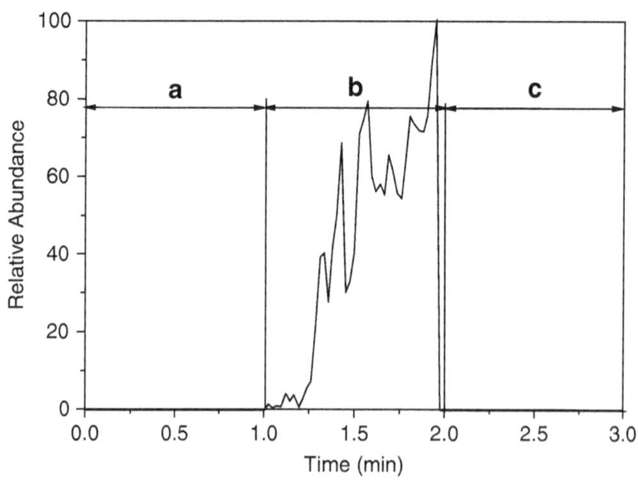

Fig. 17.3 Total ion counts obtained with (**a**) only the corona discharge on, (**b**) both corona discharge and laser on, and (**c**) only the laser on

Therefore, analyte ions can only be measured when both the laser and the corona discharge are in operation. It is obvious that the increase in the signal initiated by laser irradiation is slower than the decrease caused by stopping the corona discharge. The slow increase in the signal might be explained by the time of the heat transfer from the graphite particles to the sample molecules. The instantaneous decrease is due to the fact that the ionization process is stopped and no more ions enter the mass spectrometer. In the case when only the laser was on (c), it is reasonable to assume that the power from the diode laser was not high enough for ionization.

However, even if a few sample molecules are ionized, the absence of an electric field in the entrance region of the mass spectrometer resulting from the corona discharge reduces the transmission efficiency of ions. Therefore, these ions cannot be detected. To clarify this, the voltage of the corona discharge was decreased to the extent that the corona discharge expired, but attendant ions could still be transported into the capillary inlet. Nevertheless, there was still no measurable signal. Therefore, desorption and ionization are well decoupled in this technique.

Use of Glycerol

The purpose of glycerol is several-fold described in the literature. One of the reasons is that the viscosity of glycerol is necessary to obtain a homogenous surface (Dale et al. 1996). The signal dependence on the glycerol concentration is investigated in the following measurements. Graphite suspensions were prepared by diluting glycerol in methanol and thereafter adding 14 mg/ml of graphite powder. Besides the first proportion (60:40) of glycerol and methanol as used above, other proportions were applied as follows: (50:50), (30:70), and (0:100). Figure 17.4 shows the influence of the amount of glycerol in the mixture on the signal. With the decrease in the amount of glycerol, the signal intensity also decreases. Although protonated molecular ions of RSP could still be obtained without glycerol (Fig. 17.4d), the intensity was approximately two times weaker and the background noise increased in contrast to those with 60% glycerol (Fig. 17.4a). A high glycerol concentration is necessary for a sharp signal peak corresponding to high resolution. At a glycerol amount less than 30%, the resulting graphite layer did not have a glossy and uniform surface. This led to reduced resolution and mass calibration errors. The use of graphite is to absorb the laser power and to convert it into heat, causing rapid thermal desorption of the analyte from the TLC plate. Differing from other techniques in which pulsed lasers are used, here graphite is necessary not only for the purpose of improving the experimental results but also to obtain a signal. However, from another point of view, a tiny, economic, portable, and easy to use diode laser can be adopted to realize the function of conventional pulsed laser systems in TLC/MS. Five different graphite concentrations, 84, 42, 14, 5.6, and 2 mg/ml, were investigated to obtain an optimized result. A high concentration of

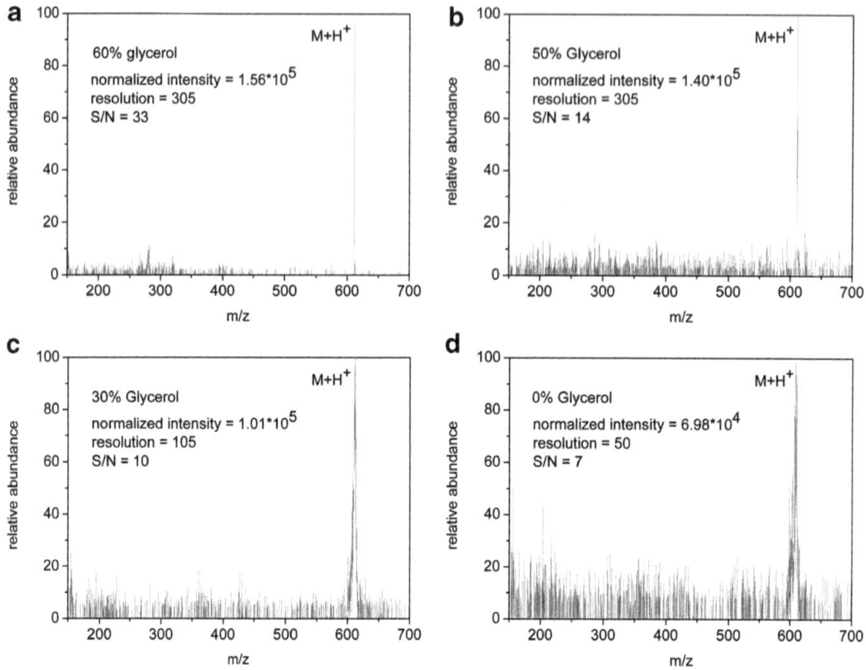

Fig. 17.4 Mass spectra of reserpine obtained by using graphite/glycerol mixture with (**a**) 60%, (**b**) 50%, (**c**) 30%, (**d**) 0% glycerol in methanol. Sample signal and spectral resolution are improved with the increase in glycerol concentration

Table 17.1 Results at different graphite concentrations in the mixture

Graphite conc. (mg/ml)	Intensity	Peak resolution	Burn
84	$1.27 \pm 0.41 \times 10^5$	60	Yes
42	$1.17 \pm 0.27 \times 10^5$	100	Yes
14	$1.56 \pm 0.29 \times 10^5$	305	No
5.6	$8.54 \pm 1.32 \times 10^4$	305	No
2[a]	1.22×10^3	610	No

[a]Average of 100 spectra

graphite leads to a high absorption of the laser power, but, on the other hand, a thick graphite layer keeps part of the analyte inaccessible to laser desorption.

As indicated in Table 17.1, the signal intensities are smaller and the resolution is lower when measured with 84 and 42 mg/ml graphite concentration than that with 14 mg/ml. Furthermore, at higher graphite concentrations the danger of igniting the graphite layer increases. Strong analyte signals and "clear" spectra were obtained with a concentration of approximately 14 mg/ml. Analysis at graphite concentrations lower than 14 mg/ml was performed as well; however, the signal intensity started to decrease rapidly because of reduced absorption of laser power.

In addition, the intensity of background signals stayed almost constant. When the graphite concentration was decreased to 2 mg/ml, only a faint gray shadow was visible on the surface of the TLC plate. Only a weak signal could be obtained in this case because the surface was not dark enough to couple the necessary laser power for desorption. The data measured with a graphite concentration of 2 mg/ml shown in Table 17.1 were obtained by averaging 100 spectra. The average intensity is approximately 100 times weaker than that at 14 mg/ml graphite concentration, but the peak resolution is higher. Sample-to-sample reproducibility is still a problem in this technique because of the inhomogeneous surface conditions for desorption. In fact, we cannot guarantee to obtain the same graphite layer after each apparent uniform sample preparation process. A better pretreatment method should be probed in future.

The choice of different graphite and glycerol proportions is a critical parameter for the desorption process. If only methanol and graphite are used, methanol penetrates quickly into the porous surface of the silica gel, and the graphite powder does not have enough time to diffuse and form a smooth layer. Glycerol keeps graphite powder in the liquid phase and results in a uniform distribution. In general, a more homogeneous surface is obtained when more glycerol is used. If the glycerol amount is small, the surface is not suitable for desorption. If the graphite concentration is less than 2 mg/ml, the resulting surface is not dark enough to absorb adequate power for desorption, and no stable sample signal can be obtained.

Dependence on Laser Power

Initially, the measurements were carried out by using the maximum output of the 1-W diode laser. We found that the analyte signal disappeared even if the laser power was decreased 10% less than 1 W. Therefore, 1 W is presumably near to the "threshold" for desorption in this method. A stronger 4 W diode laser was used to reach higher signal intensity in the following experiments. As shown in Table 17.2, experiments were performed with laser powers of 4 and 2 W. Spectra with better resolution and similar intensity were obtained at 2 W. The result at 4 W shows a marginal improvement in signal intensity, which was often canceled by poor sample-to-sample reproducibility, but resulted in notable increase in background noise and decrease in resolution. Furthermore, a very strong laser power causes much more material desorbed simultaneously including impurities from silica gel.

Table 17.2 Results at different laser powers

Laser power (W)	Intensity	Duration of signal (min)	Resolution
4	$2.20 \pm 0.58 \times 10^5$	1.5	150
2	$2.17 \pm 0.38 \times 10^5$	4.5	205
1	$1.56 \pm 0.29 \times 10^5$	6	305

Too many ions introduced into the ion trap can result in a "space charge effect," which broadens analyte peaks and causes mass errors. Laser power also has an influence on the duration of the signal. It is reasonable that lower power causes longer signal duration. The duration of the signal measured at 4 W is much shorter than that at 2 W, which means more sample molecules are vaporized in the same time range at 4 than at 2 W, but the signal intensity is nearly the same. It is possible that many ions are lost during transmission.

Full Plate Detection System

An extension of this work to a TLC plate-scanning system according to the same principle, which provides an online qualitative and quantitative full plate detection method, is also presented. Two desorption modes are tested in this study: direct LD and LD with graphite assistance (Byrdwell 1998).

We used the phospholipids, lecithin and sphingomyelin (SPM), as model analytes (for structures, see Fig. 17.5), which are the most abundant phospholipids inside living bodies. Lecithin is a group of glycerophospholipids that are involved in the metabolism of several lipid compounds; SPM is a group of phosphosphingolipids made by transferring choline from lecithin to a ceramide. SPM makes up a significant portion of the membrane lipids of the myelin sheath of nerve tissue.

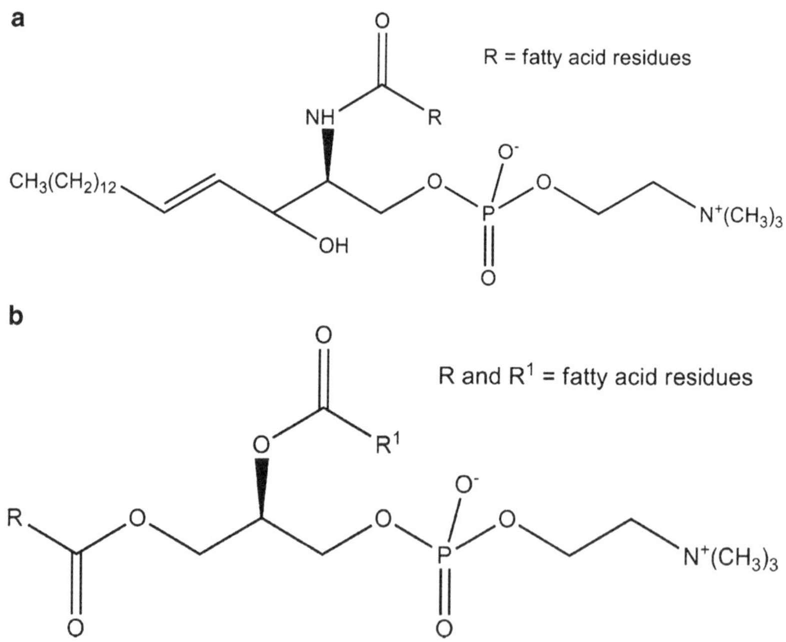

Fig. 17.5 Structures of (**a**) sphingomyelin and (**b**) lecithin

Experimental Arrangement for the Full Scanning System

To arrange a scanning system, the TLC plate can be moved linearly underneath the source of desorption or extraction or the source can be moved linearly along the plate. The first alternative was applied in the following experimental arrangement. Because of the decoupling of desorption and ionization, the ionization can be initiated before or after the transfer of the analytes, which means a transfer of ions or neutral molecules, respectively. The transfer of ions demands the modification of the mass spectrometer, whereas the transfer of molecules can lead to sample loss, which can be partially prevented by heating the transfer part. The latter system is relatively simple and easy to adapt to a wide variety of mass spectrometers. Figure 17.6 shows a sketch of the experimental arrangement used for the following experiments. The TLC plate is placed on a motor-operated xy-stage that is controlled by a waveform generator (FG-5000, WAVETEK).

A short glass tube (2.5 cm in length, 5.5 mm i.d., 7.5 mm o.d.) was installed 2 mm above the plate. An inlet hole with a diameter of 2 mm is located at the bottom side of the short glass tube for sampling the desorbed molecules. The tube is heated up to 200°C with a coiled heating wire. One end of the glass tube is closed, and the other is connected to a gas jet pump. A gas flow of 1.5 l/min of N_2 is introduced to the pump to supply a suction flow of 1.1 l/min. The gas is then transported to a transfer glass tube of 10-cm length and 7.5 mm i.d., which is positioned several millimeters in front of the capillary inlet of a LCQ Classic ion trap mass spectrometer (Thermo Finnigan, USA). A coiled heating wire maintains the temperature of the transfer tube at 350°C to reduce sample loss during transfer.

A laser beam from a diode laser (OTF 30P-40 from OPTOTOOLS) with a wavelength of 808.8 nm guided by an optical fiber is aligned through the sampling

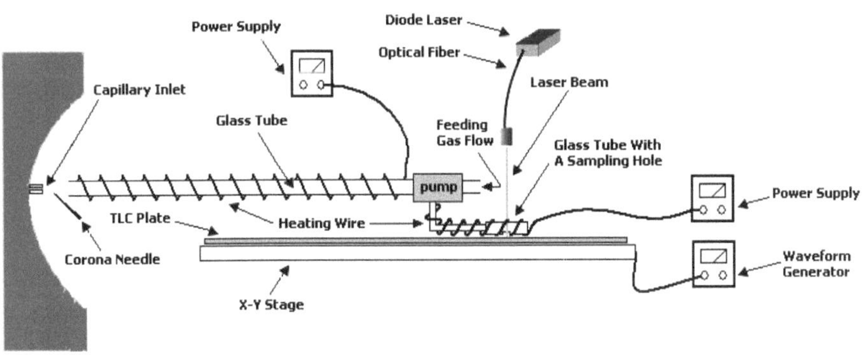

Fig. 17.6 Schematic diagram of the experimental arrangement. Sample molecules are desorbed by the laser beam from the TLC plate on a movable xy-stage. The desorbed molecules are sucked into a glass tube by a gas jet pump and then transported to the ionizing region produced by corona discharge in front of the heated capillary of the mass spectrometer

inlet and desorbs analytes from the surface of the TLC plate. The laser beam with a maximum power of 16.8 W is focused on the surface with a calculated spot diameter of 0.05 mm resulting in a maximum power density of 8.6×10^5 W/cm^2. The desorbed molecules are transferred to the area in front of the heated capillary inlet of the mass spectrometer and then ionized by a corona needle with a discharge current of 1 mA and a potential of 4 kV supplied by the LCQ system.

The capillary voltage of the mass spectrometer was set to 41.5 V, the tube lens offset voltage was set to 43 V, and the capillary temperature was maintained at 150°C. The ion injection was set to AGC (automatic gain control) mode and the full MS target was set to 3×10^6 as the synchronization between laser and mass spectrometer was not necessary because of the use of a continuous diode laser system.

Lecithin, (L-α-phosphatidylcholine, PC), from fresh egg yolk, and SPM, from bovine brain, were obtained from Sigma. The graphite powder (99.5%, 2 μm) was obtained from Fluka, chloroform (99.8%) from Aldrich, and *n*-hexane (for GC), isopropanol (for GC), methanol (for GC) and ethanol (99.8%, for chromatography) from Merck (Germany). No further treatment or purification was performed for the materials used in these experiments.

The lecithin and the SPM solutions were prepared by dissolving the powders in a mixture of chloroform/methanol/*n*-hexane (13:18:69) and ethanol/*n*-hexane (70:30), respectively. The hybrid sample solution was prepared by mixing the lecithin and the SPM solution in a ratio of 1:1. The analytes were separated on commercial silica gel TLC plates (aluminum-backed, 5×10 cm^2, Merck, Germany). The separation was carried out with a chloroform/methanol/water (20:7:1) solvent system for 15 min at room temperature. In the experiments with graphite assistance, the graphite layer was produced by spraying the graphite suspension with a sprayer on the TLC plates after development. The graphite suspension was prepared by adding graphite powder in isopropanol (10 mg/ml), followed by mixing for 5 min with an electromagnetic stirrer.

A mixture of 1 μl of PC/SPM (500 ng lecithin and 100 ng SPM) was used in the first experiments to check the feasibility of the interface. After the developed TLC plate was dry, the plate was mounted on the xy-stage. The scan was carried out with a velocity of 0.6 mm/s. In our previous work (Peng et al. 2004), a 1-W diode laser was used as the desorption source. Using this experimental setup, desorption from a white surface was not possible. However, the diode laser used in the present experiments has a maximum power of 16.8 W, which results in a power density of 10^6 W/cm^2. This power density also enables laser desorption of analytes from a white TLC plate.

A total ion chromatogram (TIC), showing two peaks from SPM and lecithin (PC) measured during the scan, and the related mass spectra are shown in Fig. 17.7. The different species of SPM or lecithin cannot be separated by conventional TLC techniques because all species in the group have nearly the same migrating distance so that they are located in the same sample spot. The mass spectra depicted in Fig. 17.3c and d show that SPM and lecithin can easily be identified. As shown in the spectra, signals of different molecular species were obtained at the same time

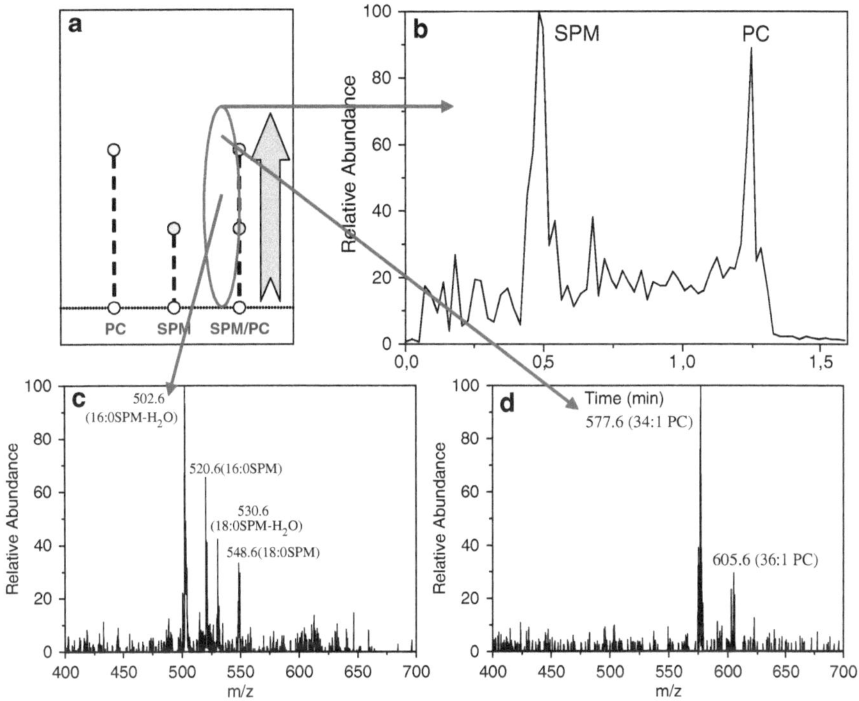

Fig. 17.7 Results from a scan on an intact TLC plate. (**a**) Schematic diagram of the experiments; (**b**) total ion chromatogram (TIC); (**c**) mass spectrum of sphingomyelin recorded during the run; (**d**) mass spectrum of lecithin recorded during the run. Only the species that give the most intensive peaks are labeled in spectra

although they were overlapping in chromatography, which in this case enhances the ability of analyzing complex samples. The peaks of species in the spectra were marked according to the amounts of carbon atoms in the fatty acid residues and the amounts of unsaturated carbon chains, such that 34:1 PC means a lecithin peak with a total of 34 carbon atoms and one unsaturated carbon chain in two fatty acid residues. The mass spectra show the loss of a phosphocholine group (head group, MW 183) from the quasi-molecular ion, a fragmentation that is regularly observed in APCI-MS for this group of compounds (Peng et al. 2005).

Influence of the Power Density

A high power density of the laser beam is not always desired because it may cause unexpected fragmentation and destroy the TLC layer. Therefore, similar experiments were carried out on surfaces covered with five different graphite particle densities. The graphite suspension was sprayed on the TLC plates, and the graphite densities were determined by measuring the gray scale of photos taken with a camera mounted on a microscope. TLC plates with five different gray scales (1–5, see Fig. 17.8) were scanned with different power densities and a scan velocity of 0.6 mm/s.

The signal area measured during these experiments is shown in Fig. 17.8 as a function of the laser power density. The SPM signal was first obtained when the laser power was higher than 500 mW focused on a spot with a diameter of 50 mm (power density: 2.5×10^4 W/cm^2) and when the TLC plate with the highest graphite density was used. Therefore, the necessary power density is approximately 30 times lower than that measured when a TLC plate was used without graphite coating. A power of 4 W (power density: 2×10^5 W/cm^2) was enough to measure an intensive signal when the plate with the gray scale value 1 was used. Increasing graphite particle density results in a reduced demand on laser power density for desorption. For the relatively "light-colored" plates with the gray scale values 1, 2, and 3, the signal area increased with higher laser power. However, for the darker TLC plates, values 4 and 5, the signal area decreased with higher laser powers. Ignited tracks were observed on these two plates at the sites where the laser beam was scanned. Obviously, these conditions caused fragmentation of the analytes and reduced signal intensity.

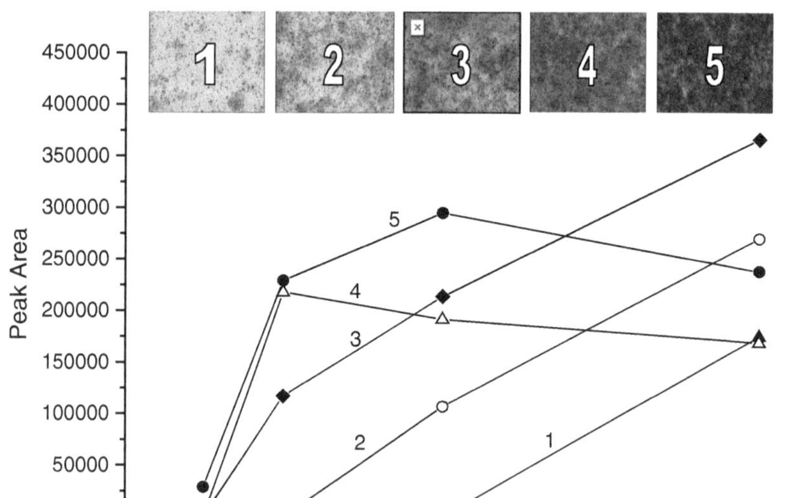

Fig. 17.8 Peak area of the SPM signal on the TLC plates covered with different graphite particle densities and scanned with different laser powers. The pictures were taken with a camera mounted on a microscope (area shown 2.0 mm × 1.5 mm)

Rapid Screening for Specific Substances

One of the important applications of the TLC technique is rapid screening for specific substances in complex samples. This can be conveniently realized with the device described here. To demonstrate the feasibility of the technique, we simulated the screening of SPM in the PC/SPM mixture. One reference sample containing only SPM and four synthetic samples containing PC/SPM mixtures were developed simultaneously on a TLC plate. The SPM concentration was identical in all five samples. First, the migrating distance of SPM can be identified by scanning the reference sample along the migrating direction. Then, the scan was continued from the center of the SPM spot in the reference sample in orthogonal direction of the development. As shown in Fig. 17.9, SPM signals were obtained from all synthetic samples at the same migrating distance of the plate. The signal areas (also shown in Fig. 17.9) of the four mixed samples were observed to be close to that of the reference sample. The average relative error is 7.0%. If the migrating distance of the analyte is known, as is usually the case for screening tasks, the overall measurement time can be quite short. For the conditions chosen here, it was approximately 1 min. However, when the scanning velocity is increased or sample displacement decreased, the time necessary for such a screening run could be even shorter.

In TLC/MS, often the quantification is performed off-line and represents a complicated and time-consuming procedure. The sample spot has to be located by a visualization method first. Then, the stationary phase containing the analyte has to be removed from the plate followed by extraction of the analyte with a solvent. Finally, the usual quantification procedure has to be carried out for the analyte solution. Another approach is the extraction of the analyte directly from the

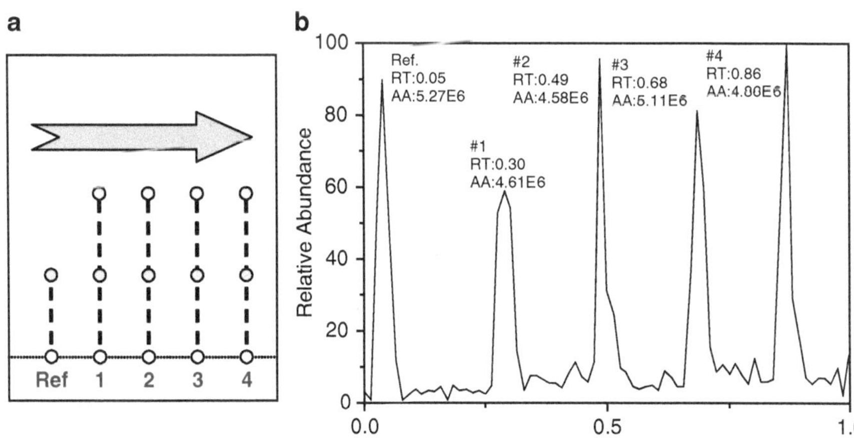

Fig. 17.9 (**a**) Schematic diagram of rapid screening; (**b**) total ion chromatogram of a SPM screening experiment. Ref. and #1, #2, #3, #4 means reference sample and four synthetic samples; RT is the retention time; AA means integral peak area

plate with a liquid (Luftmann 2004b) and the direct transport of the extraction liquid to a mass spectrometer. However, the measurement is still performed intermittently from one sample spot to another, which is not really suitable for high-throughput analysis. In contrast, the experimental setup presented here is able to provide a quick and continuous online quantitative measurement technique. Nevertheless, reliable quantification is a critical aspect using desorption techniques. Obviously, the surface condition is an important factor for laser desorption. Especially on TLC plates, the inhomogeneity of the microscopic surface condition at different locations results in differences in the desorption efficiency of analyte molecules. Consequently, the use of an internal standard is highly beneficial. The ideal internal standard would be the application of isotopically labeled analyte species (e.g., ^{13}C-labeled). However, a very few isotopically labeled biomolecules are commercially available. Another possibility is the use of structural homologues, e.g., other phospholipids, as internal standards; however, even slightly different structures can lead to a separation from the analytes during the TLC run.

Internal Standard

To overcome the problem, a suitable reference compound was added into the mobile phase. After the development of the sample, this compound was distributed homogeneously over the whole area below the solvent front on the plate. The "background" signal of this compound was then used as an internal standard. In this way, the chromatographic property of the internal standard is not important anymore and a wide range of compounds can be chosen for this purpose. Here, RSP was used as the internal standard (0.6 g/l in the mobile phase), a compound that is often used for calibration of LC/MS systems. In a normal quantification procedure, the ratio between the signal area of analyte (A_x) and the signal area of the internal standard is used to build a calibration curve. In this work, the internal standard yields a steady signal. One possibility is to use the integral area of the internal standard during the appearance of the analyte signal. However, this area is subject to change due to the chromatographic reproducibility, i.e., it will change with the width of the analyte signal although its signal area might be the same.

To reduce this kind of deviation, the average intensity of the internal standard (AI_{st}) was used, which is defined here as the integral area during the appearance of the analyte signal divided by the number of scans. The quantification of SPM was carried out based on the procedure described above. The strongest signal (m/z 502.6) was used as the basis to calculate the signal area. For each point in the calibration curve, five synthetic samples with the same SPM concentration were measured. They were applied simultaneously on one plate and scanned orthogonally to the direction of the development as described above. Less than 1 min was required for such a scan. As shown in Fig. 17.10, the ratio of the signal area of SPM and the average intensity of RSP showed a good linear correlation ($R^2 = 0.9991$) for the sample amount investigated (0–1.0 μg). The internal standard used here

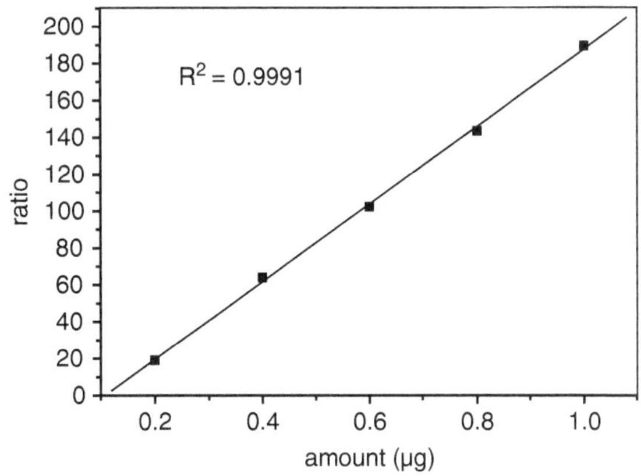

Fig. 17.10 Calibration curve for the ratio of peak area of SPM and average intensity of reserpine. Five samples were measured for each calibration point

shows no structural conformity with the analyte. In principle, isotopically labeled compounds or structural homologues could also be used as internal standards in the same way as described here, which possibly would even result in a better accuracy. However, because a relatively large amount of internal standard is necessary to provide sufficient signal intensity, this kind of internal standard application would not be suited for high-cost internal standard materials.

Notes

TLC, a novel technique, combined with diode laser-induced desorption/atmospheric pressure chemical ionization is demonstrated in this chapter. The use of a graphite suspension and the decoupling of desorption and ionization allow diode lasers to be used in TLC/MS for the first time. This will provide a possible fast and easy separation and detection method in the future. The graphite suspension plays the role of energy absorber and transfer medium and has proved to be a suitable material to couple the power of a continuous diode laser. Glycerol is used to improve signal intensity and decrease background noise, and it is important for obtaining a satisfying graphite layer.

Corona discharge is used to ionize the molecules desorbed by the diode laser that is not strong enough to achieve the ionization step. Matrix-related peaks are limited at low mass range, and no fragments are observed. Sample-to-sample reproducibility is still a problem in this technique in spite of careful sample preparation. A better sample preparation method should be developed in order to obtain a more homogenous surface of the graphite layer.

In this chapter, a novel interface to realize the coupling of TLC and MS is demonstrated. A diode laser, which is compact, easy to use, and cost effective, is used as the laser desorption source. The TLC plate is placed on a moving xy-stage to realize full plate scanning. There are two scan modes: a scan along the direction of the analyte development, which is appropriate for detecting all the components in one sample, and a scan across several sample spots orthogonally to the direction of development, which is suitable for high-throughput analysis of the same compound or compounds with the same R_f from several samples. The mass spectrometer is coupled to provide an identifying ability to the TLC technique, which is important for the analysis of unknown substances or overlapping spots, as well as to improve the reliability of the analysis. A gas jet pump connected with heated glass tubes enables transportation of the desorbed analyte molecules to the ionization region.

The measurement was carried out on both non-pretreated and graphite-coated TLC plates. Using the graphite-coated surface, the necessary laser power for desorption decreased down to 10^4 W/cm^2, approximately 30 times lower than that needed for the desorption from white surfaces. This opens up a wide range for the choice of the laser sources, including diode lasers with compact size. Although the device is mainly designed for qualitative work, analyte quantification was also carried out. An internal standard was added into the mobile phase to yield a "background" signal, which was used as a reference signal for the quantification.

References

Byrdwell WC. Dual parallel mass spectrometers for analysis of sphingolipid, glycerophospholipid and plasmalogen molecular species. Rapid Commun. Mass Spectrom. 1998; 12 (5), 256–272.

Chen Y, Shiea J, Sunner JJ. Thin-layer chromatography mass spectrometry using activated carbon, surface-assisted laser desorption/ionization. Chromatogr. A 1998; 826 (1), 77–86.

Coon JJ and Harrison WW. Laser desorption-atmospheric pressure chemical ionization mass spectrometry for the analysis of peptides from aqueous solutions. Anal. Chem. 2002; 74 (21), 5600–5605.

Coon JJ, Mchale KJ and Harrison WW. Atmospheric pressure laser desorption/chemical ionization mass spectrometry: a new ionization method based on existing themes. Rapid Commun. Mass Spectrom. 2002; 16 (7), 681–685.

Cotter R. Cationized species in laser desorption mass-spectrometry. Anal. Chem. 1981; 53 (4), 719–720.

Crecelius A, Clench MR, Richards DS, Parr V. J. Thin-layer chromatography – matrix-assisted laser desorption ionisation – time-of-flight mass spectrometry using particle suspension matrices. Chromatogr. A 2002; 958 (1–2), 249–260.

Dale MJ, Knochenmuss R, Zenobi R. Graphite/liquid mixed matrices for laser desorption/ionization mass spectrometry. Anal. Chem. 1996; 68 (19), 3321–3329.

Fanibanda T, Milnes J, Gormally J. Thin-layer chromatography mass-spectrometry using infrared-laser desorption. Int. J. Mass Spectrom. Ion. Process. 1994; 140, 127–132.

Gusev AI, Proctor A, Rabinovich YI, Hercules DM. Thin-layer chromatography combined with matrix-assisted laser-desorption ionization mass-spectrometry. Anal. Chem. 1995; 67 (11), 1805–1814.

Gusev AI, Vasseur OJ, Proctor A, Sharkey AG, Hercules DM. Imaging of thin-layer chromatograms using matrix/assisted laser desorption/ionization mass-spectrometry. Anal. Chem. 1995; 67 (24), 4565–4570.

Hercules DM. Pure Appl. Chem. 1983; 55, 1869.

Krutchinsky AN, Dolgin AI, Utsal OG, Khodorkovski AM. Thin-layer chromatography laser-desorption of peptides followed by multiphoton ionization time-of-flight mass-spectrometry. J. Mass Spectrom. 1995; 30 (2), 375–379.

Li L, Lubman DM. Resonant 2-photon ionization spectroscopic analysis of thin-layer chromatography using pulsed laser desorption-volatilization into supersonic jet expansions. Anal. Chem. 1989; 61 (17), 1911–1915.

Luftmann H. A simple device for the extraction of TLC spots: direct coupling with an electrospray mass spectrometer. Anal. Bioanal. Chem. 2004; 378 (4), 964–968.

Mehl JT, Hercules DM. Direct TLC-MALDI coupling using a hybrid plate. Anal. Chem. 2000; 72 (1), 68–73.

Mehl JT, Gusev AI, Hercules DM. Coupling protocol for thin layer chromatography matrix-assisted laser desorption ionization. Chromatographia 1997; 46 (7–8), 358–364.

Novak FP, Hercules DM. Thin-layer chromatography-laser mass-spectrometry (TLC-LMS) of triphenylmethane dyes - initial results. Anal. Lett. 1985; 18 (4), 503–518.

Oka H, Ikai Y, Kondo F, Kawamura N, Hayakawa J, Masuda K, Harada KI, Suzuki M. Development of a condensation technique for thin-layer chromatography fast-atom-bombardment mass-spectrometry of non-visible compounds. Rapid Commmun. Mass Spectrom. 1992; 6 (2), 89–94

Peng S, Ahlmann N, Kunze K, Nigge W, Edler M, Hoffmann T, Franzke J. Thin-layer chromatography combined with diode laser desorption/atmospheric pressure chemical ionization mass spectrometry. Rapid Commun. Mass Spectrom. 2004; 18 (16), 1803–1808.

Peng S, Edler M, Ahlmann N, Hoffmann T, Franzke J. A new interface to couple thin-layer chromatography with laser desorption/atmospheric pressure chemical ionization mass spectrometry for plate scanning. Rapid Commun. Mass Spectrom. 2005; 19 (19), 2789–2793.

Roger K, Milnes J, Gormaolly J. The laser desorption laser ionization mass-spectra of some methylated xanthines and the laser desorption of caffeine and theophylline from thin-layer chromatography plates. Int. J. Mass Spectrom. Ion. Process. 1993; 123 (2), 125–131.

Schurrenberg M, Dreisewerd K, and Hillenkamp F. Laser desorption/ionization mass spectrometry of peptides and proteins with particle suspension matrixes. Anal. Chem. 1999; 71 (1), 221–229.

Somsen GW, Morden W, Wilson ID. Planar chromatography coupled with spectroscopic techniques. J. Chromatogr. A 1995; 703 (1–2), 613–665.

Therisod H, Labas V, Caroff M. Direct microextraction and analysis of rough-type lipopolysaccharides by combined thin-layer chromatography and MALDI mass spectrometry. Anal. Chem. 2001; 73 (16), 3804–3807.

Wilson ID. The state of the art in thin-layer chromatography-mass spectrometry: a critical appraisal. J. Chromatogr. A 1999; 856 (1–2), 429–442.

Wu J, Chen Y. J. A novel approach of combining thin-layer chromatography with surface-assisted laser desorption/ionization (SALDI) time-of-flight mass spectrometry. Mass Spectrom. 2002; 37 (1), 85–90.

Chapter 18
HPTLC Hyphenated with FTIR: Principles, Instrumentation and Qualitative Analysis and Quantitation

Claudia Cimpoiu

Abstract In recent years, much effort has been devoted to the coupling of high-performance thin-layer chromatography (HPTLC) with spectrometric methods because of the robustness and simplicity of HPTLC and the need for detection techniques that provide identification and determination of sample constituents. IR is one of the spectroscopic methods that have been coupled with HPTLC. IR spectroscopy has a high potential for the elucidation of molecular structures, and the characteristic absorption bands can be used for compound-specific detection. HPTLC–FTIR coupled method has been widely used in the modern laboratories for the qualitative and quantitative analysis. The potential of this method is demonstrated by its application in different fields of analysis such as drug analysis, forensic analysis, food analysis, environmental analysis, biological analysis, etc. The hyphenated HPTLC–FTIR technique will be developed in the future with the aim of taking full advantage of this method.

The identification of separated compounds from a mixture in the absence of standards is one of the problems in high-performance thin-layer chromatography (HPTLC). Moreover, the quantitative analysis involves the previous identification of detected compounds. The selectivity and sensitivity of component detection and identification can be improved by coupling of two or more analytical techniques in HPTLC. For these reasons, in the last years, the coupled or "hyphenated" chromatographic techniques are becoming more common in analytical separation. These techniques have the goal of a rapid and efficient chromatographic separation and online identification of the separated compounds. The hyphenation of HPTLC separation technique with spectrometric methods in the analysis of complex mixtures represents the state of the art in modern analytical laboratories. The increased

C. Cimpoiu
Faculty of Chemistry and Chemical Engineering, Babes Bolyari University, Cluj Napoca, Romania
e-mail: ccimpoiu@chem.ubbcluj.ro

MM. Srivastava (ed.), *High-Performance Thin-Layer Chromatography (HPTLC)*,
DOI 10.1007/978-3-642-14025-9_18, © Springer-Verlag Berlin Heidelberg 2011

amount of information obtained by these hyphenated techniques is sufficient for the identification of the compound structure and for the quantification. HPTLC can be coupled with ultraviolet–visible (UV/VIS) and fluorescence spectrometry, infrared spectrometry (IR), Raman spectrometry, photoacoustic spectrometry (PA), and mass spectrometry (MS) (Gocan and Cimpan 1997; Cseharti and Forgacs 1998). Because IR spectrometry has been successfully coupled with liquid chromatography (LC) (Somsen and Visser 2003), many attempts are focused on the coupling of HPTLC and IR spectroscopy (Somsen et al. 1995). The IR spectroscopy has a high potential for the elucidation of molecular structures, and the characteristic absorption bands can be used for compound-specific detection. HPTLC and FTIR coupling approaches can be divided into two groups: indirect and direct methods. The indirect coupling involves either the transfer of the substance from the spot to a nonabsorbing IR material (KBr or KCl) or in situ measurement of excised HPTLC spots when the spectra are recorded directly from the plate (Rager and Kovar 2001). The direct methods are based on the direct hyphenated HPTLC–FTIR technique introduced by Glauninger et al. (1989). Until then, the combination of HPTLC and ultraviolet–visible (UV–VIS) spectroscopy was the only online coupling method available in planar chromatography. The information content of UV–VIS spectra is rather poor and rarely enables unambiguous identification of a substance and a chromophore is needed for UV detection.

Almost all chemical compounds give good IR spectra that are more useful for identification of unknown substances and discrimination between closely related substances (Pfeifer et al. 1996). HPTLC–FTIR spectra make possible the detection and quantification of even non-UV absorbing substances on HPTLC plates (Stahlmann and Kovar 1998). These reasons make this hyphenated technique universally applicable. The direct online coupling HPTLC–FTIR offers some advantages in comparison with other hyphenated techniques (HPTLC–Raman spectroscopy, HPTLC–PA, and HPTLC–MS) such as the simple operation and the optimized operational aspects of online coupling.

Principle, Instrumentation and Data Presentation

The principle of HPTLC–FTIR hyphenated technique depends on scanning the plate with an IR beam in a diffuse reflectance infrared Fourier transform (DRIFT) unit connected to a Bruker IFS 48 FTIR spectrometer (Fig. 18.1). The plate is fixed on to a computer-controlled x,y-stage. The special mirror arrangement was constructed to enable DRIFT measurements and to eliminate largely the specular (Fresnel) reflectance in the 3,600–1,350 cm^{-1} region. The diffuse reflectance containing the desired spectral information is collected and directed to the mercury, cadmium, and telluride (MCT) detector.

Diffuse reflectance is not a direct method of measurement. The particles of the samples scatter, remit, and absorb most of the radiation, and the intensity of the reflected radiation is the same in every direction. The desired spectral

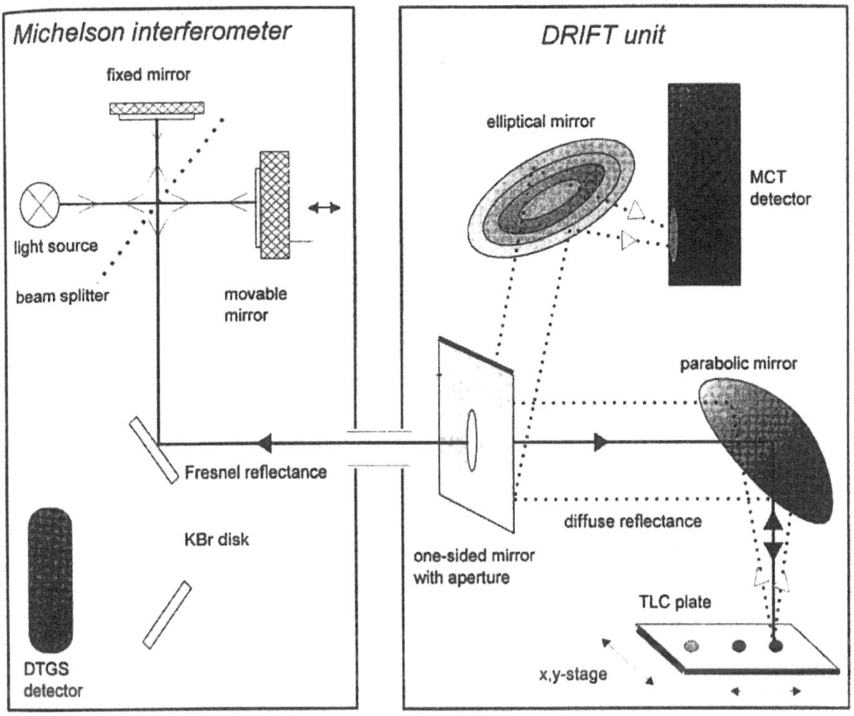

Fig. 18.1 Diagram of Bruker HPTLC–FTIR unit

Fig. 18.2 Interaction between the radiation and the samples: 1, 2 – directional Fresnel radiation; 3 – scattered Fresnel radiation; 4, 5 – diffuse radiation

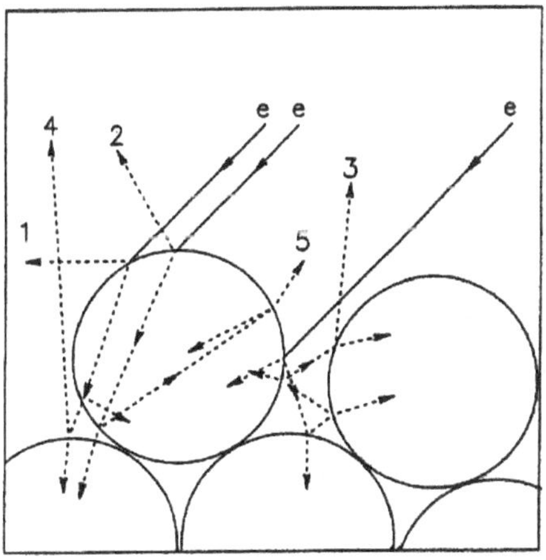

information about the sample is contained in the diffusely remitted radiation. A part of the remitted radiation, called Fresnel reflectance (Fig. 18.2), does not contain spectral information and leads to distortion and shifting of the band in the reflectance spectra.

This specular reflectance must be minimized in order to obtain the desired quality of spectral information. The diffuse reflectance follows the Lambert cosine law:

$$\frac{df/dI_r}{d\omega} = \frac{CS_0}{\pi \cos\alpha \times \cos\theta} = B\cos\theta, \tag{18.1}$$

where I_r is the remitted radiation flux in an area f (cm^2) and solid angle ω (sr); S_0, the intensity of irradiation (W/cm^2); constant C (<1), the fraction of the incident radiation flux which is remitted; α, the angle of incidence; θ, the angle of observation; and B, the radiation density or surface brightness (W/cm^2 sr) (Fig. 18.3).

A simplified solution to this equation is obtained by making several assumptions by Kubelka and Munk and is known as Kubelka–Munk function.

$$f(R_\infty) = K/S = k'c, \tag{18.2}$$

where R_∞ is the ration of the diffuse reflectance single-beam spectra of sample and reference at infinite thickness of the sample layer; K, the absorption coefficient; S, the scattering coefficient; k', a proportionality constant; and c, the concentration of an absorbing compound. This equation allows the correlation between sample concentration and the intensity of scattered light in a similar way as Bouguer–Beer law can predict for spectrometry by transmittance. The Kubelka–Munk equation is applicable to very dilute sample and can be used in the case of spot on HPTLC plates. An advantage of DRIFT technique in comparison with transmission spectrometry is that the signal-to-noise ratio diminishes only with the square root of the concentration.

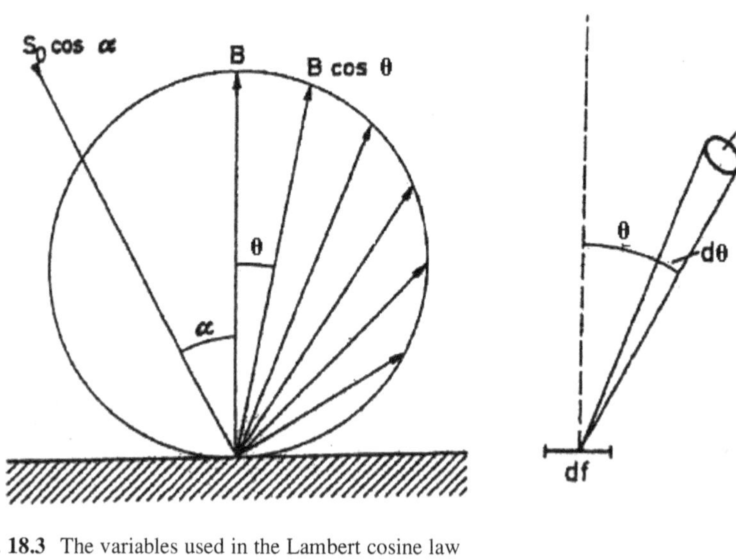

Fig. 18.3 The variables used in the Lambert cosine law

In the direct coupling HPTLC–FTIR method, some difficulties appear in comparison with the approaches involving the transfer of the substance to a nonabsorbing material. One of the most important problems is the absorption of conventional stationary phases, e.g., silica gel, which absorb strongly in the IR range. It is very difficult to obtain reliable spectra in the regions where the layer shows strong IR absorption. The "reststrahlen" effect could arise at wavelengths where the matrix presents high absorbency and refractive index. The "reststrahlen" effect consist of no diffuse reflectance occuring almost entire reflected from the surface. The silica gel, the most widely used adsorbent in HPTLC, presents absorption bands between 1,350 and 1,000 cm^{-1} and above 3,550 cm^{-1} which are superimposed on the spectra of compounds and only the region between 3,550 and 1,350 cm^{-1} can be evaluated (Bayerbach et al. 1999). Therefore, the measurements on silica gel in these regions are not possible, but it can make measurements up to 1,000 cm^{-1} on cellulose. The best results are obtained when the mixture of silica gel 60 and magnesium tungstate (1:1) is used as stationary phase. This adsorbent improved signal-to-noise ratios and enhanced the performance of the diffuse reflectance of the matrix (Bauer et al. 1998). Another problem is due to the particle sizes, particle size distribution, and the layer thickness, which affect the scattering, remitting, and absorbing of the radiation by matrix. The stationary phase with particle diameter of 10 μm, a small particle distribution, and a layer thickness of 200 μm on the glass is found to be ideal in the mid IR (Rager and Kovar 2001). Moreover, the binder or the fluorescence indicator added to the adsorbent and the mobile phase could be altered the HPTLC–FTIR spectra (Bauer and Kovar 1999).

The data generated by HPTLC–FTIR analysis are saved on the hard disk. The interferograms are then converted using commercially available GC-IR software obtaining different ways of information presentations (Fig. 18.4). The DRIFT spectra can be calculated from the interferograms using fast Fourier transformation (FFT) (Griffiths and De Haseth 1986). The spectra can be converted into normalized reflectance spectra (Fig. 18.4a, b), into quasi-absorbency units (Fig. 18.4c), or into Kubelka–Munk units (Fig. 18.4d). The quasi-absorbency units and the Kubelka–Munk units are proportional to concentration. By Gram–Schmidt vector orthogonalization, a chromatogram can be created from the interferogram data (Fig. 18.4f). The chromatogram obtained is a substance-nonspecific one, in which the total integrated absorption of the sample is measured (Kovar 1991). This procedure is one of the possibilities to locate the spots on the plates. Another procedure is using window chromatograms (Fig. 18.4e), which are the frequency-dependent chromatograms at any wavelength of interest. The first method can be used to increase selectivity, and the second method is universally applicable and independent of wave number (Morlock and Kovar 2003).

The recorded IR spectra are the base of compound identification. The IR spectrum of an organic compound provides a unique fingerprint because IR spectrum of a polyatomic molecule is based on molecular vibrations, which depend on atomic masses, bond strengths, and intra- and intermolecular interactions. The IR spectrum can be distinguished from the IR absorption patterns of other compounds including isomers. The identification can be realized by fitting the reference spectra

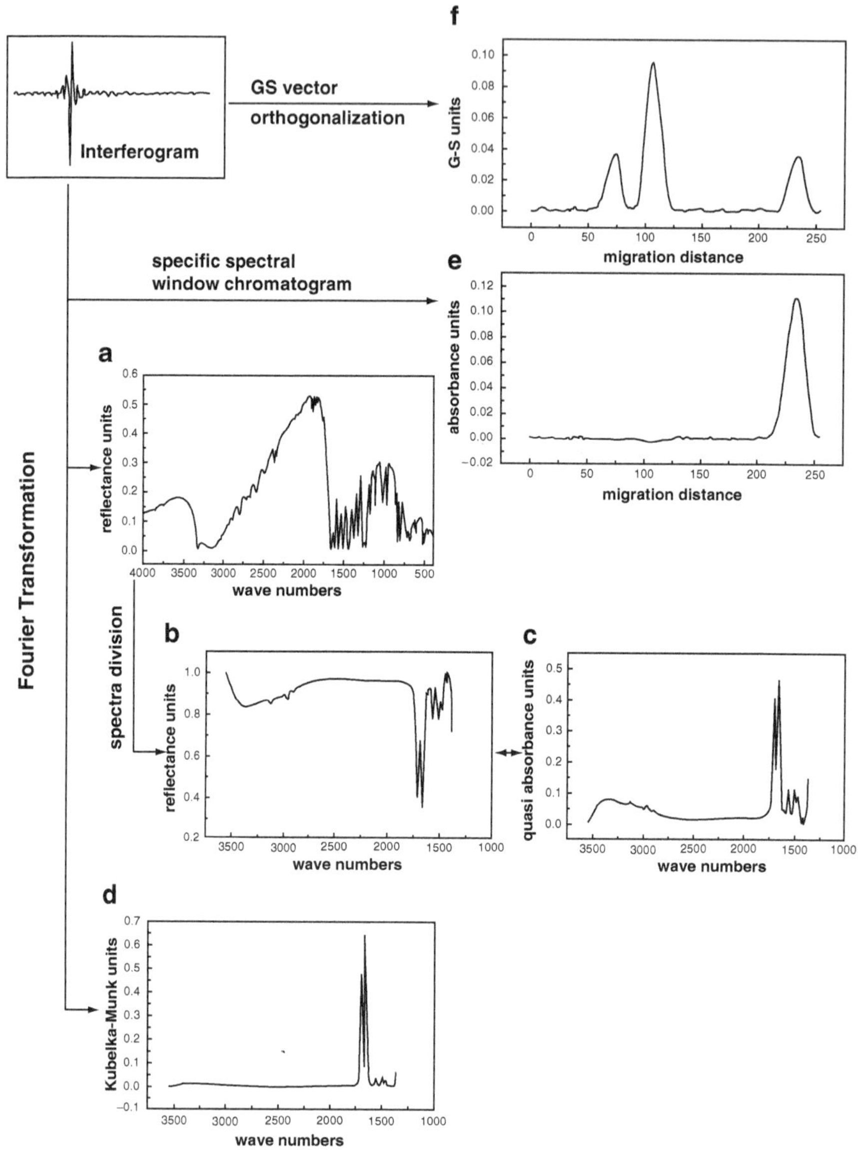

Fig. 18.4 The possibilities of data presentation

to sample spectra and visual comparison. The spectra of separated compounds are the spectra extracted at the peak maxima. A reference spectrum is needed for identification of compounds because false bands, called artifacts, arise due to the irregularities in the stationary phase. The reference spectra must be recorded in the same conditions because in the case of in situ measurements the interaction between adsorbed substance and the layer leads to significant changes in the maximum peak position. The compounds separated by HPTLC can be also identified using an HPTLC–FTIR library. The band position, width, and intensity are automatically compared and the reliability of the results is described in terms of hit quality ($\geq$800). The more detailed spectra and the increased hit quality can be obtained by postrun measurement of the located spots with higher resolution. These methods allowed unambiguous identification of separated compounds.

Qualitative and Quantitative Analysis

The identification by direct HPTLC–FTIR hyphenated technique is demonstrated by the applications in the field of drug analysis – amphetamines, morphine, and indole derivatives (Kovar et al. 1995), benzodiazepines (Kovar et al. 1991), phtalazine derivatives (Cimpoiu et al. 2003); food analysis – pigments of raisins (Cserhati et al. 2000) and pigments of chestnut sawdust (Cserhati et al. 2001); forensic analysis (Brandt and Kovar 1997; Pfeifer and Kovar 1995); and biological analysis (Pfeifer et al. 1996; Pisternick et al. 1997). The quantitative analysis by HPTLC–FTIR technique is generally applied for the substances that do not absorb in the UV/VIS range and when the precision required is not too high. The lack of precision is due to the increasing of the sample broadening with the migration distance and due to the measurement not exactly at the peak maximum. These problems are due to the circular infrared beam with small diameter. The determination of compounds is made on the basis of evaluation of the peak areas in the Gram–Schmidt trace or in the window diagram, or by the evaluation of Kubelka–Munk spectra with integration of their strongest bands. The method using the Gram–Schmidt traces indicates the changes in absorbency over the whole spectral region. From this reason this method is practical and suitable for rapid determinations. The evaluation of the peak areas in the window chromatogram is appropriate for the quantification of individual substances. An advantage of this method is a better signal-to-noise ratio, but the disadvantage is the lower precision. More precise results are obtained using the evaluation of Kubelka–Munk spectra. The limit of identification and determination is ten times higher than that obtained by densitometry (Stahlmann 1999). This method has the disadvantages of the measurement only of the fraction of the substance in the peak maxima and the additional processing step.

Salient Features of HPTLC–FTIR

In conclusion, none of these methods is perfect and appropriate for all samples. The choice of one method depends on the goals of analysis. Moreover, in the quantitative analysis, great attention must be paid to parameters influencing the application, development, and measurement to avoid the presence of the errors in the final results. Relative standard deviations between 1.3% and 6.1% were achieved for the determination of pure substances without sample preparation (Frey et al. 1993). Many components from different types of samples have been quantified by online HPTLC–FTIR method. An example of online HPTLC–FTIR application is the determination of edetic acid (EDTA) from environmental samples (Wolff and Kovar 1994a). EDTA is determined in surface water samples by online coupling of HPTLC and FTIR after the treatment of water with cobalt (II) acetate and formic acid and the enrichment of the sample by solid phase extraction (SPE). The limits of detection and determination were calculated to be 250 and 450 ng, respectively. The method has the advantages of analyzing the sample without derivatization and the discrimination against related substances.

Another example is the analysis of hexobarbital, phenobarbital, caffeine, salicylic acid, and ascorbic acid by this hyphenated technique (Wolff and Kovar 1994b). For the improvement in identification and determination limits and signal-to-noise ratio, the analysis has been carried out using automated multiple development (AMD) on water-resistant silica gel 60WRF$_{254S}$. The determined identification limits were found to be 55 ng for hexobarbital, 55 ng for phenobarbital, 30 ng for caffeine, 220 ng for salicylic acid, and 240 ng for ascorbic acid.

The potential of HPTLC–FTIR coupled method has been demonstrated by its application in forensic chemistry (Brandt and Kovar 1997). The 11-nor-Δ^9-tetrahydrocannabinol-9-carboxylic acid (THC-COOH) was determined in urine using HPTLC–UV/FTIR online coupling. The sample was enriched by SPE after alkaline hydrolysis of ester glucuronide. The derivatization with an azo dyestuff is not necessary. The detection limits for THC-COOH is 4 ng/mL for UV and 14 ng/mL for IR, and therefore, this method enables qualitative and quantitative analysis in the region of the 20 ng/mL cut-off.

Notes

HPTLC coupled with FTIR spectroscopy is a powerful technique for the identification and determination of the sample compounds. This technique is a nondestructive method that has the advantage of the elimination of the substance transfer from the spot. The IR spectra are like a fingerprint of the analyzed substances; therefore, this method enables unambiguous identification of the substances. With modern FTIR instrumentation, IR spectroscopy is suitable for molecular recognition. Due to the simplicity and robustness of HPTLC, great attention must be given to further developments of hyphenated HPTLC–FTIR method which has a great potential for identification and determination of compounds from complex mixtures. This technique together with HPTLC–UV could be used in the routine analysis in the modern analytical laboratories.

References

Bauer GK and Kovar K-A. Direct HPTLC-FTIR On-line coupling: interaction of acids and bases with the binder in precoated HPTLC plates. J Planar Chromatogr-Mod TLC 1999; 11, 30–33.

Bauer GK, Pfeifer AM, Hauck HE and Kovar K-A. Development of an optimized sorbent for direct HPTLC-FTIR on-line coupling. J Planar Chromatogr-Mod TLC 1998; 11, 84–89.

Bayerbach S, Gauglitz G, Kovar K-A and Pisternick W (1999). Dünnschichtchromatographie in memoriam Professor Dr Jork, published by InCom Sonderband.

Brandt C and Kovar K-A. Determination of 11-Nor-Δ^9-tetrahydrocannabinol-9-carboxylic acid in urine by use of HPTLC-UV/FTIR on-line coupling. J Planar Chromatogr-Mod TLC 1997; 10, 348–352.

Cimpoiu C, Miclaus V, Damian G, Puia M, Casoni D, Bele C and Hodisan T. Identification of new phtalazine derivatives by HPTLC-FTIR and characterization of their separation using some molecular properties. J Liq Chromatogr Relat Technol 2003; 26, 2687–2696.

Cseharti T and Forgacs E. Hyphenated techniques in thin layer chromatography. J Assoc Off Anal Chem 1998; 81, 329–332.

Cserhati T, Forgacs E, Candeias M, Vilas-Boas L, Bronze R and Spranger I. Separation and tentative identification of the main pigment fraction of raisins by thin-layer chromatography – Fourier transform infrared and high performance liquid chromatography – ultraviolet detection. J Chromatogr Sci 2000; 38, 145–150.

Cserhati T, Forgacs E, Morais MH and Ramos AC. TLC-FTIR of color pigments of chestnut sawdust. J Liq Chromatogr Relat Technol 2001; 24, 1435–1445.

Frey OR, Kovar K-A and Hoffmann V. Possibilities and limitations of assays by on-line coupling of thin layer chromatography and FTIR spectroscopy. J Planar Chromatogr-Mod TLC 1993; 6, 93–99.

Glauninger G, Kovar K-A and Hoffmann V (1989). In Software Entwicklungen in der Chemie 3, Proceeding of 3rd Workshop Comput. Chem., Berlin, Germany, published by Springer, Berlin.

Gocan S and Cimpan G. Compound identification in thin layer chromatography using spectrometric methods. Rev Anal Chem 1997; XVI, 1b–24b.

Griffiths PR and De Haseth JA (1986). Fourier Transform Infrared Spectrometry, published by Wiley-Interscience, New York.

Kovar K-A dan Hoffmann V. Möglichkeiten und Grenzen der direkten DC-FTIR Kopplung. GIT Spezial Chromatographie 1991; 11, 1197–1201.

Kovar K-A, Enβlin HK, Frey OR, Rienas S and Wolff SC. Die direkte DC-IR-Kopplung zur Identifizierung und Bestimmung von Substanzgemischen. GIT Spezial Chromatographie 1991; 11, 95–101.

Kovar K-A, Dinkelacker I, Pfeifer A, Pisternick W and Wössner A. Identifizierung von Suchstoffen mit Hilfe der HPTLC-UV/IR Kopplung. GIT Spezial Chromatographie 1995; 15, 19–24.

Morlock G and Kovar K-A (2003). Handbook of Thin Layer Chromatography, published by Marcel Dekker Inc., New York.

Pfeifer AM and Kovar K-A. Identification of LSD, MBDB, and atropine in real samples with on-line HPTLC-FTIR coupling. J Planar Chromatogr-Mod TLC 1995; 8, 388–392.

Pfeifer A, Tolimann G, Ammon HPT and Kovar K-A. Identification of adenosine in biological samples by HPTLC-FTIR on-line coupling. J Planar Chromatogr-Mod TLC 1996; 9, 31–34.

Pisternick W, Kovar K-A and Ensslin H. High-performance thin-layer chromatogeaphic determination of N-ethyl-3,4-methylenedioxy-amphetamine and its major metabolites in urine and comparison with high performance liquid chromatography. J Chromatogr B 1997; 688, 63–69.

Rager IOC and Kovar K-A (2001). Planar Chromatography, published by Springer Scientific Publisher, Budapest.

Somsen GW and Visser T (2003). Encyclopedia of Analytical Chemistry, published by John Wiley and Sons Ltd, Chichester.

Somsen GW, Morden W and Wilson ID. Planar chromatography coupled with spectroscopic techniques. J Chromatogr 1995; 703, 613–665.

Stahlmann SA. Ten-year report on HPTLC-FTIR online coupling. J Planar Chromatogr-Mod TLC 1999; 12, 5–12.

Stahlmann S and Kovar K-A. Analysis of impurities by high-performance thin-layer chromatography with Fourier transform infrared spectroscopy and UV absorbance detection in situ measurement: chlordiazepoxide in bulk powder and in tablets. J Chromatogr A 1998; 813, 145–152.

Wolff SC and Kovar K-A. Determination of edetic acid (EDTA) in water by on-line coupling of HPTLC and FTIR. J Planar Chromatogr-Mod TLC 1994; 7, 286–290.

Wolff SC, Kovar K-A. Direct HPTLC-FTIR measurement in combination with AMD. J Planar Chromatogr-Mod TLC 1994; 7, 344–348.

Index

Kolibri 360 GmbH
Ferdinand-Jühlke-Straße 7
99095 Erfurt, Deutschland
produktsicherheit@kolibri360.de